Pharmacological Research on Traditional Herbal Medicines

Pharmacological Research on Traditional Herbal Medicines

Edited by

Hiroshi Watanabe

Research Institute for Oriental Medicines (Wakan-Yaku)

Toyama Medical and Pharmaceutical University

Toyama, Japan

and

Takeshi Shibuya

Department of Pharmacology

Tokyo Medical University

Tokyo, Japan

Taylor & Francis

Taylor & Francis Group

LONDON AND NEW YORK

 Printed by Biddles Short Run Books, King's Lynn

Reprinted 2003
by Taylor & Francis
11 New Fetter Lane, London EC4P 4EE

Transferred to Digital Printing 2003

Taylor & Francis is an imprint of the Taylor & Francis Group

British Library Cataloguing in Publication Data

Pharmacological research on traditional herbal medicines
1. Herbs - Therapeutic 2. Medicinal plants
I. Watanabe, Hiroshi
615.3′2

ISBN 13: 978-9-0570205-44

Front cover line drawing:
Aconitum hakusanense (Ranunculaceae). Courtesy of Professor Tsuneo Namba, Professor Emeritus of the Research Institute for Oriental Medicines (Wakan-Yaku), Toyama Medical and Pharmaceutical University, Toyama, Japan.

CONTENTS

PREFACE

Traditional medicine has been practiced for millennia, resulting in a particularly long and rich heritage that continues to influence growing acceptance of the efficacy and clinical use of traditional herbal medicines. Practices of traditional medicine can be categorized into two simple groups. One category constitutes highly evolved and disciplined applications of both medical and pharmaceutical methodologies. Among some of the most developed practices are Traditional Chinese Medicine and Traditional Indian Medicine (such as Ayurveda and Sidda), which engender distinct, comprehensive and systematic principles as the foundation of their specific therapeutic practices. Traditional Chinese Medicine has been widely adopted in many neighboring regions of China, where it has been customarily modified by incorporating indigenous ingredients and/or local practices; a few specific examples include, Kampo Medicine (Japan), Han Medicine (Korea), and North Medicine (Vietnam). Ayurveda is also popularly exercised worldwide, largely advocated and exercised by Indian scholars and emigrants.

Distinctly, Folk Medicine comprises another category of traditional medicines, commonly characterized by the application of simple indigenous remedies, but often lacking in any systematized methodology needed to demonstrate genuine efficacy.

The cultural, social and legislative status of traditional medicine varies considerably around the world. In Japan, the system of traditional (Kampo) medicine was officially integrated into the healthcare system in 1976. Today many Kampo herbal medicines are commonly available as OTC or prescription drugs. Japan's pharmaceutical industry accounts for sales of approximately 6.2 trillion yen annually, of which Kampo medicines constitute around 2%, according to 1995 statistics. Nearly 65–70% of Japanese medical doctors utilize Kampo medicines in clinical practice, either exclusively or in combination with modern drugs. The nation's medical school curriculum is now being revised, to better educate physicians on the theories and uses of Kampo medicines, while until now Japanese doctors had been trained only in the applications of modern drugs.

In Germany, herbal remedies are particularly prevalent, both as self-medications and prescription drugs, constituting nearly 5.4% of all medical prescriptions and 10% of the entire drug market. In the US, however, the Food and Drug Administration has conventionally prohibited the lawful use of herbal medicines. Nevertheless, the growing interest in herbal medicines throughout the world can be ascribed to benefits of cost effectiveness, efficacy and generally reduced side-effects, of many of these treatments. The US government has recently classified herbal medicines and the products of medicinal plants as dietary supplements. However, this half-step measure lacks the regulation necessary to prevent the misuse of herbal medicines (designated as dietary supplements) which can exhibit adverse effects, and chronic or acute toxicity if over-ingested.

More recently, though, the US has substantially increased its funding allocated to 'Complimentary and Alternative Medicine' by millions of dollars, while further establishing The Office of Alternative Medicine, in order to better facilitate research and clinical trials. These are welcome measures towards the development of more practical and effective regulatory policies concerning the use of traditional herbal remedies and phytomedicines.

In the midst of the present boom in natural products and herbal medicines, this book intends to introduce some of the most current research on traditional herbal medicines, and to depict potential and functional applications of these agents in our modern healthcare systems, which continue to be practiced most widely in Asia. The contributing studies in this book also characterize state-of-the-art pharmacological methods, which are being employed to determine the pharmacological mechanisms and develop more effective formulations of traditional herbal remedies, in concert with their potential roles in our modern healthcare systems. The editors believe much more basic pharmacological research into traditional herbal medicines is essential, and that this book will assist continued research as an important reference.

Hiroshi Watanabe

PRIMARY CONTRIBUTORS

Nuntavan Bunyapraphatsara (3)
Medicinal Plants Information Center
Faculty of Pharmacy
Mahidol University
Bangkok 10400
Thailand

M. Eric Gershwin (1)
Division of Rheumatology/Allergy and Clinical Immunology
Department of Internal Medicine
University of California at Davis
TB 192
Davis, CA 95616
USA

Masao Hattori (14)
Department of Cell-Resources Engineering
Research Institute for Oriental Medicines (Wakan-Yaku)
Toyama Medical and Pharmaceutical University
2630 Sugitani
Toyama 930-0194
Japan

Nguyen Thi Thu Huong (7)
Department of Pharmacology
Research Institute for Oriental Medicines (Wakan-Yaku)
Toyama Medical and Pharmaceutical University
2630 Sugitani
Toyama 930-0194
Japan

Takeshi Miyata (9)
Department of Pharmacology
Faculty of Pharmacy
Kumamoto University
Kumamoto 862-0973
Japan

Nobuyoshi Nishiyama (6)
Department of Chemical Pharmacology
Faculty of Pharmaceutical Sciences
University of Tokyo
7-3-1 Hongo, Bunkyo-ku
Tokyo 113-0033
Japan

Han Rui (2)
Department of Pharmacology
Institute of Materia Medica
Chinese Academy of Medical Sciences
Beijing
People's Republic of China

Haruhiko Shinozaki (8)
Department of Pharmacology
The Tokyo Metropolitan Institute of Medical Science
3-18-22 Honkomagome, Bunkyo-ku
Tokyo 113-0021
Japan

Kimiyasu Shiraki (13)
Department of Virology
Toyama Medical and Pharmaceutical University
2630 Sugitani
Toyama 930-0194
Japan

Kazuo Toriizuka (10)
Oriental Medicine Research Center
The Kitasato Institute
5-9-1 Shirokane, Minato-ku
Tokyo 108-0072
Japan

Hiroshi Watanabe (5)
Research Institute for Oriental
Medicines (Wakan-Yaku)
Toyama Medical and Pharmaceutical
University
2630 Sugitani
Toyama 930-0194
Japan

Kazuo Watanabe (11)
Faculty of Pharmaceutical Sciences
Chiba University
1-33 Yayoi-cho, Inage-ku
Chiba 263-0022
Japan

Yasuo Watanabe (4)
Department of Pharmacology
Tokyo Medical University
6-1-1 Shinjuku, Shinjuku-ku
Tokyo 160-0022
Japan

Haruki Yamada (12)
Oriental Medicine Research Center
The Kitasato Institute
5-9-1 Shirokane, Minato-ku
Tokyo 108-0072
Japan

1. COMPLEMENTARY AND ALTERNATIVE MEDICINE AND ASTHMA

ROBERT M. HACKMAN[1], JUDITH S. STERN[1], SEISHI SUEHIRO[2] and M. ERIC GERSHWIN[3]

[1]*Department of Nutrition and*
[3]*Division of Allergy and Clinical Immunology, Department of Internal Medicine, the University of California at Davis, Davis, California 95616*
[2]*Nippon Zoki Pharmaceutical, Osaka, Japan*

INTRODUCTION

Over 12 million Americans suffer from asthma. It is especially prevalent in children, but affects people of all ages. Alarmingly, the incidence and mortality of asthma are increasing as the number of hospital admissions for the treatment of asthma has increased; the number of deaths attributed to asthma has nearly doubled since 1976. Moreover, despite the thoughts of William Osler that "an asthmatic does not die of asthma, but rather pants one's way into old age", the number of older Americans with reactive airway disease is also increasing even in the absence of direct tobacco exposure.[1–4] Much of our information base on diagnosis and current status of asthma treatment is based on a recent NIH DHEW review.[4] Asthma is a common lung disease characterized by reversible airway obstruction and airway inflammation.[2,3] It is characterized by paroxysmal bronchospasm, inflammation, hypersecretion of mucus, airway wall edema, and bronchial hyperreactivity. As there are a number of parallels and correlations between asthma and sinusitis and often common denominators exist, the incidence and severity of chronic sinusitis has likewise dramatically increased. The incidence of sinusitis rises with age and can be a particular problem for older Americans. The reasons for these unsettling trends are complex and successful therapeutic approaches are elusive. However, throughout the world, hundreds of unproved, but yet potentially applicable treatment programs are used which rely on complementary and alternative medicine (CAM) for the management of asthma and sinusitis. These programs rely on a variety of protocols including nutrition, botanical medicine, homeopathy, and acupuncture, to name just a few. While conventional therapy is typically directed toward reducing airway inflammation and attenuating bronchial hyperreactivity, investigation of CAM approaches toward treatment and prevention of asthma may help yield new medical modalities to address an important personal and public health problem.

Epidemiology

Asthma is a chronic condition of varying severity, affecting nearly 12 million

Americans. Asthmatics experience well over 100 million patient days of restricted activity annually and costs of asthma care exceed $4.6 billion a year.[4] The point prevalence of asthma is approximately 3 to 5% of our population.[5] This figure varies markedly with age, gender, race, smoking level, and socioeconomic status. Although the public primarily associates asthma with children, 40% of people with asthma first develop the disease after the age of 40.[6]

Asthma Mortality

Numerous epidemiological studies demonstrate increases in asthma mortality in developed countries.[7–15] Asthma mortality in the US decreased for people between the ages of 5 and 34 years from 0.4 deaths per 100,000 in 1968 to a low of 0.16 in 1977. Since 1977, mortality rate for asthma has steadily increased, rising up to 0.42 deaths/100,000 in 1987.[13] Between 1982 and 1986, *hospitalizations* for asthmatics between 5 and 34 years of age peaked during the months of September–November, whereas asthma *deaths* for this age group appeared to peak in June–August.[12] Deaths from asthma are highest among those over 65 years of age, compared to other age cohorts.[4]

Even though our knowledge regarding the pathogenesis and mechanisms of asthma is improving, existing data clearly indicate an increase in asthma mortality. Reasons for this increase may be grouped into 5 broadly recognized categories: coding artifacts,[14] epidemiological trends,[12] medication trends,[16] environmental changes,[17] and nutritional influences.[18–23] Additionally, previously unrecognized CAM practices may elucidate further insights into the causes of asthma mortality.

Physician Visits, Hospitalizations and Health Care Costs

In 1989, over 479,000 hospitalizations were recorded in which asthma was the first listed diagnosis.[4] Health care costs associated with asthma exceed $4.6 billion annually. In 1988, almost 15 million asthma-related visits were made to physicians.[4]

Racial and Ethnic Factors

In 1979, blacks of both sexes were about twice as likely to die from asthma as whites. By 1988, the asthma death rate was almost 3 times greater among blacks than whites.[4] It is not clear why prevalence of asthma and its complications are more common among black cohorts compared to whites. Certain health damaging behaviors, such as smoking or inhaling second-hand smoke, are more common in blacks than in whites. Blacks also have less access to health care. In addition, an underlying genetic predisposition to asthma may influence the higher incidence in blacks.[9]

Age Considerations

Children

According to the 1988 US National Health Interview Survey on Child Health,[24] asthma is the leading cause of school absence in the US. Childhood asthma is a major reason for health care usage, totaling over 3.4 million patient visits in 1980 to 1981[25] and 149,000 hospitalizations in 1987.[26]

Seniors

The incidence of asthma in those 65 yrs and older is 5.2%, and 40% of people with asthma first develop the disease after the age of 40. Greater disability among seniors has been linked to a history of asthma. A number of factors appear to influence the relatively high incidence of asthma among seniors. In 10% of asthmatics an acute inflammatory response to aspirin and related nonsteroidal anti-inflammatory agents precipitates an attack.[27] Seniors are the most common users of aspirin and related nonsteroidal anti-inflammatory agents, thus increasing their risk of aspirin-induced asthma. A second factor which may precipitate asthma, particularly among seniors, is sinusitis. Sinusitis is proposed to be a leading, and commonly undiagnosed, cause of asthma.[28]

Risk Factors for Asthma

We have shown that when appropriate quantitative genetic models are applied to data from large-scale twin studies, overwhelming support exists for the notion that genetic factors are important in the etiology of asthma.[29] The importance of non-additive genetic factors in asthma is less clear. We have also reviewed the primary environmental factors that trigger genetically predisposed individuals to develop asthma.[29] The most significant environmental variables include allergen exposure, air pollution, cigarette exposure, respiratory viral infections and oxygen radical damage. Perhaps increased exposure of susceptible individuals to such triggers helps explain the increase in asthma incidence and mortality.

Asthma Pathogenesis

The mast cell and eosinophil continue to be regarded as central participants in asthma pathogenesis. Appreciation of asthma as an inflammatory disease process has led to recognition of the importance of other inflammatory cells and their respective products in a complex immunologic response. Recently, the role of cytokines and the potential role of neuropeptides have been explored. Evidence for eosinophils' role in asthma is strong.[30,31] Data support ongoing eosinophil and mast cell degranulation in symptomatic asthmatics. Recruitment and activation of eosinophils is markedly inhibited by corticosteroids. Therapeutic success of corticosteroids in blocking late-phase reactions and modifying bronchial hyperresponsiveness is further support of a proinflammatory role for eosinophils in asthmatic airways.[32]

Substantial data suggest that in asthmatics' airways, T lymphocytes are activated and encourage interaction with eosinophils, contributing to ongoing airway inflammation and bronchial hyperreactivity. The major mechanism by which lymphocytes interact with other immune/inflammatory cells is through cytokine production. Cytokines act via specific receptors, and form a network by which interleukins, interferons, lymphokines, monokines, and a variety of "factors" are related.[32]

Reactive Oxygen Species (ROS) and Asthma

Exposure to certain environmental hazards, such as cigarette smoke and chemical fragrances, can play a significant role in potentiating asthma because they act as triggers.[17] Common triggers include viral infections, cold air, sinusitis, dust, pollens, animal dander and foods. Many of these aggressions may result in the release of ROS[43] by inflammatory cells. Oxygen free radicals generated by neutrophils,[34] eosinophils,[35] and other cells[36] are implicated in asthma by a variety of mechanisms, including bronchoconstriction, induction of mucus secretion, and microvascular leakage.[37,38] Reactive oxygen species can also cause an autonomic imbalance between muscarinic-receptor-mediated contraction and beta-adrenergic-mediated relaxation of pulmonary smooth muscle.[37,39] Neutrophils from asthmatics produce increased amounts of superoxide anion compared with those from normal subjects after stimulation by either *N*-formyl-methionyl leucyl-phenylalanine or phorbol-myristate acetate. Selenium levels in sera from asthmatics have been found to be lower than normal, although the extent to which the antioxidant capability of the circulating blood cells is compromised is not clear.[40]

Aspirin and Related Nonsteroidal Anti-Inflammatory Agents as Precipitating Factors in Asthma

The association of aspirin sensitivity, asthma, and nasal polyps was first described by Widal and colleagues in 1922.[41] This clinical entity, later named the aspirin triad, was further studied in the late sixties with a clearly described clinical course of the syndrome.[42] Adverse reactions to aspirin and other nonsteroidal anti-inflammatory drugs (NSAID) may have different clinical presentations and different pathogenesis.[43]

The disease course and clinical picture are characteristic.[42–46] For the majority, first symptoms appear during the third or fourth decade of life; many of these patients have a negative family history. A typical patient starts to experience intense vasomotor rhinitis characterized by intermittent and profuse watery rhinorrhea. Over a period of months, chronic nasal congestion appears and physical examination reveals nasal polyps. Bronchial asthma and intolerance to aspirin develop during subsequent stages. Intolerance presents a unique picture: Within 1 hour following ingestion of aspirin, acute asthmatic attacks develop, often accompanied by rhinorrhea, conjunctival irritation, and scarlet flush of the head and neck. These reactions are dangerous and can be life threatening. A single therapeutic dose of aspirin or other anticyclooxygenase agent can provoke violent bronchospasm, shock, unconsciousness, and respiratory arrest.[27] In patients with aspirin-induced asthma, not only aspirin but all other anti-inflammatory drugs that inhibit cyclooxygenase precipitate bronchoconstriction.

The Cyclooxygenase Theory of Aspirin-Induced Asthma (AIA)

The possibility that AIA might result from specific inhibition in the respiratory tract of a single enzyme, namely cyclooxygenase,[47] has received increasing attention. The hypothesis stimulated a number of studies on the mechanism of bronchoconstriction, all of which operate within the frame of the cyclooxygenase theory. The major assumption is that inhibition of cyclooxygenase triggers specific biochemical reactions which can lead to open asthma attacks. One explanation for AIA is that arachidonic acid is shunted from the generation of prostaglandins to the biosyntheis of leukotrienes. It is not clear in which cells of the respiratory tract alterations in arachidonic acid metabolism might occur during AIA. Leukocytes, especially eosinophils, present in large amounts in nasal and bronchial tissue of aspirin-sensitive asthmatics, could be considered as a source of leukotrienes, and might be activated during aspirin-precipitated asthma attacks.[48]

Relationship Between Upper and Lower Airways

For many years the upper and lower airway of humans have been regarded as anatomically and fundamentally separate with little or no relationship. Diseases of the upper and lower airway may coexist. For example, 80% of patients with asthma have rhinitis symptoms, and 5 to 15% of patients with perennial rhinitis have asthma. A number of relationships between the upper and lower airway can be identified. The nose serves as an important filter of inspired air. Relatively large particles are captured by the hairs within the nostrils and other noxious substances are trapped in mucus. Nasal obstruction or a failure in the filter function would therefore increase the allergen/irritant burden to the lower airway, thus potentiating lower airway hyperresponsiveness.

Heating and humidification of inspired air are important functions of the nose. Such functions are largely provided by a highly vascularized mucosa of the turbinates and septum. If inspired air bypasses warming and humidification provided by the nose, cooler, dryer air is delivered to the lungs, which can potentiate a phenomenon referred to as exercise-induced asthma (EIA). Exercise is an important trigger for bronchial asthma and is thought to be initiated by loss of water and heat from the lower airway. Reduction in severity of EIA can be achieved by breathing through the nose rather than through the mouth during exercise.[49]

Increased Lower Airway Reactivity During Upper Respiratory Infections

In both children and adults with asthma, certain upper respiratory tract viral infections (URIs) provoke wheezing. Respiratory syncytial virus is most common in young children, with rhinovirus and influenza virus being more prevalent in older children and adults.[50] Obstructive changes in small airway function of the lung associated with viral illness may persist for up to 5 wks even after clinical illness has resolved. Viral URIs also cause airway hyperreactivity.

A number of mechanisms may explain how viral URIs and sinusitis contribute to airway reactivity, wheezing, and the pathogenesis of asthma[28]:

1. Exfoliation of bronchial epithelium whereby a viral URI inflames the bronchial epithelium. This will sensitize rapidly adapting sensory vagus fibers located primarily in epithelium of large airways. Exposure of sensitized fibers to an irritant, such as histamine, causes reflex bronchospasm.
2. Increased permeability of mucous membranes may result in inflammation of bronchial epithelium which allows for increased permeability of antigen.
3. Viral URIs may block β-adrenergic responsiveness in asthmatics.
4. Leukocytes of asthmatics release increased amounts of histamine when incubated with respiratory viruses. This enhanced basophil mediator release appears to be associated with production of interferon by the viruses.
5. Some respiratory viruses, in particular respiratory syncytial virus and parainfluenza, stimulate production of allergic antibody IgE to virus.[51] IgE attaches to exfoliated respiratory mast cells and reacts with virus to increase release of histamine, which causes wheezing. Enhanced histamine release causes the virus to stimulate production of interferon, which in turn enhances release of histamine.

Nasal-Sinus-Bronchial Reflex

Studies have been performed in both experimental animals and humans which indicate that stimulation of receptors in the upper airway result in bronchoconstriction. As early as 1870, a substantial increase in lower airway resistance was reported by stimulating the nose of cats with either ether or sulphur dioxide.[52] Electrical stimulation of the nose was found to result in increased lower airway resistance in cats.[53] These findings were extended to demonstrate that sectioning of the vagus blocked changes in lower airway resistance.[53] Even physical intervention, specifically correction of nasal septum deviation, restores normal lower airway function.[54]

Histamine challenge studies on patients with perennial rhinitis caused a significant drop in forced expiratory volume in one second.[55] The frequent association of paranasal sinus disease and bronchial asthma has been noted for many years. A number of clinical studies as early as the 1920s and 1930s emphasized the importance of sinusitis as a trigger for asthma in many patients.[56–58]

A recent study indicated that 75% of patients admitted with status asthmaticus had abnormal sinus X-rays.[59] Although more objective evidence that sinusitis triggers or exacerbates asthma would be helpful in further clarifying this issue, the data suggest that patients who present with difficult-to-control asthma will improve when coexistent sinusitis is cleared by medical and/or surgical treatment. This can be considered as strong suggestive evidence for an etiologic role of sinusitis in lower airway disease.[28]

The eosinophil appears to play an important role in mediating injury to bronchial epithelium in chronic asthma. Recently, the role of the eosinophil in chronic inflammatory disease of paranasal sinuses was investigated with tissue from patients who underwent surgery for chronic sinusitis. Sinus tissue from patients with sinusitis who also had chronic asthma and/or allergic rhinitis was extensively infiltrated with eosinophils. In contrast, sinus tissue from patients with chronic sinusitis alone had no eosinophils.

Another proposed mechanism for sinusitis as an aggravator of asthma is local stimulation of irritant receptors by inflammatory mediators with resultant reflex bronchospasm. Leukotrienes, prostaglandin (PG) D_2, and histamine levels were measured in maxillary sinus lavage fluid obtained during surgery for chronic sinusitis. The levels of leukotrienes, histamine, and PGD_2 in sinus fluid were significantly elevated and in the range associated with local inflammation and irritant receptor stimulation.[86]

COMPLEMENTARY AND ALTERNATIVE THERAPY IN ASTHMA

Conventional management of asthma is based on use of a limited number of drugs, special techniques such as hyposensitization, and environmental adjustment. Each of these current approaches is the outcome of a broad collection of practices, some of which make good sense and are based on reasonable science, and many of which illustrate a history of qualified success. However, the annals of asthma therapy are replete with the alternative and the mystical, with unproven remedies, and with reliance on a basic group of useful drugs that are progenitors of many of those that are still in favor today.[61–62] A comprehensive review of unconventional asthma therapies of today and for the past five millennia is beyond the scope of this article, and further detailed discussions can be found in articles by Dr. Irwin Ziment.[63–66]

Oriental Remedies

Chinese knowledge of botanical medicine dates back to the discovery of the herbal therapeutic *Ma huang* around 3000 BC. It was initially used as a stimulant, but was also used for respiratory afflictions and other diseases.[67] The active ingredient in *Ma huang* was subsequently found to be ephedrine, an effective bronchodilator. *Ma huang* has been central to asthma treatment in traditional forms of Chinese medicine. Numerous composite preparations that are used in treatment of asthma rely on this botanical. For example, *Ma huang* is combined with gecko lizard tails in "Ge Jie Anti-Asthma Pill"; other traditional mixtures are prescribed under names such as Minor Blue Dragon Combination.[64,67,68] Some preparations contain a compound with disodium cromoglycate-like activity, but many other favorite traditional Chinese remedies without *Ma huang* still remain to show proven value. Some traditional antiasthma agents, such as *bupleurum* and *pinellia*, and *magnolia* have not been found to be effective.[68]

Extracts of *Gingko biloba* have been shown to inhibit platelet-activating factor (PAF), thus offering a scientific explanation for the use of this herb for treatment of asthmatic coughs.[69] Other traditional agents are known to antagonize PAF including various fungal derivatives such as gliotoxin from a wood fungus, and kadsurenone from *haifentenga*, a Chinese medication found in *Piper futokadsura*, and an extract of *Tussilago farfara*. This latter herb is known in the West as coltsfoot, and has been long used in both the West and in China as a mild antitussive.

Traditional Chinese medicine may have introduced steroid therapy in a crude fashion through use of fetal and placental extracts and urine of pubescent children, practices sometimes used in the treatment of asthma. While none of Chinese herbal medications used to treat asthma have been found to have corticosteroid activity, it is quite possible that many components have anti-inflammatory activity. For example, licorice has been extensively studied and does have such activity. *Ledebouriella seseloides, Rehmannia glutinosa,* and *Paeonia lactflora* also have anti-inflammatory activity.

In Japan, a form of traditional practice, Kampo, has its origins in traditional Chinese medicine. Saiboku-To, a mixture of herbs that has been used for over 200 years for treatment of asthma, is one such Kampo formula. Recent analytical studies suggested possible mechanisms for its effectiveness in treating asthma.[70,71] Another modern use of Saiboku-To in asthma aims at reducing the dose of steroids in steroid-dependent severe asthma.[72]

Indian Remedies

Several non-Western forms of medical practice in India exist, including Ayurveda, naturopathy, unani, and homeopathy.[64] Most Ayurvedic medications used for asthma are still relatively unknown to Western scientists, but one plant, *Tylophora indica asthmatica,* has been a standard remedy in the treatment of respiratory disorders in which mucus accumulation is a symptom.[65] In 1964, a noted Indian physician, Dr. Shivpuri, demonstrated that chewing leaves of *Tylophora indica* was as effective in relieving asthmatic bronchospasm as was epinephrine. The traditional drug, vasaka, derived from *Adhatoda vasica* or the malabar nut tree, is smoked in cigarettes for asthma and tuberculosis. *Adhatoda vasica* is the source of bromhexine and ambroxol, two mucoregulating drugs used in Europe.

Leaves of *Datura stramonium.*, called d'hatura in India, contain a bronchodilator. Traditional use of this plant as a tobacco remedy for asthma originated in India, and *D. stramonium* cigarettes eventually became popular for treating asthma and bronchitis in Europe and North America.

Several other CAM practices developed in India are used in Western alternative treatment of asthma. These include yoga breathing and relaxation exercises, transcendantal meditation, and vegetarian diets.

Homeopathy

Homeopathy is widely practiced in India, although this treatment originated in Germany and was initially expanded to Europe and America. The homeopathic physician claims to analyze a constellation of physical, psychological and emotional domains in order to devise an individualized therapeutic program which uses very diluted extracts of botanicals and other compounds. Because each person is treated individually, and because asthma is viewed as secondary to underlying body imbalances, it is difficult to identify specific homeopathic remedies which work in a large, heterogeneous population.

African Remedies

Desmodium adscendens is a traditional botanical medicine used in Ghana as a treatment for asthma and other diseases associated with excessive smooth muscle contraction. Plant extracts inhibit contraction of guinea pig ileum caused by electrical field stimulation[73] and contractions of sensitized guinea pig airway smooth muscle induced by arachidonic acid/or leukotriene D_4.[74,75] This botanical contains several different elements which aid in relaxation of smooth muscle, acting to inhibit NADPH-dependent monooxygenase pathway of arachidonic acid metabolism.[75] Recently, *D. adscendens* has been shown to contain triterpenoid glycosides which function as high-affinity activators of calcium-dependent potassium channels.[76] As such, this ethnomedicine contains the most potent known potassium channel opener yet discovered.

Complementary and Alternative Western Remedies

Many CAM remedies currently employed in the US and Europe stem from folk remedies previously used in Europe. Most European practices actually originated in the Middle East and were derived over centuries from work of physicians in ancient Akkadia, Sumeria, Mesopotamia, and Egypt.[64,65] Originally rooted in the concept of magical energy in plants, systematic investigations by physicians in ancient Greece and Rome eventually developed a fairly rational pharmacopeia by the first century AD.

Non-botanical Greco-Roman approaches were also used throughout Europe, including cupping, leeching, scarification and venesection, supplemented by massage, exercises, baths, inhalations, fomentations, poultices, emetics, cathartics, and clusters.[64] Dietary prescriptions were routinely employed in the management of all diseases, and for asthma, various herbs, spices, animal extracts and food eliminations or additions were widely advocated by health experts of the time.

Moses Maimonides, the noted 13th century physician, prescribed an integrated lifestyle for the asthmatic patient. His chief dietary remedy was a spicy, herbal mixture of chicken broth which contained herbs such as fennel, parsley, oregano, mint, and onion.[77,78] It is now well known that hydration of the upper respiratory tract is helpful in alleviating some asthmatic symptoms, and robust consumption of broth is one way to accomplish such hydration. The role of vegetable and herbal extracts in soup broth remains to be examined, although onions and garlic have some protective effect against allergic reactions.

North American plant derivatives were used in European respiratory therapeutics beginning in the 16th century. Important botanicals included guaiac wood, ipecacuanha, tobacco, and chili peppers. As discussed previously, cigarettes containing *D. stramonium* were used, and often also included horehound, mullein, coltsfoot, lobelia, cubebs, seaweed, and marijuana. Glyceryl guaiacolate (guaifenesin), which has mucus-loosening, expectorant properties, was eventually derived from guaiac wood. The mechanism of action of guaiac is the same as that produced by subemetic doses of ipecacuanha, which cause activation of the gastropulmonary mucokinetic reflex, thereby resulting in secretion of low viscosity mucus.[63] Capsaicin, an active ingredient in cayenne and chili peppers, stimulates the mucokinetic reflex and releases and then depletes Substance P stored in nonadrenergic noncholinergic nerves. The subsequent release of mucus, and possible bronchospasm, may help to decrease mucus production and prolong dilation of bronchial muscles.

In 18th century Europe, strong coffee was widely used for treatment of asthma.[66] The subsequent discovery that caffeine functions as an effective bronchodilator helped verify this practice. Although drinking tea was not particularly favored as an asthma treatment, tea leaves were the original soource of theophylline, whose name actually means tea-leaf. Another 19th century European drug used to treat asthma was saltpeter, or potassium nitrate. Saltpeter was inhaled for the muscle-relaxing nitrate vapor. Potassium nitrate was also used in combination with *D. stramonium* and other herbs in tobacco preparations.

In recent years, isocyanates found in pungent vegetables, such as onion, have been shown to be protective against allergic reactions in the guinea pig.[77] Mandrake and related solaneous plants, a source of atropinic agents, offer benefits similar to those of *D. stramonium.* The related *Atropa belladonna* was long recognized in Europe to be a potent pharmacologic agent, and appears to have been favored by observant lay healers as an antiasthma remedy hundreds of years before it became popular in the 19th century medical community.

Nutrition

Dietary adjustments have been made throughout the ages to control asthma.[78] Unusual additions and rigorous elimination of articles of diet have had their advocates. In more recent years, various dietary practices developed, such as the macrobiotic diet based on concepts of balancing yin and yang. Elimination and rotation diets are popular with some patients, who avoid items such as seafoods (especially shellfish), chocolate, eggs, nuts, milk, cheeses, and certain liquors. In some cases, a true food allergy can be avoided, but many patients subscribe to a limited diet without any clear evidence that the avoided foods are harmful. Undoubtedly, certain foods or additives can make asthma worse, but it is a challenge to construct a realistic elimination diet unless an obvious relationship between specific foods and an individual's asthma exists. For example, avoidance of restaurant salads or various packaged foods could be of importance in patients with sulfite sensitivity. Food coloring has been considered as a possible cause of asthma in patients with marked allergy. A more recent concept has emerged that excess dietary salt may increase airway hyperresponsiveness. There may be reason for avoiding salty products if further scientific testing verifies this concept.

We have suggested other possible links between diet and asthma.[80] A potentially important dietary variable in regard to asthma is the amount and composition of polyunsaturated fatty acid (PUFA) intake. Much attention has been focused on the omega-3 class of fatty acids, such as eicosapentanoic acid (EPA). EPA competitively inhibits formation of prostaglandins and leukotrienes derived from omega-6 PUFA, arachidonic acid, including LTC_4, LTD_4, and LTE_4, which activate ion transport,[81] mucus secretion,[82] smooth muscle contraction and LTB_4, which stimulates neutrophil chemotaxis.[83] In addition to being a competitive inhibitor of arachidonate, eicosapentanoic acid is itself a substrate for the biosynthesis of leukotrienes as well as prostaglandins.

The drive to remove animal fats and cholesterol from the diet to prevent atherosclerosis has resulted in replacement of animal fats with vegetable oils in manufactured food, fast-food frying, and home food preparation. In addition, margarine, mayonnaise, salad dressings, ready-to-eat frozen foods, and 1/4 to 1/3 of American meals eaten outside the home have dramatically altered the composition (but not amount) of our fat intake.[84] This has resulted in a doubling of linoleic acid content, precursor to arachidonate, in human adipose tissue from 8% to about 15% in the last 30 years.[85] In addition, arachidonic acid that is eaten directly from the diet (predominantly from meats, eggs, and dairy fats) enriches the phospholipid pool with arachidonate.[86] Changes in the diet of Americans in the last three decades, while beneficial with respect to heart disease, may have resulted in alterations in body lipid composition that are partly responsible for the increase in asthma mortality seen over this period. We do not suggest that a high dietary intake of omega-6 fatty acids has a primary role in pathogenesis of asthma, but believe that it may amplify clinical severity in the case of an acute asthma attack, thus resulting in increasing mortality.

Zinc is a key nutrient with respect to modulation of immune function. Marginal dietary zinc intake may occur in many Americans. The trend of reducing red meat intake also reduces one of the best dietary zinc sources. Inadequate intake may modulate the severity of asthma, as we have recently proposed.[23]

Another potential link between nutrition and asthma involves magnesium. The modern American diet is low in magnesium,[87] with 58 percent of women aged 19 to 50 getting less than 70 percent of the Recommended Dietary Allowance for this essential mineral.[89–91] Intravenous and aerosolized magnesium hyposulfite have been demonstrated to be antianaphylactic in humans since the 1920s.[121–123] More recently, intravenous magnesium administered during an acute severe bronchial asthma attack improves FEV_1 as well as clinical signs.[92–94] The bronchodilating effect of magnesium may be due to interference with calcium handling by bronchial smooth muscle cells. An intriguing question for CAM research is whether dietary magnesium supplementation can attenuate severity of asthma symptoms.

Probiotics may also be a useful area of investigation in regard to nutritional influences on asthma. Consumption of microorganisms in yogurt and other beneficial bacteria is associated with a reduction in susceptibility to a variety of diseases. In recent studies, we have found that daily consumption of 450 g of yogurt with live cultures is associated with a five-fold increase in interferon-gamma by stimulated lymphocytes.[95] Since interferon-gamma inhibits of IgE synthesis, and since hyperproduction of IgE is implicated in asthma, probiotic supplementation may attenuate this sequence of events and reduce the severity of asthmatic symptoms.

REFERENCES

1. M.E. Addy, Some secondary plant metabolites in Desmodium adscendens and their effects on arachidonic and metabolism. *Prostaglandins and Leukotriene in Essential Fatty Acids* **47**, 85–91 (1992).
2. M.E. Addy and J.F. Burka, Effect of Desmodium adscendens fractions on antigen- and arachidonic acid-induced contractions of guinea pig airways. *Canadian Journal of Physiology and Pharmacology* **66**, 820–825 (1988).
3. M.E. Addy and J.F. Burka, Effect of Desmodium adscendens fraction 3 on contractions of respiratory smooth muscle. *Journal of Ethnopharmacology* **29**, 325–335 (1990).
4. H.L. Alexander, A historical account of death from asthma. *Journal of Allergy* **34**, 305–317 (1963).
5. J.P. Arm, C.E. Horton, J.M. Mencia-Huerta, F. House, N.M. Eiser and T.T. Clark, *et al.*, Effect of dietary supplementation with fish oil lipids on mild asthma. *Thorax* **43**, 84–92 (1988).
6. N. Barnes, P.J. Piper and J. Costello, The effect of an oral leukotriene antagonist L-649, 923 on histamine and leukotriene D_4-induced bronchoconstriction in the normal man. *Journal of Allergy and Clinical Immunology* **79**, 816–821 (1987).
7. P.J. Barnes, Reactive oxygen species and airway inflammation. *Free Radical Biological Medicine* **9**, 235–243 (1990).
8. P.J. Barnes, K.F. Chung and C.P. Page, Platelet-activating factor as a mediator of allergic disease. *Journal of Allergy and Clinical Immunology*, 751–769 (1988).
9. A.S. Buist and W.M. Vollmer, Reflections on the rise in asthma morbidity and mortality. *JAMA* **264**, 1719–1720 (1990).
10. S.S. Bullen, Incidence of asthma in 400 cases of chronic sinusitis. *Journal of Allergy* **4**, 402–408 (1932).
11. P. Burney, A diet rich in sodium may potentiate asthma. Epidemiologic evidence for a new hypothesis. *Chest* **91** (Suppl), 143S–148S (1987).
12. P.G.J. Burney, J.E. Neild and C.H.C. Twort, *et al.*, Effect of changing dietary sodium on the airway response to histamine. *Thorax* **44**, 36–41 (1989).
13. W.E. Busse, The precipitation of asthma by upper respiratory infections. *Chest* **87** (Suppl.), 44–48 (1985).
14. C.M. Carlson and M.E. Sachs, Is alternative medicine an alternative for the treatment of asthma? *Journal of Asthma* **31**, 149–151 (1994).
15. T.B. Casale and S.J. Smart, Pathogenesis of asthma: mediators and mechanisms. In M.E. Gershwin and G.M. Halpern, eds., *Bronchial Asthma: Principles of Diagnosis and Treatment*, Humana, Totowa, NJ, pp. 1–22.
16. P. Chanez, G. Dent, T. Yukawa, P.J. Barnes and K.F. Chung, Generation of oxygen free radicals from blood eosinophils from asthma patients after stimulation with PAF or phorbol ester. *European Respiratory Journal* **3**, 1002–1007 (1990).
17. C.C. Chang, S.D. Phinney, G.M. Halpern and M.E. Gershwin, Asthma mortality: another opinion – is it a matter of life and ... bread? *Journal of Asthma* **30**, 93–103 (1993).
18. M. Cluzel, M. Damon, P. Chanez, J. Bousquet, A. Cratas de Paulet, F.B. Michel and P. Godard, Enhanced alveolar cell luminol-dependent chemiluminescence in asthma. *Journal of Allergy and Clinical Immunology* **80**, 195–201 (1987).
19. D.W. Cockcroft, R.E. Ruffin, J. Dolovich and F.E. Hargreave, Allergen-induced increase in non-allergic bronchial reactivity. *Clinical Allergy* **7**, 503–513 (1977).
20. M.P. Cosman, A feast for aesculapius: historical diets for asthma and sexual pleasure. *Annual Reviews in Nutrition* **3**, 1–33 (1983).
21. G.K. Crompton, Beta-2-agonists in asthma. *Lancet* **337**, 43–46 (1991).
22. S. Cunningham-Rundles, R.S. Bockman and A. Lin, *et al.*, Physiological and pharmacological effects of zinc on immune response. *Annals of the New York Academy of Sciences* **587**, 113 (1990).
23. J.G.R. De Monchy, H.F. Kauffman, P. Venge, G.H. Koeter, H.M. Jansen, H.J. Sluiter and K. De Vries, Bronchoalveolar eosinophilia during allergen-induced late asthmatic reactions. *American Reviews in Respiratory Diseases* **131**, 373–376 (1985).
24. J.C. Delaguente, Nutrients and immune responses. *Rheumatic Disease Clinics of North America* **17**, 203–220 (1991).
25. R. DiToro, G. Galdo Capotorti, G. Gialenella, M. Miraglia del Giudice, R. Moro and L. Perrone, Zinc and copper status of allergic children. *Acta Paediatrica Scandinavia* **76**, 612–617 (1987).

26. W.E. Dixon and T.G. Brodie, The bronchial muscles, their innervation and the action of drugs upon them. *Journal of Physiology* **29**, 93–97 (1903).
27. R. Dodge, M.G. Cline and B. Burrows, Comparisons of asthma, emphysema and chronic bronchitis in a general population sample. *American Review of Respiratory Diseases* **133**, 981–986 (1986).
28. R.R. Dodge and B. Burrows, The prevalence and incidence of asthma and asthma-like symptoms in a general population sample. *American Review of Respiratory Diseases* **122**, 567–575 (1980).
29. C.J. Doelman and A. Bast, Oxygen radicals in lung pathology. *Free Radical Biological Medicine* **9**, 381–400 (1990).
30. J. Dolovich, F.E. Hargreave, M. Jordana and J. Denberg, Late-phase airway reaction and inflammation. *Journal of Allergy and Clinical Immunology* **83**, 521–524 (1989).
31. W. Dorsch, O. Adam, J. Weber and T. Ziegeltrum, Antiasthmatic effects of onion extracts—detection of benzyl and other isothiocyanates (mustard oils) as antiasthmatic compounds of plant origin. *European Journal of Pharmacology* **107**, 17–24 (1985).
32. R. Evans, R.W. Wilson, H.M. Rosenburg, F.M. Chevarley and M. Feinleib, National trends in the morbidity and mortality of asthma in the US. Prevalance, hospitalization and death from asthma over two decades: 1965–1984. *Chest* **91** (Suppl), 65S–74S (1987).
33. M.G. Fowler, M.G. Davenport and R. Garg, School functioning of US children with asthma. *Pediatrics* **90**, 939–944 (1992).
34. C. Fuller, W. Richards, V. Gilsanz and M.D. Nelson, Sinusitis in status asthmaticus. *Journal of Allergy and Clinical Immunology* **85** (abstract), 222 (1990).
35. M.S. Gottlieb, Relation of intranasal sinus disease in the production of asthma. *Journal of the American Medical Association* **85**, 105–109 (1925).
36. G.M. Halpern, K.G. Vruwink, J. Van de Water, C.L. Keen and M.E. Gershwin, Influence of long-term yogurt consumption in young adults. *International Journal of Immunotherapy* **7**, 205–210 (1991).
37. M. Homma, K. Oka, T. Yamada, T. Niitsuma, H. Ihto and N. Takahaski, A strategy for discovering biologically active compounds with high probability in traditional Chinese herbal remedies: an application of Saiboku-To in bronchial asthma. *Analysis of Biochemistry* **202**, 179–187 (1992).
38. R. Jackson, M.R. Sears, R. Beaglehole and H.H. Rea, International trends in asthma mortality: 1970 to 1985. *Chest* **94**, 914–918 (1988).
39. A.B. Kay, Asthma and inflammation. *Journal of Allergy and Clinical Immunology* **87**, 893–910 (1991).
40. C.L. Keen and M.E. Gershwin, Zinc deficiency and immune function. *Annual Reviews in Nutrition* **10**, 413–431 (1990).
41. I. Kratchmer, Physiologic relationships between nasal breathing and pulmonary function. *Laryngoscopy* **76**, 30–35 (1966).
42. D.J. Lane, What can alternative medicine offer for the treatment of asthma? *Journal of Asthma* **31**, 153–160 (1994).
43. T.H. Lee, R.L. Hoover, J.D. Williams, R.I. Sperling, J. Ravalese and B.W. Spur, *et al.*, Effect of dietary enrichment with eicosapentaenoic and dicosahexanoic acids on in vitro neutrophil and monocyte leukotriene generation and neutrophil function. *New England Journal of Medicine* **312**, 1217–1224 (1985).
44. G.D. Leikauf, I.F. Ueki, J.H. Widdicombe and J.A. Nadel, Alteration of chloride secretion across canine tracheal epthelium by lipoxygenase products of arachidonic acid. *American Journal of Physiology* **250**, F47–F53 (1986).
45. A. Lumiere, L'hyposulfite de mg dans l'instabilite humorale et dans l'anaphylaxie. *Program of Medicine* (Paris), **55**(3), 110–113 (1931).
46. S. Makino, Preventive therapy in Japan. In J. Morley, ed., *Preventive Therapy in Asthma.* London: Academic, pp. 230–52.
47. Z. Marom, J.H. Shellhamer, M.K. Bach, D.R. Morton and M. Kaliner, Slow reacting substance, leukotriene C_4 and D_4, increase the release of mucus from human airways *in vitro. American Reviews in Respiratory Diseases* **126**, 449–451 (1982).
48. O.B. McManus, G.H. Harris, K.M. Giangiacomo, P. Feigenbaum, J.P Reuben, M.E. Addy, J.F. Burka, G.J. Kaczorowski and M.L. Garcia, An activator of calcium-dependent potassium channels isolated from a medicinal herb. *Biochemistry* **32**, 6128–6133 (1993).
49. S.M. Meltzer, B. Goldberg, P. Lad and J. Easton, Superoxide generation and its modulation by adenosine in the neutrophils of subjects with asthma. *Journal of Allergy and Clinical Immunology* **83**, 960–966 (1989).

50. A.H. Mendis, T.J. Venaille and B.W. Robinson, Study of human epithelial cell detachment and damage: effects of proteases and oxidants. *Immunology of Cell Biology* **68**, 95–105 (1990).
51. E.A. Mitchell, Is current treatment increasing asthma mortality and morbidity? *Thorax* **44**, 81–84 (1989).
52. E.A. Mitchell, Racial inequalities in childhood asthma. *Social Science in Medicine* **32**, 831–836 (1991).
53. M.T. Murray and J.E. Pizzorno, *An Encyclopedia of Natural Medicine.* Rocklin, CA: Prima Publishing.
54. National Heart, Lung and Blood Institute, Asthma statistics, data fact sheet. Bethesda, MD: National Institutes of Health, May (1992).
55. National hospital discharge survey: annual summary, *Vital and Health Statistics*, Series 13, No. 99. Hyattsville, MD: National Center for Health Statistics; 1989: 30. US Department of Health and Human Services publication (PHS) 89-1760 (1987).
56. F.P. Nijkamp and P.A. Henricks, Receptors in airway disease. Beta-adrenoceptors in lung inflammation. *American Reviews in Respiratory Diseases* **141**, S145–S150 (1990).
57. J.H. Ogura and J.E. Harvey, Nasopulmonary mechanisms: experimental evidence of the influence of the upper airway upon the lower airway. *Acta Otolaryngology* **71**, 123 (1971).
58. H. Okayama, T. Aikawa, M. Okayama, H. Sasaki and S. Mue, *et al.*, Bronchodilating effects of intravenous magnesium sulphate in bronchial asthma. *Journal of the American Medical Association* **257**, 1076–1078 (1987).
59. Oriental Healing Arts Institute, *How to Treat Yourself with Chinese Herbs.* Los Angeles, CA (1980).
60. Patterns of ambulatory care in pediatrics; the national ambulatory medical care survey, US, January 1980–December 1981 (data from the National Health Survey). *Vital and Health Statistics*, Series 13, No. 75. Hyattsville, MD: National Center for Health Statistics; 1983: 30. US Department of Health and Human Services publication (PHS) 84-1736.
61. L.F. Paulozzi, J.J. Coleman and A.S. Buist, A recent increase in asthma mortality in the northwestern United States. *Annals of Allergy* **56**, 392–395 (1986).
62. S.D. Phinney, A.B. Tang, S.B. Johnson and R.T. Holman, Reduced adipose 18:3w3 with weight loss by very low calorie dieting. *Lipids* **25**, 798–806 (1990).
63. F. Ramond and P. Ramond, Le mg et son role en therapeutique. Seminars in *Hopitale Paris* **2**(25), 89–93 (1946).
64. G. Rolla, C. Bucca, E. Cavia, W. Arossa, M. Bugiani and Cesanol, *et al.*, Acute effects of intravenous magnesium sulfate for airway obstruction of asthmatic patients. *Annals of Allergy* **61**, 388–391 (1988).
65. M. Samter and R.F. Beers, Jr, Aspirin Triad. Clinical studies and consideration of its pathogenesis. *Annals of Internal Medicine* **68**, 975–983 (1968).
66. R. Shturman-Ellstein, R.J. Zeballos, J.M. Buckley and J.F. Souhrado, The beneficial effect of nasal breathing on exercise induced bronchoconstriction. *American Reviews in Respiratory Diseases* **118**, 72–77 (1978).
67. K. Sladek and A. Szczeklik, Mast cell (MC) and eosinophil (EOS) activation during aspirin-provoked leukotriene (LTs) release in aspirin-induced asthma (AIA). *Journal of Allergy and Clinical Immunology* **89**(abstract 354), 233 (1992).
68. R.G. Slavin, Chronic sinus disease and asthma. In M.E. Gershwin and G.M. Halpern, eds., *Broncial Asthma: Principles of Diagnosis and Treatment.* Humana, Totowa, NJ, pp. 549–61 (1994).
69. S.L. Spector, Common triggers of asthma. *Postgraduate Medicine* **90**, 50–52; 55–9 (1991).
70. A.M. Stephen and N.J. Wald, Trends in individual consumption of dietary fat in the United States, 1920–1984. *American Journal of Clinical Nutrition* **52**, 457–469 (1990).
71. D.D. Stevenson, Diagnosis, prevention and treatment of adverse reactions to aspirin and nonsteroidal anti-inflammatory drugs. *Journal of Allergy and Clinical Immunology* **74**, 617–622 (1984).
72. B.D. Stone, J.W. Georgitis and B. Matthews, Inflammatory mediators in sinus lavage fluid. *Journal of Allergy and Clinical Immunology* **85**(abstract), 22 (1990).
73. J. Stone, L.J. Hinks, R. Beasley, S.T. Holgate and P.A. Clayton, Reduced selenium status of patients with asthma. *Clinical Science* **77**, 495–500 (1989).
74. A. Szczeklik, Analgesics, allergy and asthma. *Drugs* **32**(Suppl 4), 148–163 (1986).
75. A. Szczeklik and K. Sladek, Aspirin, related nonsteroidal anti-inflammatory agents, sulfites and other food additives as precipitating factors in asthma. In M.E. Gershwin and G.M. Halpern, eds., *Bronchial Asthma: Principles of Diagnosis and Treatment*, Humana Press, Totowa, NJ, pp. 563—580 (1994).
76. A. Szczeklik, R.J. Gryglewski and G. Czerniawska-Mysik, Relationship of inhibition of prostaglandin biosynthesis by analgesics to asthma attacks in aspirin-sensitive patients. *British Medical Journal* **1**, 67–69 (1975).

77. A. Szczeklik, R.J. Gryglewski and G. Czerniawska-Mysik, Clinical patterns of hypersensitivity to nonsteroidal anti-inflammatory drugs and their pathogenesis. *Journal of Allergy and Clinical Immunology* **60**, 276–284 (1977).
78. A. Szczeklik, C. Virchow and M. Schmitz-Schumann, Pathophysiology and pharmacology of aspirin-induced asthma. In C.P. Page and P.J. Barnes, eds., *Handbook of Experimental Pharmacology. Pharmacology of Asthma*, Springer-Verlag, Berlin-Heidelberg, pp. 98, 291–314 (1991).
79. I.L. Tamarin, Le sulfate de magnesium dans l'asthme bronchique. *Vrachebaya Gaz* **4**, 1811–1817 (1927).
80. S. Toda, M. Kimura, M. Ohnishi and K. Nakashima, Effects of the Chinese herbal medicine "Saiboku-To" on histamine release from and the degranulation of mouse peritoneal mast cells induced by compound 48/80. *Journal of Ethnopharmacology* **24**, 303–309 (1988).
81. USDA (US Department of Agriculture), Nationwide Food Consumption Survey: Continuing Survey of Food Intakes of Individuals. Women 19–50 years and Their Children 1–5 years, 4 Days. *Report No. 86-3.* Hyattsville, MD: Nutrition Monitoring Division, Human Nutrition Information Service (1986).
82. N.G. Waller, S. Teuber and M.E. Gershwin, The genetics and epidemiology of asthma. In M.E. Gershwin and G.M. Halpern, eds., *Bronchial Asthma: Principles of Diagnosis and Treatment*, Humana, Totowa, NJ, pp. 1–22 (1994).
83. F.L. Weille, Studies in asthma XIX. The nose and throat in 500 cases of asthma. *New England Journal of Medicine* **215**, 235–238 (1936).
84. K.B. Weiss, Seasonal trends in the US asthma hospitalizations and mortality. *Journal of the American Medical Association* **263**, 2323–2328 (1990).
85. K.B. Weiss and D.K. Wagener, Changing patterns of asthma mortality. Identifying target populations at high risk. *Journal of the American Medical Association* **264**, 1683–1687 (1990).
86. R.C. Welliver, Upper respiratory infection in asthma. *Journal of Allergy and Clinical Immunology* **72**, 341–346 (1983).
87. R. Whang, Magnesium deficiency: pathogenesis, prevalence, and clinical implications. *American Journal of Medicine* **82**, 24–29 (1987).
88. M.F. Widal, P. Abrami and J. Lermoyez, Anaphylaxie et idiosyncrasie. *Presse Medicine* **30**, 189–193 (1968).
89. World Health Organization. Manual for the International Statistical Classification of Diseases, Injuries, and Causes of Death. Geneva: World Health Organization (1977).
90. K. Yan and C. Salome, The response of the airways to nasal stimulation in asthmatics with rhinitis. *European Journal of Respiratory Diseases* **128**(Suppl.), 105–109 (1983).
91. I. Ziment, *Respiratory pharmacology and therapeutics.* W.B. Saunders, Philadelphia, PA (1978).
92. I. Ziment, Five thousand years of attacking asthma: an overview. *Respiratory Care* **31**, 117–136 (1986).
93. I. Ziment, The management of common respiratory diseases by traditional Chinese drugs. *Oriental Healing Art International Bulletin* **13**(2), 133–140 (1988).
94. I. Ziment, Historic overview of mucoactive drugs. In P.C. Braga and L. Allegra, eds., *Drugs in Bronchial Mucology*, Raven, New York, pp. 1–33 (1989).
95. I. Ziment, Unconventional therapy in asthma. In M.E. Gershwin and G.M. Halpern, eds., Bronchial Asthma: Principles of Diagnosis and Treatment. 3rd ed., Humana Press, Totowa, pp. 413–42 (1994).

2. TRADITIONAL CHINESE MEDICINE AND HERBAL MEDICINE: CURRENT RESEARCH AND THEIR CLINICAL APPLICATION FOR THE TREATMENT AND PREVENTION OF CANCER

HAN RUI

Department of Pharmacology, Institute of Materia Medica, Chinese Academy of Medical Sciences, Beijing, People's Republic of China

I. INTRODUCTION

Traditional Chinese medicine represents the enormous accumulation of daily experiences in medical practice by the Chinese people since ancient times. It has a unique theory which includes theory of the pathogenesis, therapeutic methodologies, prescriptions and drugs. According to the theory of traditional Chinese medicine there are five visceral organs, namely the heart, liver, spleen, lung and kidney. Although the terms used for these organs are the same as those in Western medicine, their implications are quite different in those two medical systems. We must be aware that in traditional Chinese medicine the term Kidney is not merely an organ in anatomy, but represents a functional concept which sometimes means the endocrine systems. Likewise, the term Heart is not merely the organ in the modern sense but sometimes represents the functional state of the central nervous system. Therefore, in order to apply traditional Chinese medicine rationally we have to comprehend its theory carefully.

Chinese herbal medicine is also a good resource to explore. The huge territory of China, which encompasses subtropical, temperate and cold zones is good for the growth of a variety of plants. According to a general survey of medicinal plants in China, the number of these plants in general usage amounted to more than four thousand.

Shen Nong Ben Cao Jing and Ben Cao Gang Mu are the most famous classical books of Chinese Materia Medica. Their descriptions of therapeutic usages are very informative even today. For example, *Angelica sinensis* ameliorates menstruation disorders and *Artemisia annua* is effective against malaria. All these therapeutic effects have been confirmed by contemporary scientific research. Chemical, pharmacological and clinical studies published in recent years have demonstrated that Chinese Materia Medica and traditional Chinese medicine are a great treasure house of knowledge.[3, 4]

It is a general feature that cancer patients are diagnosed by the techniques of Western medicine in terms of pathological, biochemical, X-ray and/or other examinations and treated by a combination therapy including traditional Chinese medicine. In particular, traditional Chinese medicine is widely used to ameliorate the side effects of radiation and/or chemotherapy. It is also used for the promotion of recovery from surgery as a tonic or biological response modifier in clinical practice in China. The major side effects of radiation and/or chemotherapy are exemplified as disturbance of the gastrointestinal system, hemopoesis depression, weakness and inflammation in some organs. For the treatment of these dysfunctions and symptoms, the following herbal drugs are used frequently.[6,7]

1. **For the detoxification and extinguishing of fever:** *Lonicera japonica* Thunb., *Forsythia viridissima* Lindley, *Sophora subprostrata* Chun et T. Chen, *Belamcanda chinensis* (L) DC, *Scutellaria baicalensis* Georg.
2. **For the anorexia and gastrointestinal side effects:** *Cyperus rotundus* L., *Atractylodis* Rhizoma, *Magnolia obovata* Thunb., *Pinellia ternata* Breitenbach, *Citrus unshiu* Peel
3. **For the treatment of weakness and hematopoiesis dysfunction:** *Astragalus membranaceus* Bunge, *Panax ginseng* C. A. Meyer, *Angelica sinensis* (Oliv.) Diels, *Lycium chinense* mill, *Rehmannia glutinosa* (Gaerth) Lib.

II. ANTINEOPLASTIC DRUGS DEVELOPED FROM TRADITIONAL CHINESE MEDICINE AND HERBAL MEDICINE

The experience of traditional Chinese medicine and folk medicine affords a valuable approach to the development of new antineoplastic drugs.[1,2,5]

1. *Indigofera tinctoria* and Indirubin

Guided by the therapeutic principle of "purging the hepatic substantial fire" of traditional Chinese medicine, a group of hematologists from the Institute of Hematology, Chinese Academy of Medical Sciences, has demonstrated the efficacy of a prescription called Dan Gui Lu Hui Wan for the treatment of chronic myelocytic leukemia. This prescription consists of eleven Chinese medicinal herbs, namely, *Angelica sinensis* (Oliv) Diels, *Aloe vera* L., *Gentiana scabura* L., *Saussurea lappa* Clarke, *Scutellaria baicalensis* Georgi, *Phillodendron chinensis* Schneid, *Coptis chinensis* Franch, *Gardenia jasminoides* Ellis, *Rheum palmatum* L., *Indigofera tinctoria* L. and *Moschus moschiferus* L. Pharmacological studies demonstrated that *Indigofera tinctoria* is active in the inhibition of Walker carcinosarcoma 256 in rats. Phytochemical studies showed that the minor component from *Indigofera tinctoria*, indirubin (Figure 1), is the active principle of this herb. Clinical studies showed a positive result in the treatment of chronic myelocytic leukemia. In view of the fact that the indirubin content of the herb is very low, our chemists have succeeded in the total synthesis of this compound. On the basis of structure-activity relationship studies of these compounds, the second generation of the drug, called N-methylisoindigotin, has been developed recently (Figure 1). Clinical studies indicated that it is more effective than the parent drug with fewer side effects.

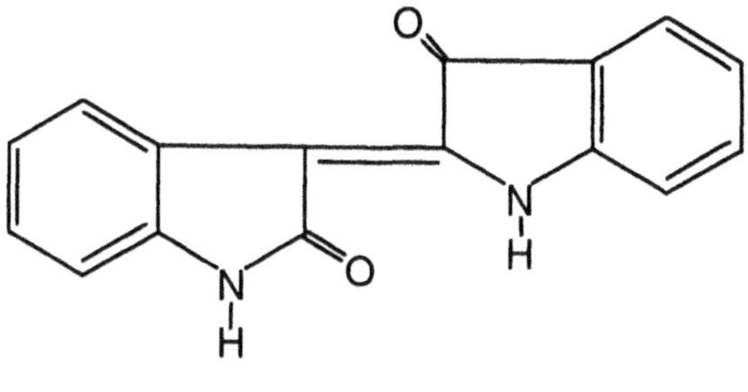

Indirubin

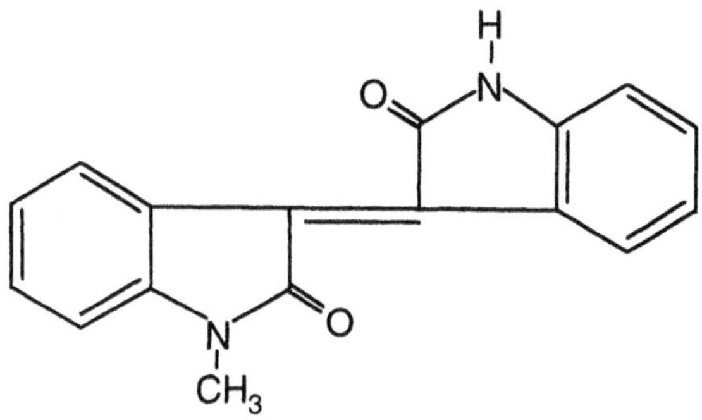

***N*-methyl isoindigotin**

Figure 1 Chemical structure of indirubin and N-methyl isoindigotin.

2. *Iris pallasii* and *irisquinone*

The seeds of *Iris pallasii* Fischer var. chinensis have been used in folk medicine in China for the treatment of malignancies. The chemists of the Tianjin Institute of Materia Medica have isolated an active principle called irisquinone from the seeds of this plant. Chemically, irisquinone is 6-methoxy-2-cis-heptadecenyl-1,4-phenylquinone (Figure 2). Experimental therapeutic studies have shown that it is effective against transplantable tumors in mice. At the moment it is undergoing phase I clinical studies.

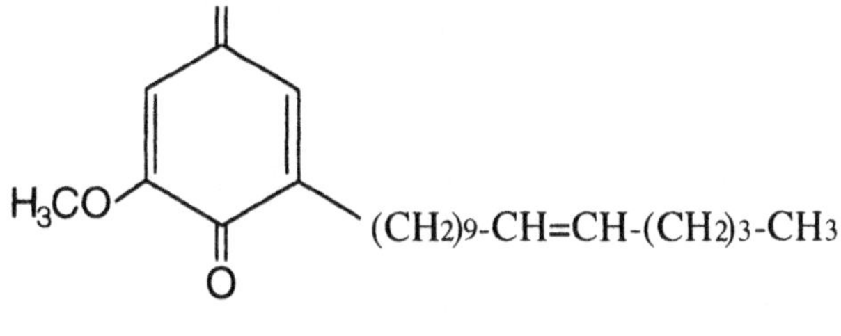

Figure 2 Chemical structure of irisquinone.

3. *Cephalotaxus hainanensis*, Harringtonine and Homoharringtonine

Cephalotaxus fortunei Hook F. was used by folk practitioners in Fujian Province, China, for the treatment of neoplastic diseases. Powell *et al.* reported that the active principles of *Cephalotaxus harringtonia* (Forbes) Koch are active against rodent leukemias. On the basis of these reports, we have been working on a systemic investigation of *Cephalotaxus hainanensis* Linn, which is indigenous to China, in terms of resource, chemistry, pharmacology and toxicology studies. So far, more than 16 alkaloids and two lactones have been isolated from the stem bark of this tree and characterized by infrared spectrometry, mass spectrometry, nuclear magnetic resonance spectrometry and X-ray diffraction analysis. Their chemical structures are shown in Figure 3. All these compounds have been screened on rodent tumor systems. It was found that the major component of these alkaloids called cephalotaxine is inactive but the esters of this alkaloid, homoharringtonine, harringtonine, isoharringtonine and deoxyharringtonine, are active against rodent leukemias.[8, 9] To explore the mechanism of action of harringtonine, we studied its effect on the incorporation of radioactive precursors of DNA, RNA and protein. Experiments demonstrated that the incorporation of ^{3}H-leucine into the protein of L-1210 cells was inhibited very quickly and severely by harringtonine, followed by the inhibition of ^{3}H-thymidine incorporation into DNA. It has no effect on RNA synthesis. On the basis of preclinical and toxicological studies, harringtonine and homoharringtonine were recommended for clinical trial. Clinical studies indicated that harringtonine and homoharringtonine are active against acute myelocytic leukemia. Through multiple center studies, these drugs are widely used in China for the treatment of acute myelocytic leukemia.[10,11] In a HOAP Protocol (H:

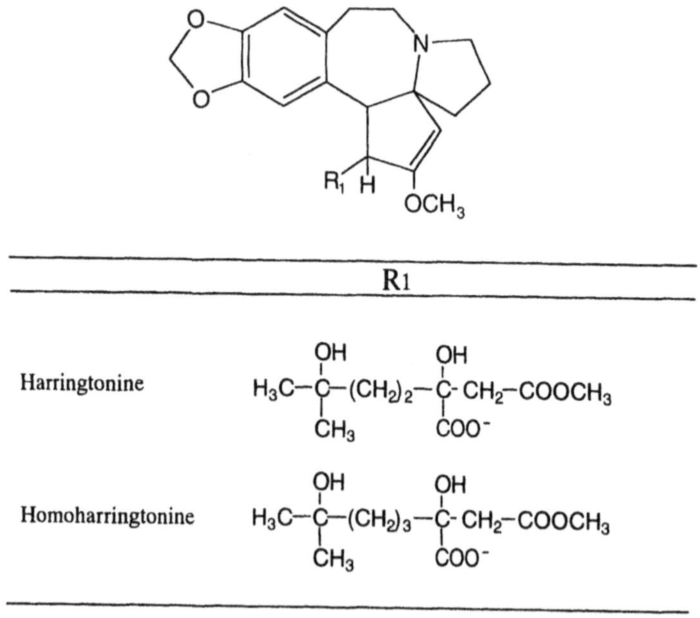

Figure 3 Chemical structure of harringtonine and homoharringtonine.

homoharringtonine, O: oncovine, A: cytarbin or AraC, P: prednisone) of combination chemotherapy developed in China for acute myelocytic leukemia, the complete remission rate is as high as 50%. It is interesting to find that harringtonine and homoharringtonine are good inducers of the programmed death of human acute promyelocytic leukemia cells *in vitro* at a concentration of $1–2 \times 10^{-7}$ mol/L in terms of typical apoptosis and the DNA "ladder".

4. *Taxus chinensis* and Taxol

Originally taxol was isolated from the stem bark of *Taxus brevifolia* by Wani *et al.* Horowitz *et al.* reported that it has a unique mechanism of action in terms of the promotion of tubulin polymerization. On the basis of a field survey of the resources of Taxus species in the mainland of China, *Taxus chinensis* (Pilg.) Rehd. was found to be the most interesting species. So far, more than 14 chemical compounds, including cephalommanine, taxol and baccatin III, have been isolated and characterized in the Insitute of Materia Medica, Chinese Academy of Medical Sciences in Beijing (Figure 4). Pharmacological studies demonstrated that taxol and cephalommanine are the two major active principles for the inhibition of

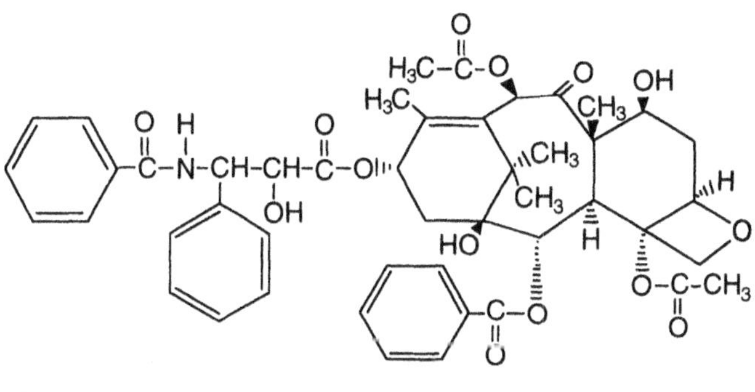

Taxol

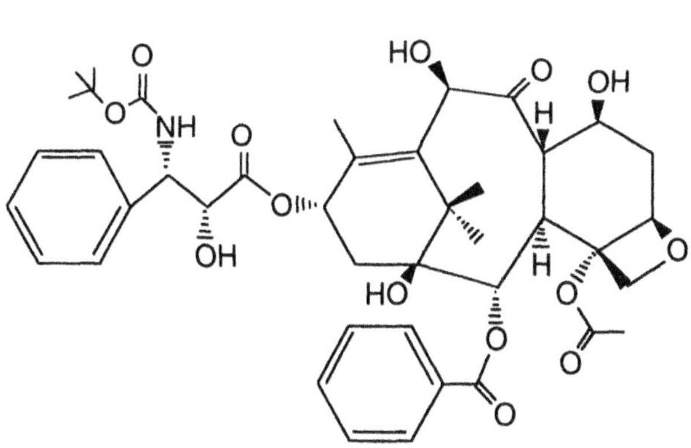

Taxotere

Figure 4 Chemical structure of taxol and taxotere.

cancer. Experimental therapeutic experiments indicated that taxol exhibited remarkable inhibition on a variety of human cancer cell lines. *In vivo* studies showed that taxol is effective against rodent tumors. Biochemical and pharmacological studies demonstrated that taxol promotes microtubule assembly and reduces the lag time for the assembly *in vitro.* Flow cytometry has shown that taxol decreased the cell population in the Gl phase at a concentration of 25 nmole/L, while the cell population in the G2+M phase was increased significantly. In addition, a polyploidy cell population with 8C DNA content appeared on the DNA histogram. The action of taxol is time dependent as well as concentration dependent.[12–14] The home-made taxol is in phase II studies in China now.

5. *Camptotheca accuminata* and Camptothecin

Camptotheca accuminata Decne is indigenous to the mainland of China and has been used in folk medicine in Zhejiang Province for the topical treatment of psoriasis. Wall *et al.* reported that the active principle of the stem bark from *Camptotheca accuminata* is camptothecin (Figure 5) which is active for the inhibition of tumor

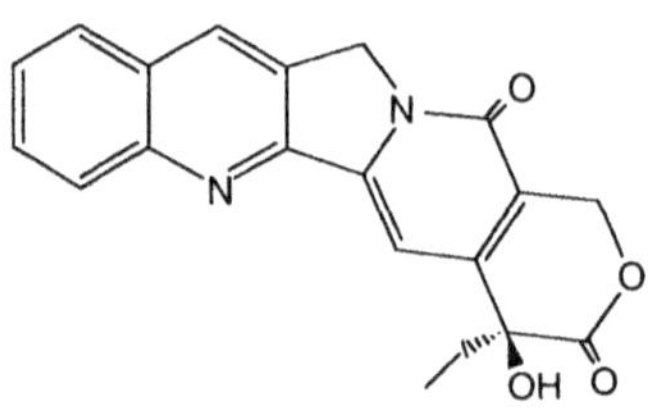

Camptothecin (CPT)

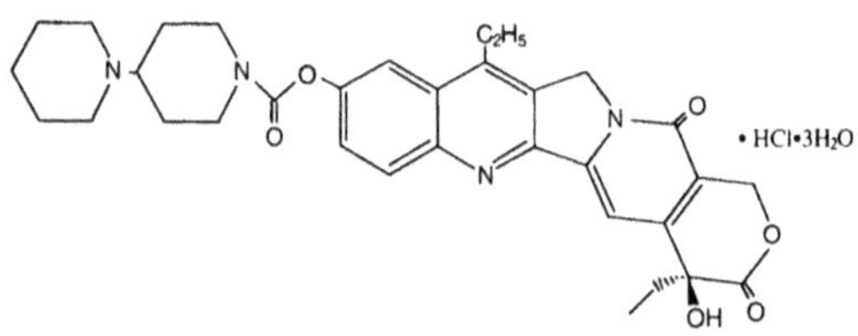

Irinotecan Hydrochloride (CPT-11)

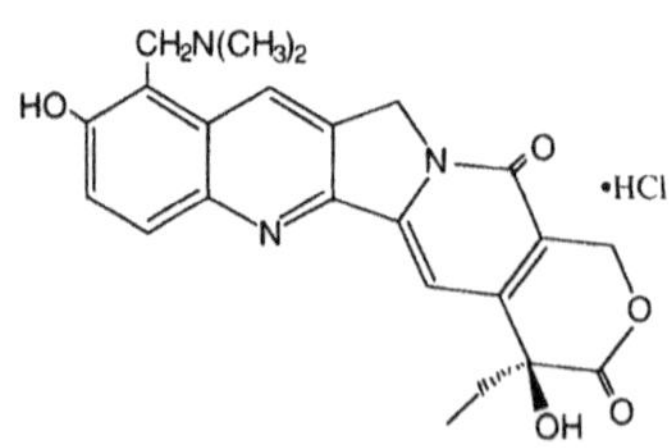

Topotecan (TPT)

Figure 5 Chemical structure of camptothecin, irinotecan and topotecan.

growth in animals. The scientists in Shanghai have dedicated themselves to chemical and pharmacological studies of this plant. It was reported that camptothecin and 10-hydroxycamptothecin exhibited a wide antitumor spectrum in rodent tumors and 10-hydroxycamptothecin is more active and less toxic than camptothecin.[15] Recently a group of Japanese scientists has succeeded in the development of new camptothecin derivatives called CPT-11 which offer very good prospects for the treatment of lung cancer.

III. INDUCERS OF CELL DIFFERENTIATION OF CANCER FROM TRADITIONAL CHINESE MEDICINE

Cancer can be considered as a disorder of cell differentiation which is readily illustrated by hematologic neoplasms. Oncogenic conversion, defined as the arrest of differentiation without loss of proliferation capacity, can occur at any of the intermediate maturation steps. Consequently the maturation-arrested cells continue to proliferate and a population of immature "cancer" cells emerges and gives rise to adverse clinical manifestations. It is well known that the traditional target of cancer chemotherapy is to kill the cancer cells. The historical achievements of cancer chemotherapy demonstrated that this approach is fruitful, but the drawback of this approach is poor selectivity and severe side effects, which limited its further development. In addition, the bad news is that cytotoxic anticancer drugs do very little to the bulk of cancers, for example large bowel cancer, pancreatic cancer, liver cancer, soft tissue sarcomas, stomach cancer, non-small cell lung cancer, head and neck cancer, melanoma and cervical cancer. Therefore, it is the general view that further major therapeutic advance is unlikely to come from empirical studies of new candidate agents developed by the traditional cytotoxic approach.

Recently, it was found that some neoplastic cells can be induced to differentiate to normal or normal-like cells *in vitro*, for example teratoma, neuroblastoma, leukemia and melanoma. In particular, the finding that human acute promyelocytic leukemia could be induced to complete remission by all-trans-retinoic acid opened a new frontier — the so-called differentiation therapy of cancer. This approach was regarded as a new modality of cancer treatment.

We have been working for more than ten years to try to find new inducers of cancer cell differentiation from natural products. So far, more than one thousand samples have been screened by the human acute promyelocytic leukemic cell line (HL-60) and several interesting inducers from traditional Chinese medicine have been found.

1. *Pueraria lobata* and daidzein

Pueraria lobata (wild) Ohwi has been used for the treatment of hypertension and angina pectoris for many years in traditional Chinese medicine. It was reported that the two major active principles are daidzein and puerarin (Figure 6). There was no

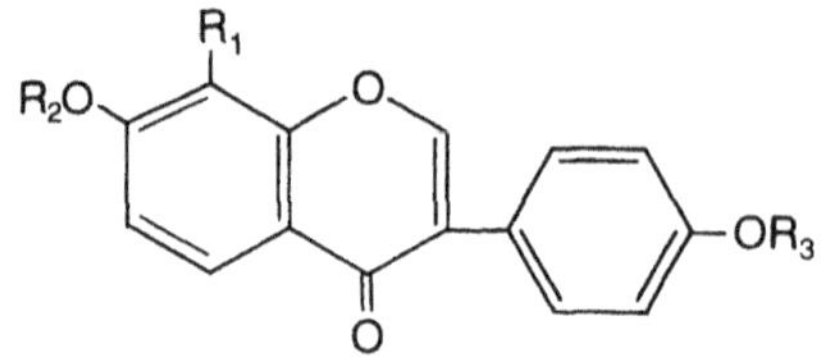

	R_1	R_2	R_3
Daidzein	**H**	**H**	**H**
Daidzin	**H**	**glucose**	**H**
Pueralin	**glucose**	**H**	**H**

Figure 6 Chemical structure of daidzein and its derivatives.

information on the activity of these compounds on cancer cell differentiation. Our experiments demonstrated that daidzein was active in the induction of cell differentiation of HL-60 cells in terms of morphological change, NBT reduction activity and flow cytometry. It was shown that at a concentration of 18 and 22 μg/ml daidzein induced the cells to differentiate along the myelocyte lineage. Flow cytometry demonstrated that daidzein exhibited a Gl arrest in cytokinetics of HL-60 cells. At a concentration of 14 μg/ml of daidzein, 63.6% of the cells were arrested in the Gl phase.[16] It is interesting to note that the expression of oncogene c-myc and c-myb was inhibited by daidzein but the expression of oncogene c-fos was enhanced by this compound which coincided with the differentiation of HL- 60 cells.

Interestingly, there is a report that the majority of flavones extracted from *Citrus aurantium nobilis* pericarpium and fruit peel of *Citrus reticulata* Blanco exhibited differentiation activity in HL-60 cells.[17]

2. *Boswellia carterii* and Boswellic Acid

Boswellia carterii Birdw. has been used for many years in traditional Chinese medicine for activating blood circulation and relieving pain. Liu *et al.* reported that one of the active principles of the resin of *Boswellia carterii* (alpha-boswellic acid acetate, BC4) is a potent inhibitor of DNA topoisomerase I and II. In searching for new inducers of cell differentiation of HL-60 cells, it was found that the active principle of this plant, BC4 (Figure 7), can induce cell differentiation of HL-60 cells in terms of NBT reduction, phagocytosis and morphological change. At a concentration of 10 μg/ml of BC4, 80% of the cells become NBT positive and 90% of the cells showed phagocytosis of polysterene latex beeds.

3. *Panax ginseng* and ginsenoside Rh_2

Panax ginseng is one of the most famous drugs used since ancient times. It was

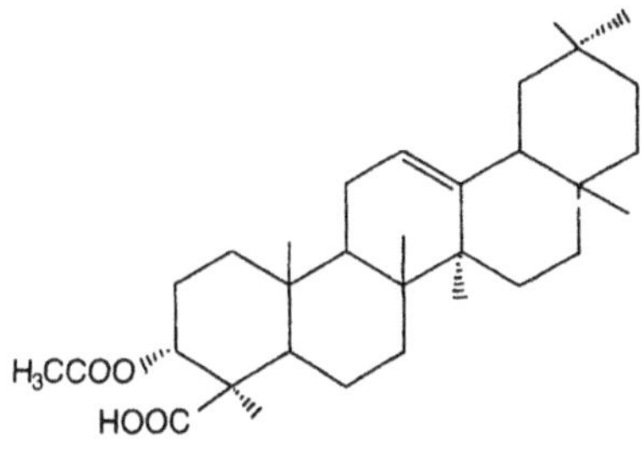

Bc-4-I

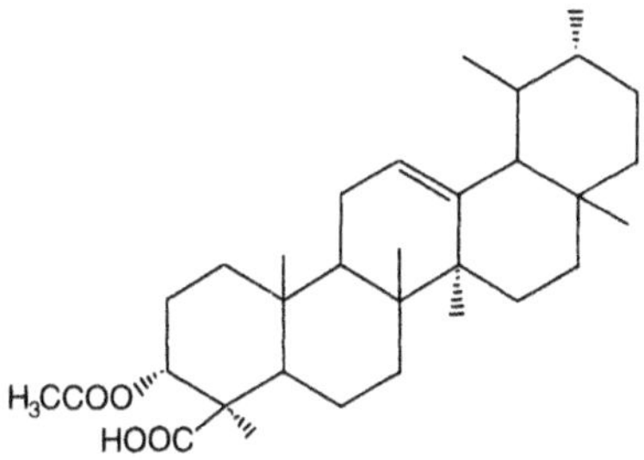

Bc-4-II

Figure 7 Chemical structure of acetylboswellic acid.

ranked as the "best drug" in *Shen Nong Ben Cao Jing*. Pharmacological studies have demonstrated that *Panax ginseng* is able to strengthen resistance to harmful stress of the host and stimulate phagocytosis in normal and tumor-bearing mice. It is interesting to note that Odashima *et al.* have reported that ginsenoside Rh_2 (Figure 8), one of the characteristic components of red ginseng, could exhibit a significant induction of cell differentiation of melanoma B-16 cells *in vitro*.[18] We have found that at a concentration of 8 μg/ml it inhibited cell growth of B-16 melanoma completely, and the morphology of these cells changed so that they could be differentiated. Flow cytometry showed that most of the melanoma cells were arrested in the Gl phase.[21]

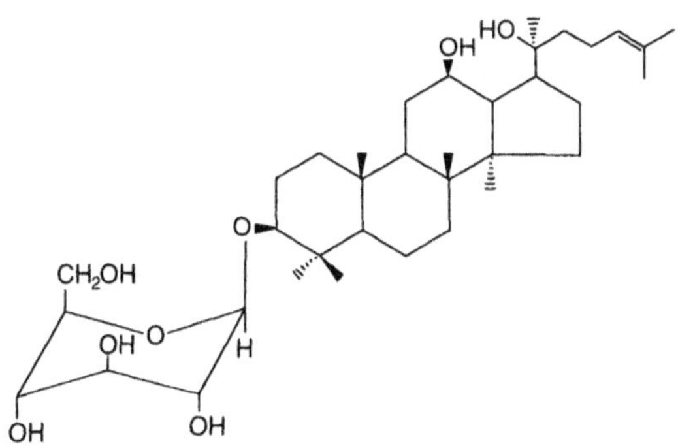

Figure 8 Chemical structure of ginsenoside Rh_2

IV. CANCER CHEMOPREVENTION AND CHEMOPREVENTIVE AGENTS

It is generally agreed that the ultimate purpose of cancer research is to lower the mortality and incidence of cancer. In the earliest ancient classical book of traditional Chinese medicine, called *Huang Di Nei Jing*, a famous principle of medicine, that the role of a good doctor is to treat before sickness, was described. In spite of the great success of early diagnosis and treatment of cancer, the total incidence of cancer per year is going up. Therefore, how to reduce the incidence of cancer is a big challenge for the medical community and the pharmaceutical community as well. The aim of cancer chemoprevention is to use natural products or chemical compounds to reverse or inhibit the malignant transformation of cells and prevent the metastasis of cancer.[19–29] In the ideal case, chemoprevention would be the less painful, more economic and more reasonable approach for the control of cancer. According to the point of view of traditional Chinese medicine, "medicine and food are from the same source". Many edible plants and medicinal plants have been described in the classics of Chinese Materia Medica. They could provide a potential resource for the research and development of chemopreventive agents of cancer.[27–29] In recent years the antimutagenic activity of plants became of considerable interest. It is now clear that higher plants contain a variety of preformed secondary metabolites that represent a structurally diverse array of antimutagenic compounds. Japanese scientists have reported that an extract of green tea decreases the mutagenic activity of carcinogen MNNG and tumor promotion induced by phorbol. Several groups of Chinese scientists have confirmed these results.

We have been interested in searching for chemopreventive agents of cancer since the late 1970s. On the basis of established methodologies of experimental chemoprevention *in vitro* and *in vivo*, we have been engaged in the screening of chemopreventive agents for cancer in terms of antimutagenesis, antipromotion and anticarcinogenesis. It was found that in addition to the retinoids synthesized by our Institute, some traditional Chinese medicine showed very promising results.[22–26]

1. Red Ginseng

Traditional Chinese medicine has been a source of nourishing tonics which were also called Fu Zheng drugs, that is, drugs for reinforcing body resistance to pathogens or internal pathological factors. Red ginseng is a processed *Panax ginseng* which was regarded as a typical tonic in traditional Chinese medicine. Our studies demonstrated that red ginseng extract can inhibit skin papillomas induced by carcinogen DMBA at doses of 100, 200 and 400 mg/kg, p.o. Under the action of red ginseng extract, the incidence of papilloma decreased significantly and the number of papilloma per mouse also decreased dramatically. These results were coincident with the epidemiological observation of Korean scientists in terms of chemopreventive activity of cancer in humans.

2. *Curcuma longa* and Curcumin

Curcuma longa L. is a drug of choice for "eliminating stagnation and activating the blood" in traditional Chinese medicine. It is used to relieve abdominal pain and also as a food additive. It has antibacterial activity *in vitro* and the major active principle of this plant is curcumin (Figure 9). Experiments have been done in our laboratory to explore the chemopreventive potential of curcumin for cancer.[30] Experiments indicated that curcumin can inhibit the His revertants induced by methyl methane sulfonate in Salmonella/microsome plate incorporation assay. It also decreased the micronucleus formation induced by cyclophosphamide. As an antipromotor of cancer it decreased the ear edema caused by croton oil in mice.

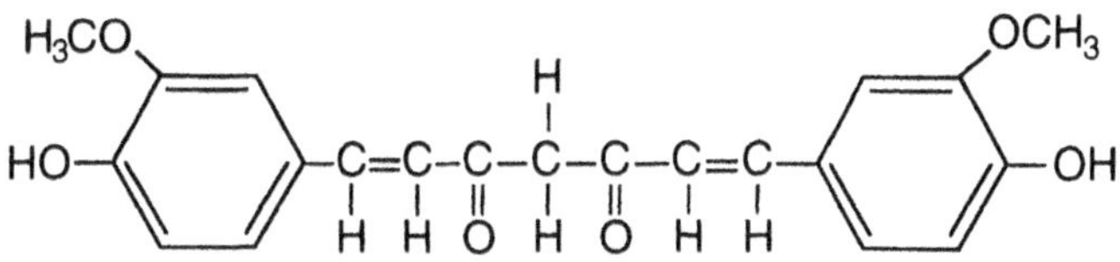

Figure 9 Chemical structure of curcumin.

3. *Glycyrrhiza uralensis* and Glycyrrhetinic Acid

Glycyrrhiza uralesis Fisch. is widely used as an antidote in traditional Chinese medicine. It is also used as a food additive. Nishino *et al.* have reported that one of the predominant components, glycyrrhetinic acid, exhibited antipromotion activity in a two-stage model of chemical carcinogenesis and antimutagenic activity in Salmonella typhimurium TA 98 and TA 100 systems. Our experiments demonstrated that glycyrrhetinic acid inhibited the ear edema induced by croton oil in mice. It also inhibited the activity of epidermal ornithine decarboxylase at a dose of 50–200 mg/kg for three days. It is interesting that the rapid DNA damage induced by carcinogen benzopyrene was significantly protected by this compound.[31]

4. *Rhus chinensis* and Chinese Gallotannin

Rhus chinensis and its major component, Chinese gallotannin, is widely used as an astringent in traditional Chinese medicine as well as in Western medicine. Recently it was found that Chinese gallotannin has a variety of biological activity in terms of antitumor, antivirus and antifungal activity. Pharmacological studies in our laboratory demonstrated that Chinese gallotannin inhibited the malignant transformation of V79 cells induced by benzopyrene *in vitro* at a concentration of 5–20 μg/ml.[32] It is interesting to note that under the action of Chinese gallotannin, bezopyrene-induced pulmonary adenoma in A/J mice was inhibited profoundly. The incidence of tumors and the number of tumors per mouse in the treated group were significantly lowered. Other experiments indicated that Chinese gallotannin also inhibited DMBA/croton oil-induced skin papilloma in mice. The latent period of the tumor was prolonged and the incidence of the tumor was also decreased in a dose-dependent manner. In the transplantable tumor system, Chinese gallotannin exhibited a significant antitumor activity in sarcoma 180, Lewis lung carcinoma and colon carcinoma 26 bearing mice. Further studies showed that Chinese gallotannin is a strong free radical scavenger in terms of O_2^- and DPPH. Interestingly, it also inhibited the metabolic activation of benzopyrene by inhibiting the P-450 Al and AHH enzyme activities. Chinese gallotannin showed a strong antimutagenic activity *in vitro* and protected the chromosome damage induced by cyclophosphamide.

5. Recipe "Antitumor B"

Recipe "Antitumor B" is a complex recipe of traditional Chinese medicine which consists of six medicinal herbs, *Sophora subprostrata* Chua et Chen, *Patrinia villosa* Thunb., *Dictamnus dasyearpus, Discorea bulbifera* L., *Purimura vugalis* L. and *Polygonum historia* L. Lin *et al.* have reported that the "Antitumor B" recipe exhibited a dramatic inhibition of chemical carcinogenesis of the esophagus induced by a carcinogen nitrosamin in rats.[33] Wei *et al.* following Lin *et al.* have reported that recipe "Antitumor B" has a significant inhibitory effect on DMBA-induced buccal carcinoma in hamsters. Fan *et al.* published that the recipe also inhibited the incidence of bladder cancer induced by N-butyl-(4-hydoxybutyl) nitrosamide (BBN) in rats. On the basis of toxicological studies, this recipe was recommended for clinical trial in a high-risk area for esophaged cancer in Linxian County, Henan Province, China. It was reported that during long-term administration of this recipe to a high-risk population for five years in the above-mentioned area, the incidence of esophaged cancer was reduced significantly.[34] In addition, there were no severe side effects. Although it is a preliminary field study which needs further confirmatory study, it is a good start to using traditional Chinese medicine in preventing cancer.

CONCLUSION

The general feature of cancer treatment in China is that cancer patients are usually diagnosed by Western medicine and treated by a combination therapy which includes surgery, radiotherapy, chemotherapy and traditional Chinese medicine. It is well recognized that traditional Chinese medicine is useful for the amelioration of side effects caused by radiation and/or chemotherapy. It is also useful for the promotion of recovery from surgery in cancer treatment.

The experience of traditional Chinese medicine and folk medicine affords a valuable approach for the research and development of new antineoplastic drugs as illustrated by the example of indirubin from Dan Gui Lu Hui Wan and irisquinone from *Iris pallasii.* It seems that there is a good opportunity to explore some chemopreventive herbs for cancer and some new inducers of cancer cell differentiation.

The application of chemotaxonomic principle of related species of plants in reference to the experience of traditional Chinese medicine and folk medicine, has furnished an alternative approach which has been used extensively and successfully for the development of new anticancer drugs. The studies of *Cephalotaxus hainanensis* and *Taxus chinensis* provide two typical examples in this respect. The continuous search for new drugs from plants will be an important frontier in the treatment and prevention of cancer.

REFERENCES

1. R. Han, Highlight on the studies of anticancer drugs derived from plants in China. *Stem Cells,* **12** (1), 53–63 (1994).
2. R. Han. *Chemical Prevention and Pharmacological Treatment of Cancer* (R. Han, ed.), Beijing Medical University and Peking Union Medical College Press, Beijing (1993).
3. B.K. Holland, Prospecting for drugs in ancient texts. *Nature,* **369**, 702–703 (1994).
4. T.T. Hien, and N. J. White, Qinghaosu. *The Lancet* **341**, 603–607 (1993).
5. I.H. Krakoff, Progress and prospects in cancer treatment: the Karnofsky legacy. *Clin. Oncol.* **12**, 432–438 (1994).
6. D.Z. Zhang, Trends in the prevention and treatment of side-effects of chemotherapy and radiotherapy by traditional Chinese medicine. *Cancer* **4** (4), 240–242 (1985).
7. H. Inui, Ameriolation of side effects of lung cancer chemotherpy by Hochu-ekki-to. *Kampo and Newest Therapy,* **2** (1), 56–60 (1993).
8. R. Han and X.-J. Ji, The antitumor effects and pharmacolocal action of harringtonine. *Chinese Medical J.* **3** (2), 131–136 (1977).
9. R. Han, and J. Roboz, Quantification of harringtonine and homoharringtonine in serum by high performance liquid chromatography. *J. Chromatog.* **233**, 203–211 (1982).
10. L. Li, L.J. Xia, C. Jiang and R. Han, Induction of apoptosis by harringtonine and homoharringtonine in HL-60 cells. *Yao Hsueh Hsueh Pao Acta Pharmaceutica Sinica.* **29** (9), 667–672 (1994).
11. T.C. Chou, F.A. Schmid, F.S. Philips and R. Han. Uptake, initial effects and chemotherpeutic efficacy of harringtonin in murine leukemic cells sensitive and resistant to vincristin and other chemotherapeutic agents. *Cancer Res.* **43**, 3074–3079 (1983).
12. R. Han, Development of new anticancer drug Taxol. *Chinese Tumor Clin.* **19** (2), 223–226 (1993).
13. W.M. Chen, Chemical constituents and biological activity of plants of the Taxus genus [Review] [Chinese]. *Yao Hsueh Hsueh Pao — Acta Pharmaceutica Sinica.* **25** (3), 27–40 (1990).
14. R. Han, Z.R. Li, *et al.* Pharmacological studies of new anticancer drugs, Taxol. *Tumor Treatment (Bulletin of National Tumor Prevention and Treatment Meeting).* 959–968 (1989).
15. W.J. Slichenmyer, E.K. Rowinsky, R.C. Donehower and S.H. Kaufmann. The current status of camptothecin analogues as antitumor agents. *J. Natl. Cancer Inst.* **85** (4), 271–291 (1993).
16. R. Han, L. Jiao and H.Y. Liu. Effects of S86019, an active component from Pueralia lobata on cell differentiation and cell cycle traverse of HL-60 cells. *Chinese J. Cancer Res.* **285**, 51–53 (1990).
17. Y.K. Jing, K. Nakaya and R. Han. Differentiation of promyelocytic leukemia cells HL-60 induced by daidzein *in vitro* and *in vivo. Anticancer Res.* **13**, 1049–1054 (1994).
18. S. Odashima, Y. Nakayabu, N. Honjo, H. Abe and S. Arichi. Induction of phenotypic reverse transformation by ginsenosides in cultured Morris hepatoma cells. *Europ. J. Cancer* **15**, 885–892 (1979).
19. R. Han, Trends of antitumor drugs. In: *Prevention and Treatment of Tumor development* (Li Bing, ed.), Beijing Medical University and Peking Union Medical College Press, pp. 81–93 (1994).
20. T. Kinoshita, U. Sankawa *et al.* Induction of differentiation in murine erythroleukemia cells by flavonoids. *Chem. Pharm. Bull* **41** (4), 714–719 (1993).
21. L. J. Xia and R. Han, Induction of cell differentiation of melanoma B-16 by ginsenoside Rh_2. *Acta Pharmaceutica Sinica.*
22. R. Han, L. Jiao *et al.* Evaluation of N-4 (carboxyphenyl) retinamide as a cancer prevention agent and a cancer chemotherapeutic agent. *In Vivo* **4**, 153–160 (1991).
23. R. Han Chemical prevention of carcinogenesis new way for preventing and treating tumors. *Chinese Tumor Clin.* 224–227 (1991).
24. S. M. Lippman, S. E. Benner and W. K. Hong. Chemoprevention: Stragtegies for control of cancer: *Cancer* (suppl.) **72** (3), 984–990 (1993).

25. W. Bollag and E.E. Holdener. Retinoids in cancer prevention and therapy. *Annals of Oncology 3*, 513–526 (1992).
26. L.W. Wattenberg. Prevention–therapy–Basic sciences and the resolution of the cancer problems : Presidential Address. *Cancer Res.* **53**, 5890–5896 (1993).
27. J.M. Cassady. Natural products as a source of potential cancer chemotherapeutic and chemoprevention agents. *J. Natural Products*, **53** (1), 23–41 (1990).
28. Jui Han (Han Rui). Highlights of the cancer chemoprevention studies in China. *Preventive Medicine* **22**, 712–722 (1993).
29. S.M. Lippman, W. N. Hittelman, R. Lotan, U. Pastorino and W.K. Hong. Recent advances in cancer chemoprevention. *Cancer Cella* **3** (2), 59–65 (1991).
30. X.G. Chen, Z.D. Fu and R. Han, Cancer chemopreventive activity of curcumin. *Chinese Medical Sciences J.* (to be published).
31. X.G. Chen and R. Han Effect of glycyrrhetinic acid on DNA damage and unscheduled DNA synthesis induced by benzopyrene, *Chinese Medical Sciences J*, **4** (1), 16–19 (1994).
32. X.G. Chen and R. Han. Studies on the cancer chemopreventive activity of 9201 from Rhus chinensis and its mechianism of action. *Chinese Medical Sciences J.* (to be published).
33. P.Z. Lin, J.S. Chang, S.G. Cao and R. Han. Second line prevention from esophageal cancer inhibitory therapy to block precancerou lesions. *Chinese J. Cancer Res.* **1** (1), 37–46 (1988).
34. P.Z. Lin. Studies on medicamentous inhibitory therapy for esophageal precancerous lesion—3- and 5-year inhibitory effects of antitumor-B, retinamide and riboflavin. *Proceedings of the Chinese Academy of Medical Sciences and the Peking Union Medical College* **5** (3), 121–129 (1990).

3. RATIONAL USE OF MEDICINAL PLANTS IN PRIMARY HEALTH CARE

NUNTAVAN BUNYAPRAPHATSARA

Medicinal Plants Information Center, Faculty of Pharmacy, Mahidol University, Bangkok 10400, Thailand

HISTORICAL BACKGROUND

Medicinal plants have been important components of Thai traditional medicine since this was introduced into Thailand 700–1,000 years ago. The Ayurvedic system was adapted to Thai culture and tradition to create a distinct Thai traditional medicine. However, the recipes and methods of preparation have been transferred through generations within the family and only a few written documents exist. The oldest is *Tamraa Phra Osot Phra Naraai* (the medical text of King Naraai). This text, which contained prescriptions for medicines presented to the king (1657–1688), was probably compiled during the reign of King Nippan Thaai Sa (1709–1733) or of King Borommakot (1733–1758).[1,2]

From the mid-eighteenth century, the practice of traditional medicine was interrupted, because most of the texts and recipes were lost or destroyed during wars. The attempt to compile and revive traditional medicine was made by King Rama I and King Rama II. The old recipes and medical knowledge were restored and displayed on the wall and stone plaques at Wat pho or Watphotharam, the temple which is now known as Watprachetuphonwimolmangklaram Rajworawihaan. Later, in 1871, the royal texts were collected, revised and published in Samut Khoi*. The recipes from all these texts and stone plaques serve as the basis for the texts used today.

The use of medicinal plants declined after the traditional medical program at Siriraj Hospital, the first medical school established by King Rama V (1887), was eliminated in 1903 by H.S.H. Krom Phrachainaatnarainthorn, the director of the school. At present, we in Thailand rely almost totally on imported drugs, which leads to many problems, as follows:

1. Large amounts of foreign exchange are lost.
2. The cost of drugs is too high for people with low incomes.
3. Western medicines are available in various dosage forms and are easy to take. The ease of using modern medicines allows uneducated people to abuse the drugs and to take them improperly and unnecessarily.
4. A shortage of drug supplies may arise in some situations, such as wartime.

* Samut Khoi is a long concertina-like book made from a strip of paper and pleated so that it can be opened like a book. This paper is made from the bark of Khoi (*Streblus asper* (Retz.) Lour, Moraceae.

5. The government cannot provide equal health care for rural and urban areas, especially in terms of drug supplies, owing to the high cost of imported drugs.

All these problems prompted the Thai government seriously to consider revitalizing the use of medicinal plants, which are valuable local resources. The first development plan was set up in 1982. A study by the Faculty of Pharmacy, Mahidol University, suggested three approaches:

1. Development of medicinal plants for primary health care.
2. Development of medicinal plants for the pharmaceutical industry.
3. Development of medicinal plants for export.

Among these three approaches, the development of medicinal plants for primary health care showed the highest potential.[3]

The Thai government follows the guideline set by the World Health Organization that developing countries should consider herbal medicine as an alternative method for health care.[4] It is not only sensible economically, but this medicine is accessible to practitioners at all times. However, the information needed for the development plan is inadequate owing to the loss of texts during wars and the tendency to keep and transmit the knowledge within the family. The Thai government started the development program with the available information.

PRINCIPLES

Briefly, the principles of Thai traditional medicine are as follows: the human body consists of four primary elements, namely earth, water, air or wind, and fire. Ether, which is the fifth element of Ayurveda, is seldom mentioned in the literature published on Thai traditional medicine. There are 42 main body elements, 32 being substantial. Of the 32 substantial elements, 20 are classified as earth elements, meaning that their chief quality is that of earth, or solidity. The other 12 are liquids, or water elements. The remaining 10 are six elements of wind, having the quality of movement, and four of fire, having the quality of heat.

The main causes of disease are abnormalities of body elemenets arising from changes in the season of the year, the age of the patient, the time of day when the person became ill and the place where the patient lives.

Medicinal plants and massage are the important tools in treatment. The word "drug" in Thai traditional medicine refers to any recipe containing at least two ingredients that come from animals, plants or minerals. These raw materials are called "haesatwatthu". The role of drugs in traditional medicine is to correct the disorder of the four elements so that the patients become well again. Each ingredient is added for a specific purpose; however, only the practioners know this purpose and they kept it secret.

The activities of Phaesatwatthus or ingredients are related to their "tastes". Three major tastes are cold, hot and astringent. Medicinal plants that possess cold, hot and astringent tastes are used for abnormally of fire, wind and water elements, respectively. The practioners compound the complex recipes for each patient based on medical interviews and using the above knowledge. The patient is usually asked to come back after a few days. If the improvement is not satisfactory, the practitioner will modify or change the recipe. Complex recipes are commonly prescribed; single plant prescriptions are used only rarely.

SINGLE PLANT VERSUS COMPLEX RECIPES

There was a long debate among the government committee members about whether the development of medicinal plants for primary health care should start with single plant or complex recipes. If we consider the fact that medicinal plants and traditional medicine have played an important role in the survival of mankind for centuries, complex recipes should be the best choice. The difficulty is selecting the best recipe for each disease or symptom. The survey indicated that hundreds of different recipes were used for one disease or symptom, according to the practitioner. Moreover, no clinical records are available for evaluation. The preparation of the drug is also a problem; most of the recipes are family secrets and details of the selection of raw materials and the amounts used are not available.

The advantages of using single plants to start the program are as follows:

1. It is easier to clarify or trace the cause of any abnormal incidents that might occur.
2. Scientific evidence for confirming the efficacy of the medicinal plant is easily integrated.
3. Preparation of the drug is simpler.
4. It is easier to carry out clinical research for further development.

For all these reasons, the Thai government decided to start the program with single plant recipes. Once herbal medicines have been adopted by doctors and health workers, the use of complex recipes will be promoted.

INTEGRATION OF SCIENTIFIC EVIDENCE

In countries like Thailand, where western medicine is the major health care system, traditional medicine is opposed by doctors and health care personnel. For traditional medicine to be revitalized, the efficacies claimed for its application need to be proven scientifically. Pharmacological data can be obtained through searchers of various databases. The plants for which there is scientific evidence of beneficial activity are selected. Even though many plants are reputed to possess healing properties, only those with pharmacologically confirmed activity are recommended for use in primary health care.

The reliance on scientific data is not only to confirm the efficacy but also for safety. The plants selected should have passed toxicity tests, including mutagenic and teratogenic tests, or they should be commonly consumed as food. It should be noted that in the case of external use the evaluation of toxicity data is not as strict as in the case of internal use.

Let us look at some examples. Guava leaves (*Psidium guajava* Linn) have been used to mask the smell of alcohol on the breath,[5–9] for wound healing,[10,11] the treatment of diarrhea,[5–14] dysentery,[5,6,10,12,15,16] gum inflammation,[11] chronic wounds,[17] infectious wounds,[15,17] ingrowing nails,[8] skin diseases,[18] mucous bloody stools,[5,6,12,15] and allergy to mosquitoes;[14] they have also been applied as a pus and lymph absorber[15,17,19,20] and as a deodorant.[19–24]

Pharmacological studies of guava leaves showed antispasmodic properties that were due to inhibition of acetylcholine activity.[25,26] It was found later that quercetin and quercetin-3-arabinoside were the active components.[26] Moreover, the leaf extract exhibited antibacterial activity against Shigella dysenteriae.[27] In a clinical trial of patients infected with Vibrio cholera, guava leaves showed higher potency than tetracycline.[28] All these confirm the efficacy of guava leaves as an antidiarrheal agent. Although extensive toxicity data are not available, the clinical trial showed no toxic effects. Young guava fruit contain tannin which possessse antidiarrheal activity;[29] this is also recommended for the treatment of diarrhea.

Another example is *Andrographis paniculata* (Burm. f.) Nees, which is used to promote health, for the treatment of sore throat,[30] abscess, herpes zoster and herpes simplex. It is also used as an antipyretic,[31] antidysenteric, antidiarrheal and anti-inflammatory agent.[30] A search for pharmacological data found evidence of antispasmodic activity of the alcohol extract,[32,33] which is due to 14-deoxy-11,12 dihydroandrographolide, andrographolide, neoandrographolide[34–36] and deoxy-andrographolide.[35,36] The alcoholic extract, andrographolide, deoxyandrographolide and neoandrographolide showed antibacterial activity.[37–39] Consequently, diterpenes from *A. paniculata* were tested in a clinical trial in patients with infectious diarrhea and Vibrio cholera. Satisfactory results were obtained in the diarrhea cases but the leaf powder was not as effective as tetracycline against V. cholera.[40] The acute and subchronic toxicity were investigated and no toxicity was observed.[41,42–44] All this evidence suggests that *A. paniculata* should be promoted as an antidiarrheal agent.

A. paniculata also showed antiinflammatory[32,45,46] and immunostimulatory activity.[47] It was recommended for the treatment of sore throat but only for non-infectious cases: it should not be used for infectious sore throat because it fails to exhibit systemic antibacterial activity.

The scientific evidence obtained in this way is sufficient for the application of these herbal medicines in primary health care, which is the health system intended for treatment of the people by the people. Only simple symptoms or diseases that individuals can diagnose themselves are included in the program. Further development for industrial purposes requires more scientific support, as recommended by the WHO. Appropriate dosage forms and quality control should also be established before these herbal remedies can be developed as "drugs" in the sense understood by modern medicine.

PREPARATION OF DRUG AND QUALITY CONTROL

The target groups of the development of medicinal plants for primary health care are villagers or people with low incomes. The preparation of the drug should be simple enough for them, therefore the traditional methods are recommended. There are 25 methods of preparation in Thai traditional medicine; the common methods are as follows.

Decoctions

The plant material is folded or cut into pieces and put into a bundle that can be held in one hand. If the material is too hard to fold, it should be cut into pieces 5–6 inches long and 1/2 inch wide before being put into the bundle. The bundle of plant material is placed in a dry pot, then 3–4 glasses of water are added to cover the bundle. If the recipe contains a small amount of plant material, only one glass of water is needed. Then the decoction is boiled for 10–30 minutes. The supernatant is decanted and it should be taken while it is warm.

Infusions

The fresh plant material is cut into thin pieces, which are dried. To overcome unwanted scent, the plant material may be roasted before use. The infusion is prepared by using one part of plant material and two parts of water. After the boiling water is poured over the drug, the infusion should be left in a closed container for 15–20 minutes before administration.

Tinctures

The plant material is ground to a coarse powder then loosely wrapped with muslin cloth. The arrack is added and left for 7 days before use.

Pils

The plant material is cut, dried and powdered. Two parts of plant materials are thoroughly mixed with one part of honey or syrup. The mixture is left for 2–3 hours, while it dries sufficiently to prevent the material sticking to the hands. The pills are made by hand and are 1 cm in diameter. They are dried in the sun. To prevent fungal contamination the pills should be redried frequently.

Squeezed juices

In some preparations, the fresh plant material is bruised, water may be added, then it is squeezed and the juice obtained is taken orally. Some rhizomes, such as *Boesenbergia*, may be roasted before bruising, as specified in the recipe.

Poultice

The fresh plant material is coarsely bruised then applied over the wounded area. Water or alcohol may be needed in some recipes. If the drug begins to dry, water may be added and the drug should be changed three times a day.

It should be noted that:

1. The plant used in the prescription should be fresh unless it is specified that dried plants should be used.
2. For internal medicines, if the method is not specified, the preparation is boiled.
3. External medicines are usually applied as a poultice unless the method is specified.
4. For oral prescription without specific instruction, it is understood that the drugs will be taken three times a day.
5. The dose varies according to the type of preparation: one half to one glass (250 ml) for decoctions, one half to 1 tablespoonful for tinctures or plant juices, 1–2 teaspoonfuls for powdered drugs, 1–2 pills and 1 glass for infusions.

Quality control of the preparation is essential for consistency in the treatment. Traditional doctors control their raw materials by specifying the time of harvest, the point of the compass that the plant faces, the part used and the appropriate growth stage. However, not all these are recorded. It is necessary for the development program to use a scientific approach to quality control. First, the active constituent(s) has to be identified. Then, the dose and form in which the herbal medicine is used should be determined. It should be emphasized that the effective dose of a pure compound is different from when it is used in the crude extract. This difference is due to the influence of other compounds on the absorption of the active compound. Some compounds may exert synergistic or antagonistic effects on the active constituents. After the effective dose has been clarified, quality control can be done by translating the measurement into a traditional form of measurement. It should be noted that this measurement is not accurate, so any symptoms or diseases that require accurate doses and control should not be treated in this way.

When the active constituent is not known, it is recommended that the chromatographic patterns or fingerprint be recorded. However, there is a possibility that the chromatographic pattern will not show the active constituents. In the case of volatile oils, a gas chromatographic fingerprint is appropriate, even when the active compound is unknown, because most of the compounds in the volatile oil are separated by gas chromatography. In other cases, if the group of the active compound is known, e.g. alkaloids, saponins, a chromatographic fingerprint is also applicable.

OTHER PRECAUTIONS

Problems may arise from misdiagnosis by a non-professional or because of inadequate clinical information. The following guidelines are recommended.

1. Prolonged use of herbal medicines should be avoided. If within a few days no improvement is observed, the patient should consult the physician. The problem may be due to wrong diagnosis.
2. If any abnormal symptoms are observed after use of a herbal medicine, the patient should stop using it and consult the physician.
3. Patients should follow the instructions of traditional doctors; any changes in the preparation have to be tested for efficacy and safety.
4. Prolonged boiling of the drug may make the extract too concentrated and cause a toxic effect.
5. The dose mentioned in the traditional text is for adults; the doses should be reduced for children.
6. Medicinal plants contaminated with bacteria or fungi should be discarded. The fungi or bacteria may destroy the active constituent or produce their own toxins.

CONCLUSION

Medicinal plants have played an important role in Thailand for centuries but the knowledge is transferred within the family, which makes the information available at present incomplete. Economic constraints and the lack of self-reliance in drug supplies forced the govemment to revitalize herbal medicine. Even though complex recipes are commonly used in Thai traditional medicine, single plants are carefully selected for clinical trials, because it is easier to clarify or trace the cause of any abnormal incidents that occur. Plants with proven scientific activity are selected if they have passed toxicity tests, including mutagenic and teratogenic tests, or are consumed as food. The integration of scientific evidence with traditional medicine helps to convince doctors and health personnel who are opposed to herbal medicine.

Quality control is essential for consistency in the treatment. A scientific approach is recommended to ensure the quality of raw materials. However, the quality control used in primary health care is not accurate; any symptoms or diseases that require accurate doses should not be treated with this method.

Misdiagnosis and inadequate clinical information may cause problems, therefore precautions should be taken when using herbal medicine.

REFERENCES

1. H.S.H. Prince Damrong Rajanubhap, *Tamraa Phra Osot Phra Naraai*, Bangkok: Sophonphiphat Press (1967).

2. The Committee on the Promotion of National Identity, The Ministry of Education, *Tamraa yaa jaroek Wat Raja-oros and Phra Osot Phra Naraai* (A Textbook on Traditional prescription engraved on the stone plaques at Wat Raja-oros and the medical texts of King Naaraai), Bangkok: The Secretariat Office of the Prime Minister Press, p. 5 (1980).
3. N. Bunyapraphatsara, J. Satayavivad and P. Saralamp, *et al.*, *Development strategies for medicinal plants in Thailand*, Bangkok: Faculty of Pharmacy, Mahidol University, p. 228 (1980).
4. R.H. Bannerman, WHO's programme in traditional medecine. *World Health Organization* (November), 4–6 (1977).
5. Medicinal Plant Trade Association, *Thai traditional medicine, traditional pharmacy*, Bangkok: Khunthin Aksornkit Press (1983).
6. S. Phongboonrod, *Mai thed mueong Thai, Folkloric uses of Thai and foreign medicinal plants*, Bangkok: Chaiwat Press Co. (1976).
7. A. Panjaphan, *Digging for gold in our land*, Bangkok: Anongsilp Press (1981).
8. *Proceeding of seminar on folkoric medicines of Lanna Thai*, Chiangmai: Welfare Association of Thailand (1979).
9. Yeeyuan, *Yaa chao baan klet chao moeng*, Bangkok: Asia Press (1977).
10. S. Aajsalee, *Traditional medicine for the household*, Bangkok: Pitayakaan Press (1979).
11. S. Aajsalee, *Herbal medicine*, Bangkok: Pitayakaan Press (1981).
12. B. Iamsomboon, *Plants from the forests*, Bangkok: Roongroengtham Press (1975).
13. W. Silapa-acha, *Folkloric medicines*, A pamphlet distributed by Medicinal Plants for Self-Reliance Project.
14. A. Panjaphan, A long-lasting culture of Thailand, Bangkok: Jaroenkij Press (1979).
15. P. Ponphamorn, *Traditional medicine*, Bangkok: Amnuay Saan Press (1979).
16. P. Boonyachot, *Traditional medicine*, vol. 3, Bangkok: Kasembanakil Press (1972).
17. C. Uthokapaak, *The principle of the use of medicinal plants for curing diseases*, Bangkok: Praepittaya Press (1981).
18. Medicinal Plants for Self-Reliance Project, *Treatment with medicinal plants*, newsletter number 7, Bangkok: Edison Press Products (1975).
19. S. Kitikajorn, *Use of medicinal plants, the traditional medicines*, Bangkok: Aksornthai Press (1983).
20. P. Sujamnong, *Thai-Chinese medical text book, folkloric medicines, herbal medicines and traditional medicines in Thailand*, Bangkok: Praewitaya Press (1981).
21. Pratepwimolmolee, *Folkloric drugs*, Bangkok: Makotrajwitayalai Press (1981).
22. Chewakakomarapach, *Extraordinary traditional recipes*, Bangkok: Chaimongkol Press (1974).
23. S. Kaewtanang, *Magic medicine and traditional recipes from various places*, Bangkok: Prasertsiri Press (1977).
24. A. Panjaphan, *Thailand at the thirteenth month*, Bangkok: Jaroenkil Press (1947).
25. X. Lozoya, M.M. Abou-Zaid, C. Nozzolillo and J.T. Arnason, Spasmolytic effect of the methanolic extract of Psidium guajava. *Planta Med.* **56**, 686 (1990).
26. G.D. Lutterrode, Inhibition of gastrointestinal release of acetylcholine by quercetin as a possible mode of action of Psidium guajava leaf extracts in the treatment of acute diarrheal disease. *J. Ethnopharmacol.* **25**(3), 234–47 (1989).
27. S. Praserdsook and M. Sukchotiratana, Effect of some medicinal plant extracts on the growth of dysenteria. *Symposium on science and technology*, 12th Bangkok, Thailand, October 20–22 (1996).
28. P. Thanangkul, C. Chaichanthipyuth, S. Phothidanuch, K. Naakprakob and R. Paanon, *A clinical study of Psidium guajava L. in acute diarrhoea.* Report Submitted to Primary Health Care Office, Ministry of Public Health, Thailand.
29. Martindale: The extra pharmacopoeia. J.E.F. Reynolds, ed., London: The Pharmaceutical Press, p. 779 (1989).
30. Medicinal Plants for Self-Reliance Project, *Andrographis paniculata, Medicinal Plant News* **20**, 13 (1974).
31. Anon, Do you believe in traditional medicine? *Folk Doctor Magazine* **4**(35), 103–5 (1982).
32. K. Sawasdimongkok, U. Permpipat and N. Kiatyingungsulee, *et al.*, Pharmacological study of Andrographis paniculata Nee. *Symposium on Andrographis paniculata*, National Institute of Health, Bangkok, Thailand (1990).
33. S. Thamaree, C. Pachotikarn, M. Tankeyoon and C. Itthipanichpong, Effects on intestinal motility of thirty herbal medicines used in treatment of diarrhoea and dysentry. *Chula Med. J.* **29**(1), 39–51 (1985).

34. P. Dhumma-upakorn, C. Chaichantipyuth, P. Pongjunyakul, V. Sangalungkarm and S. Chaimongkol, Spasmolytic activity of some active substances from Andrographis paniculata. *JSPS-NRCT Seminar: Pharmacological active substance from natural source.* Faculty of Pharmaceutical Science, Chulalongkorn University, Bangkok, Thailand, p. 21 (1992).
35. P. Dhama-upakorn and C. Chaichanthipayud, *et al.*, Effect of andrographolide, neoandrographolide and 14-deoxy-11,12-didehydroandrographolide on the contraction of isolated stomach and ileums. *The 8th Conference of Faculty of Pharmacy*, Chulalongkorn University, Bangkok, Thailand, December 15, 1989.
36. L.L. Garcia, Q.L. Kintaner, F.R. Fojus, F.M. Sison, N.G. Chua and B.A. Villanueva, Pharmacologic studies on the leaves of Andrographis paniculata Nees plant grown in the Philippines. *Acta Med. Philipp.* **16**(2), 59–68 (1980).
37. M. George and K.M. Pandalai, Investigations on plant antibiotics Part IV. Further search for antibiotics substances in Indian medicinal plants. *Indian J. Med. Res.* **56**, 81–4 (1949).
38. T. Pleumjai and N. Sithisomwongse, Antimicrobial activity of Andrographis paniculata Nees. *Symposium on Andrographis paniculata*, National Institute of Health, Nonthaburi, Thailand, October 22, 1990.
39. J. Sindemsuk, The antibacterial activities of the pure extract compounds from Andrographis paniculata on predominate pathogenic enteric bacilli in Thailand. *Bull. Dept. Med. Serv.* **18**, 394–400 (1993).
40. S. Gupta, M.A. Choudhary and J.N.S. Yadava, *et al.*, Antidiarrheal activity of diterpenes of Andrographis paniculata (Kal-Megh) against Escherichia coli enterotoxin in *in vivo* models. *Int. J. Crude Drug Res.* **284**, 273–283 (1990).
41. V. Chantasutra and S. Limpapanichkul, Acute antiinflammatory activity of Andrographis paniculata Nees in rats. *The eighth Conference*, Faculty of Pharmacy, Chulalongkorn University (1989).
42. M. George and K.M. Pandalai, Investigations on plant antibiotics part IV. Further search for antibiotic substance in Indian medicinal plants. *Indian J. Med. Res.* **24**, 49–53 (1949).
43. P. Dhamma-upakorn and C. Chaichantipyuth, Acute and subcronic toxicity studies of Andrographis paniculata in rats and mice. Abstr. *The 8th symposium*, Faculty of Phamacy, Chulalongkorn University Bangkok, Thailand, December, 1989.
44. N. Sithisomwongse, J. Phengchata and S. Cheewapatana, *et al.*, Acute and chronic toxicity of Andrographis paniculata Nee. *Th. J. Pharm. Sci.* **14**(2), 109–17 (1989).
45. C. Chaicharntipyuth and P. Thanangkul, Andrographis paniculata; drugs for treatment of amoebic dysentery and acute diarrhoea. *The 8th symposium*, Faculty of Pharmacy, Chulalongkorn University (1989).
46. A, Dutta and N.C. Sukul, Filaricidal properties of a wild herb, *Andrographis paniculata. J. Helminthol.* **56**, 81–4 (1982).
47. W. Deng, R. Nie and J. Liu, Comparison of pharmacological effect of Andrographolides. *Yao Hsueh T'ung Pao* **174**, 195–198 (1982).

4. CHINESE HERBAL MEDICINES: NEUROPSYCHIATRIC DISORDERS AND THE DYNAMICS OF CYTOSOLIC Ca^{2+}

YASUO WATANABE*, GEORGE F. LAWLOR, JING LI, JUN ZHANG, LI LIU and TAKESHI SHIBUYA

Department of Pharmacology, Tokyo Medical University, 6-1-1 Shinjuku, Shinjuku-ku, Tokyo 160-0022, Japan

INTRODUCTION

Natural product discovery and use in therapeutic medicine has had a distinguished history in China, dating back more than 3000 years. Many of the pharmacological effects of Chinese herbal remedies have been well evaluated and Chinese herbal medicines have long been prescribed for the treatment of several brain dysfunctions, although many of the pharmacological mechanisms of these agents in the central nervous system (CNS) are still not defined (Dictionary of Traditional Medicines, 1985; Lin, 1995; Sugaya, 1988).

More than two thousand years ago, the Chinese pharmacological bulletin, *Jin Gu Yao Lue*, described the differential diagnosis between apoplexy and paralysis, and identified therapeutic medicines for apoplexy (Figure 1). Around 700 BC, another Chinese pharmacological bulletin, *Zhu Bin Yuang Hou Lue*, defined diagnoses for CNS disorders. Natural products, including Chinese herbal remedies, have a long history in the treatment of CNS disorders such as anxiety, affective disorders, insomnia and amnesia. Some agents have been reported to have memory-enhancing, neuroprotective, anti-epileptic and anti-convulsant, anti-cholinesterase and hypothermic and sedative properties (Chinese Herbal Medicines in CNS, 1996; Chen and Gao, 1989; Sugaya, 1988; Zhang *et al.*, 1993).

Specifically, *bis*-benzylisoquinoline alkaloids, including daurioline, tetrandrine and berbamine (isolated from *Menisperium dauricum, Stephania tetrandra* and *Berberis soulieana*, respectively), exhibit potent pharmacological and therapeutic uses. These agents interact at voltage-operated Ca^{2+} and K^{+} channels, and N-methyl-D-aspartate (NMDA) receptors, which play fundamental roles in the neuroprotective and anti-ischemic properties of many Chinese herbal remedies (Lin, 1995).

The therapeutic use of Chinese herbal medicines (CHM) has traditionally employed crude plant and natural product extracts, while current medicinal practice favors the development of homogeneous and purified pharmacologically active agents. Modern chemical and pharmacological techniques are increasingly

* To whom correspondence should be addressed

《金匱要略 (Jin Gu Yao Lue)》 A.D. 200

第5章 中風歴節病 (Zhong Feng Li Jie Bin, Section 5)：

『夫風之爲病、当半身不遂、或但臂不遂者、此爲痺；脉微而数、中風使然』
"In cases of Zhong Feng (apoplexy), the hemiplegia is characteristically manifested. But in case of dysfunction of only arm and hand, the symptom is diagnosed as Bi (dyskinesia / paralysis). If the pulse is weak and rapid, then the diagnosis is Zhong Feng (apoplexy)."

Prescribed Remedies; ***Xiu-Min-Tang, Da-Chai-Hu-Tang, Tao-He-Cheng-Qi-Tang, Gui-Zhi-Fu-Ling-Wan, Dang-Gui-Shao-Yao-San, etc.***

《諸病源候論 (Zhu Bin Yuang Hou Lue)》 B.C. 700

This work provides detailed descriptions of neurological disorders.

Figure 1 Classic Chinese works on the therapeutic uses of Chinese herbal medicines.

being applied for the isolation, purification and identification of the active constituents in herbal medicines, and for determining the pharmacological mechanisms of centrally acting agents.

New research techniques are continually being developed for *in vivo* and *in vitro* experimentation, utilizing methods in behavioral pharmacology, molecular biology, neuropharmacology and electrophysiology, and employing novel animal models, specific biological markers and indicators, and cultured brain tissues. In particular, methods for the detection of intracellular Ca^{2+} levels greatly contribute to the survey of roles played by brain inorganic ions in neurochemical processes, mainly because of the important functional roles of intracellular Ca^{2+} in regulating neurological and psychiatric functions in the CNS (Grynkiewicz *et al.*, 1985; Peterson *et al.*, 1986).

Investigations employing Ca^{2+} specific fluorescent dyes show that intra- and extracellular Ca^{2+} in the CNS is an important secondary messenger for the transduction of neurochemical events (Grynkiewicz *et al.*, 1985; Choi, 1988; Komulainen and Bondy, 1988). Concentrations of intracellular levels are regulated by several kinds of Ca^{2+} channels and Ca^{2+} pumps. Ca^{2+} channels in the CNS are mainly classified as voltage-operated (sensitive) Ca^{2+} channels (VOC) and receptor-operated Ca^{2+} channels (ROC). The distinction between these two types of ion channels has been well established by molecular biological techniques, which have also shown that intracellular Ca^{2+} levels in neuronal systems regulate many brain functions (Soong *et al.*, 1993; Fasolato *et al.*, 1994; Dunlap *et al.*, 1995; Varadi *et al.*, 1995).

In this chapter we discuss mainly the pharmacological actions and mechanisms of two Chinese herbal medicines, used in the treatment of psychiatric and neurological disorders, with respect to the dynamics of cytosolic Ca^{2+} in the neuronal system, employing behavioral methods and neurochemical techniques.

SIGNIFICANCE AND PROPERTIES OF NEURONAL VOLTAGE OPERATED Ca^{2+} CHANNELS

Several metal ions common to biological systems are important for regulating physiological and psychological conditions of the brain. In neuronal systems, alterations in the extra- and intracellular ionic balance influence the quantum of neurotransmitter released and the induction of neuronal signal transduction (Dubovsky and Franks, 1983; Gibson and Peterson, 1987; Colvin *et al.*, 1991; Etcheberrigaray *et al.*, 1994; Siesjo *et al.*, 1995).

It has been demonstrated that the regulation of intracellular Ca^{2+} levels is one of the primary factors for maintaining neuronal homeostasis and controlling neuropsychiatric conditions (Chou *et al.*, 1991; Grove *et al.*, 1993; Yoshimura *et al.*, 1993); and the regulation both of intra- and extracellular Ca^{2+} concentrations is critical for maintaining neuronal functions (Dubovsky *et al.*, 1986; Miller, 1987; Shibuya and Watanabe, 1992). Extracellular Ca^{2+} concentrations are normally 10,000 to 100,000 fold higher than those of intracellular Ca^{2+}, and cytosolic Ca^{2+} levels are mainly raised by the stimulation of ion channels and receptors. For instance, the stimulation of voltage-operated channels, e.g. L-type Ca^{2+} channels that have been well characterized in pharmacological and physiological studies, induces high concentrations of Ca^{2+} to permeate into the neuronal cell (Figure 2). Under normal physiological conditions, large influxes of Ca^{2+} can be regulated by protein kinase A (PKA) through its β subunit, so-called Ca^{2+} buffering. Ions permeating into the cell bind to Ca^{2+}-binding proteins and then induce physiological responses. When receptors are stimulated by ligands, the Ca^{2+} activated enzyme, protein phospholipase C (PLC), links to specific receptors which catalyze the breakdown of phosphatidylinositol-4,5-bisphosphate and yields inositol-1,4,5-triphosphate (IP_3). IP_3 then stimulates the IP_3-sensitive cytosolic Ca^{2+} pool and induces the liberation of Ca^{2+}. The transient rise in Ca^{2+} can be regulated by Ca^{2+}-binding proteins and Ca^{2+}-ATPase in the cytosolic Ca^{2+} pool. These regulatory mechanisms help maintain the concentration of Ca^{2+} in the pool at around 1 mM. Intra- and extracellular Ca^{2+} concentrations are controlled predominantly by Ca^{2+} buffering and the capacity of the cytosolic Ca^{2+} pool, and are closely linked to the physiological condition of the cell (Figure 2).

In preclinical studies, neuronal dysfunction induced by the neurotoxin, paraquat, has been implicated with excessive levels of cytosolic Ca^{2+} (Komulainen and Bondy, 1988; Yoshimura *et al.*, 1993). Furthermore, in rats, particularly spontaneous hypertensive rats (SHR), extraordinary hyperactivity is exhibited in stressful environments (e.g., stress model applying 85 dB noise); under these conditions of "fear", frontal cortical and hippocampal Ca^{2+} channel activities are potentiated. Treatment with Ca^{2+} antagonists improves both potentiated activities and abnormal conditions (Chou *et al.*, 1991). Clinical reports have diagnosed hyper- (>5.5 mEq/l Ca) and hypo- (<4.0 mEq/l Ca) calcemia patients with mental disorders, such as depression and irritability (Dubovsky *et al.*, 1986). The implications of intracellular Ca^{2+} modulation are therefore important elements in evaluating the pharmacological mechanisms of neuropsychiatric and other drugs acting on the CNS. It is

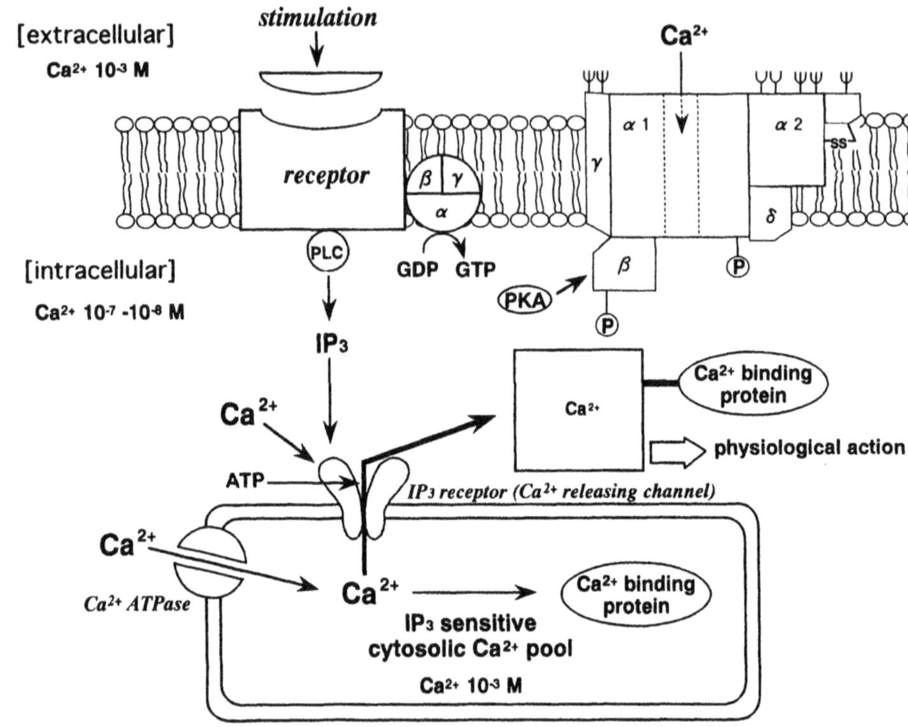

Figure 2 Significance and properties of voltage-operated channels.

well known that alterations in cytosolic Ca^{2+} levels are controlled by Ca^{2+} influx through ion channels and pumps. This influx is mainly implemented by VOC, and the activation of VOC directly influences not only Ca^{2+} release from intracellular storage, but also the turnover rate of the cytosolic Ca^{2+} cascade (Soong *et al.*, 1993; Varadi *et al.*, 1995; Randall and Tsien, 1995).

Molecular biological studies indicate that DNA clones complementary to VOC express six $\alpha 1$ subunit genes and a large number of auxiliary $\alpha 2/\delta$ and β subunit gene transcripts. The $\alpha 1$ subunits facilitate the permeation of Ca^{2+} into the cell in a voltage-dependent manner and have binding sites for VOC sensitive ligands. The $\alpha 1$ subunits can be classified into six additional subtypes by their encoded gene sequences, classes A–E and S. The β subunit maybe associated with the regulation of intracellular Ca^{2+} released through phosphorylation, and this subunit is encoded by a minimum of four separate genes, classes 1–4. $\beta 3$ and $\beta 4$ subunits are primarily found in brain VOC (Soong *et al.*, 1993).

Moreover, six different types of VOC (L, T, N, P, Q, and R-types) have been identified in peripheral tissues and the brain (Dunlap *et al.*, 1995), and their biochemical and physiological properties examined (Akaike and Watanabe, 1994; Grantham *et al.*, 1994; Wheeler *et al.*, 1994). Specific ligands for L, N and P-types include dihydropyridine (DHP) Ca^{2+} antagonists, ω-conotoxin GVIA (venom of the fish-hunting cone snail, *Conus geographus*) and ω-Aga IVA (venom of the funnel web spider, *Agelenopsis aperta*), respectively (Grove *et al.*, 1993; Grantham *et al.*, 1994). Among the identified VOC, P, Q, and R-types are primarily found in brain neuronal systems, cerebellar Purkinje and granule cells (Soong *et al.*, 1993; Randall and Tsien, 1995). The specific physiological functions of N, P, and Q-type VOC are related to the regulation of transmitter release (Wheeler *et al.*, 1994). L-type VOC, which mediate long-lasting Ca^{2+} currents, have been well characterized as DHP-sensitive channels with a heteropentametric complex subunit, since L-type VOC-sensitive antagonists have been studied in clinical fields, such as cardiovascular disease (Grove *et al.*, 1993). Some L-type VOC-sensitive antagonists pass through the blood-brain barrier, and can act as antagonists in cases of neuro-psychiatric disorders caused by high levels of cytosolic Ca^{2+} (Dubovsky *et al.*, 1986; Shibuya and Watanabe, 1992).

CHINESE HERBAL MEDICINES FOR PSYCHIATRIC DISORDERS

The therapeutic uses of various CHM for psychiatric disorders have a long tradition in China and many other Asian cultures. In some instances the combined treatment of modern drugs together with Chinese herbal remedies can help reduce potent and unwanted side effects often associated with the high dosages of contemporary, modern medicines prescribed (Tanaka, 1982; Chen and Gao, 1989; Narita, 1990). However, newer and more current approaches to clinical CHM applications have been precluded by the development of highly selective and efficacious modern drugs. In recent preclinical studies, the relationships between cytosolic Ca^{2+} dynamics and the pharmacological mechanisms of some CHM on mental disorders have been compared to those of modern psychotropic drugs (Hagino, 1990; Fujiwara, 1990; Watanabe *et al.*, 1991).

The pharmacological efficacy of the Chinese medicine, *Suan-Zao-Ren-Tang* (SZRT), on psychiatric disorders has been elucidated in behavioral studies, and has been used as an anxiolytic and hypnotic in China for over 100 years. SZRT is comprised of the five components (indicated as concentrations): Sansonin (*Zizyphi Semen*) 60.0%; Bukuryo (*Poria*) 7.5%; Chimo (*Anemarrhenae Rhizoma*) 12.5%; Kanzo (*Glycyrrhizae Radix*) 7.5%; and Senkyu (*Ligustici chuaxiong Rhizoma*) 12.5%. A recent clinical study has shown the pharmacological properties of SZRT to be similar to those of benzodiazepines, e.g., diazepam (DZP), although the pharmacological potency of SZRT is weaker than that of DZP (Shibuya *et al.*, 1986; Shibuya *et al.*, 1987). Compared to DZP, SZRT elicits lesser drug tolerance and liver toxicity (Shibuya *et al.*, 1986). While SZRT may operate by mechanisms similar to those of DZP, it is speculated to be a potentially safer drug, with fewer of the deleterious side effects associated with DZP.

As anxiolytic agents, comparative studies of SZRT and DZP on blood pressure have been performed using spontaneously hypertensive rats. Dosages of SZRT in excessive of 500 mg/kg, p.o., have been demonstrated to significantly reduce blood pressure, and the magnitude of decrease after treatment with 1 g/kg SZRT is the same as with 5 mg/kg DZP (Shibuya *et al.*, 1986). Nonetheless the same dosage of SZRT does not induce the changes in heart rate experienced with DZP. These findings suggest that the pharmacological mechanism of SZRT on blood pressure may be the result of dilation of the peripheral blood vessels or neuronal regulation. The pharmacological effects of single and concomitant treatments of SZRT and DZP on the central nervous system have also been evaluated. Experiments investigating the hypnoidal action of SZRT show that neither single (500 mg/kg, p.o.) nor repeated (500 mg/kg, p.o., for 5 days) administrations of SZRT affect the onset or duration of ethanol (EtOH) induced sleep (Shibuya *et al.*, 1986). Concomitant treatment with SZRT and DZP, however, markedly enhances EtOH-induced sleep, and diminishes the acute drug tolerance associated with repeated DZP administration. The combined administration of both SZRT and DZP could therefore serve as a better hypnotic agent, rather than the exclusive administration of DZP, in the treatment of insomnia.

The anxiolytic action of SZRT has also been examined utilizing olfactory bulbectomized (OB) rats, which exhibit hyperactive emotional behaviour (Shibuya *et al.*, 1987). The effects of SZRT and DZP on inhibiting emotional activity in OB rats are most potentiated with the concomitant use of both drugs. While the single, acute administration of SZRT does not produce the notable changes in emotional activity observed with the single administration of DZP, repeated treatment with SZRT for a period of 5 days attenuates emotional disorders in OB rats. Concomitant SZRT and DZP administration significantly diminishes hyperactive emotional activity in this animal model. The inhibitory effects of SZRT on hyperactivity are markedly more apparent in animal models, such as OB rats, than in normal animals. In anti-conflict behavioral studies employing OB rats, repeated SZRT treatment, and the concomitant administration of SZRT and DZP, exhibit efficacious results, while the single, acute administration of SZRT has been determined ineffective (Shibuya *et al.*, 1987). These results suggest that SZRT possesses anxiolytic properties, and that this agent may be most effectual when administered simultaneously with DZP. Furthermore, some Ca^{2+} channel antagonists show anxiolytic activity by modulating neuronal Ca^{2+} levels, and the anxiolytic action of SZRT can be dose-dependently enhanced by pretreatment with Ca^{2+} channel antagonists such as flunarizine and nitrendipine. Individually administered, flunarizine and nitrendipine exhibit lesser anxiolytic effects, compared with their potentiated activities when combined with SZRT. It is therefore postulated that the mechanism by which SZRT alleviates symptoms of anxiety may be attributed to the inhibition of neuronal cell Ca^{2+} channel hyperactivity.

The pharmacological effects of *Suan-Zao-Ren-Tang* on CNS disorders are compared in Table 1 to those of the anxiolytics, diazepam and tandospirone, and the Ca^{2+} antagonists, nitrendipine and verapamil. Both tandospirone, which is a serotonin receptor ($5\text{-}HT_{1A}$) partial agonist, and DZP block hyperactivity occurring at Ca^{2+} channels (Watanabe and Shibuya, 1994). These two anxiolytics and the Ca^{2+} antagonists depicted in Table 1 have similar pharmacological effects of: prolonged EtOH-induced sleeping time, attenuation of pentylenetetrazol induced seizures, and anxiolytic properties. SZRT also exhibits many of the same pharmacological effects seen with the administration of these anxiolytics and Ca^{2+} antagonists.

Table 1 Relative pharmacological effects of Chinese herbal medicine and modern medicines on CNS disorders

	Sleeping time induced by EtOH	*Aggressive behavior*	*Drug-induced seizure*	*Anxiety*
Suan-Zao-Ren-Tang	↑	↓	—	↓
Diazepam	↑	↓	↓	↓
Tandospirone	↗	⇕	⇕	↓
Nitrendipine	↗	⇕	↓	↓
Verapamil	↗	⇕	—	↓

↑: potentiation of response, ↗: less potentiation of response, ↓: attenuation of response, ⇕: inconsistency of response, –: no determination (larger symbols denote more marked response.

CHINESE HERBAL MEDICINES FOR NEUROLOGICAL DISORDERS

Extreme increases in Ca^{2+} levels in the neuronal cell are caused by large releases of glutamate (Glu), inducing strong hyperactivity at postsynaptic receptors, such as NMDA receptors (Choi, 1988; Rosenberg, 1991). This event is followed by tremendous Ca^{2+} influx from outside the cell, leading to the loss of signal transduction regulation (Landfield, 1994; Sen *et al.*, 1993; Peterson *et al.*, 1986). One possible mechanism for neuronal cell death, including delayed neuronal death, may be caused by Glu-induced increases in cytosolic Ca^{2+} levels. After brain ischemia, excessively high concentrations of released Glu stimulate specific receptors on postsynaptic neurons, such as NMDA and α-amino-3-hydroxy-5-methyl-4-isoxazole-propionate (AMPA)/kainate (KA) receptors, followed by the stimulation of VOC, facilitating a large influx of extracellular Ca^{2+} into the cytosol through both these receptors and ion channels, as depicted in Figure 3. Although AMPA/KA receptor-gated ion channels do not normally conduct Ca^{2+} into the cell, Na^+ influx depolarizes the cell membrane, thereby allowing Ca^{2+} influx mainly through NMDA and/or subsidiary AMPA/KA receptor-gated ion channels, as well as through VOC (Pigott *et al.*, 1991; Sen *et al.*, 1993). With increased intracellular levels, Ca^{2+} ions bind to calmodulin-dependent protein kinase II (CaMKII). The subsequent enhanced protein kinase activates nitric oxide synthetase (NOS), precipitating nitric oxide (NO) production. This pathway occurs in the acute phase of ischemia followed by neuronal death (Price, 1990). One current hypothesis explaining delayed neuronal death is the presence of free radicals, which can alter lipid peroxidation and protein oxidation, thus evoking hyperactivity among mitochondrial functions. Mitochondrial hyperactivity induces megapore opening and sustained, massive release of Ca^{2+} ions from the mitochondria to initiate delayed neuronal death (Duchen *et al.*, 1993; Liu, 1993; Crompton and Costi, 1988; Siesjo *et al.*, 1995).

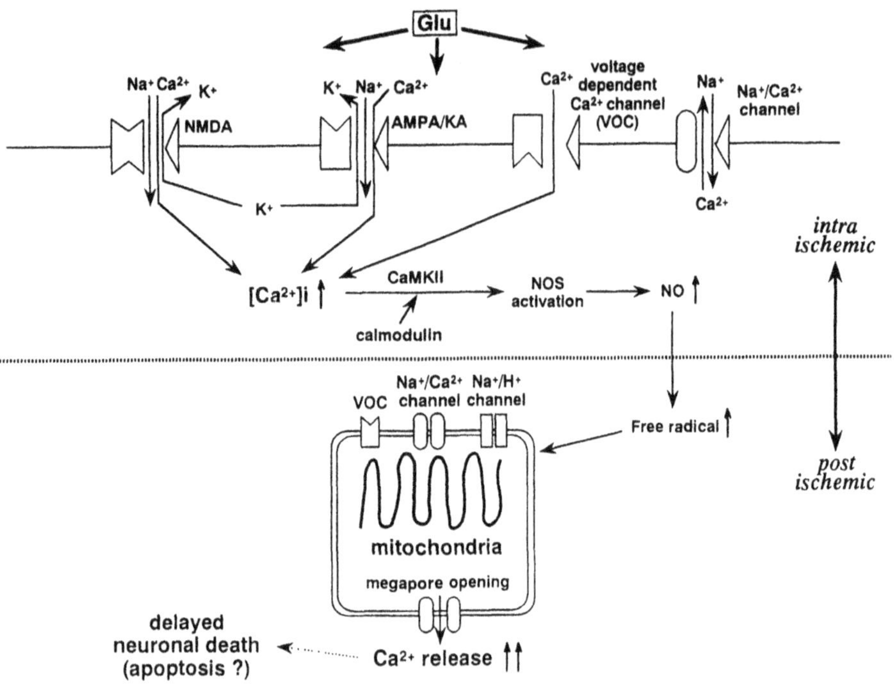

Figure 3 Glutamate-induced mechanism of neuronal cell death.

In recent reports, Toki-Shakuyaku-San, comprised of six Chinese herbs (indicated as concentrations), Toki (*Angelicae sinensis Radix*) 12.5%, Shakuyaku (*Paeoniae Radix*) 25.0%, Bukuryo (*Hoelen*) 16.7%, Sojutsu (*Atractylodis lanceae Rhizoma*) 16.7%, Takusha (*Alismatis Rhizoma*) 16.7%, and Senkyu (*Ligustici wallichii Rhizoma*) 12.5%, has been shown to ameliorate some neuronal dysfunctions, such as Alzheimer's disease, senile dementia and memory loss, as well as providing neuroprotective properties against glutamate-induced neuronal death (Liu *et al.*, 1995; Watanabe *et al.*, 1995; Hagino, 1990; Fujiwara, 1990). To investigate the central mechanism of this herbal remedy, Ca^{2+} antagonist functions have been examined using cultured cerebellar granule cells, focusing on the effects of TSS and its six Chinese herbal components on K^+ (50 mM KCl) and glutamate-evoked Ca^{2+} influx. The inhibitory effects of this medicine on both voltage-operated Ca^{2+} channels and receptor-operated channels were evaluated together with the antagonistic effects of flunarizine and nicardipine (Watanabe *et al.*, 1995). Both Ca^{2+} antagonists were found to block VOC in a dose-dependent manner, although diminished ROC activity is also exhibited as a result of reduced VOC functions (Zhang *et al.*, 1993).

Toki-Shakuyaku-San is also known to block both types of Ca^{2+} channels in dose-dependent manners, displaying IC_{50} values for VOC and ROC of 300 μg/ml and 420 μg/ml, respectively. Furthermore, studies on the six herbal components of TSS show that the inhibitory effects of Toki, Senkyu, Sojutsu, Bukuryou and Takusha, on both channel types, occur at nearly the same dosages. Shakuyaku, on the other hand, exhibits a markedly weaker inhibitory effect on VOC. The range of IC_{50}

values for the six herbs on each type of Ca^{2+} channel was between 230 and 450 $\mu g/ml$, except for Shakuyaku. Many Chinese herbs, including TSS, strongly block the activities of glutamate-related ROC, while nicardipine exhibits weaker inhibition on these channels (Watanabe *et al.*, 1995).

Brain neuronal functions can be controlled by neuronal cell Ca^{2+} levels, whereas cell death is also determined by cellular levels of glutamate (Figure 3) and other growth factors. A number of CHM block glutamate-related Ca^{2+} channels, implicating their potential therapeutic uses in treating brain disorders and neuronal damage, and providing further evidence that many Chinese herbs are most efficacious when utilized in combination rather than administered individually or as purified agents.

Extreme Ca^{2+} level increases in neuronal cells are caused by copious releases of glutamate, followed by Ca^{2+} influx via NMDA receptors. Such excessive Ca^{2+} ion levels can interfere with signal transduction, therefore the identification of new and more effective drugs to control cellular Ca^{2+} influx is important in the development of better therapies to treat brain-related disorders.

CONCLUSION

We have discussed herein two Chinese herbal medicines exhibiting clinical effects on neuronal regulation and impaired brain functions, although the mechanisms of their actions are still not well defined. Yet two distinct and inherent characteristics of CHM, and other traditional herbal remedies, complicate the evaluation of their pharmacological mechanisms. Principally, many CHM have multiple functions and therapeutic effects; a given herbal remedy may be used in treating various and unrelated diseases and symptoms. Secondly, most of these remedies constitute complex, heterogenous mixtures of various compounds.

It has long been debated that insufficient research has been conducted on the active agents in crude natural product extracts and herbal formulations, which must initially be separated, identified and purified, in order to determine the specific modes of action of each operative compound. Nonetheless, even as purified substances, establishing the precise pharmacological effects of some CHM on the brain may still be equivocal, as numerous brain and neurological functions remain unclear.

Increasingly, conventional dogma recognizes that a variety of neurological and psychiatric disorders are related to or are controlled by cytosolic Ca^{2+} dynamics in the neuronal cell (Watanabe and Shibuya, 1994). This has been well substantiated by contemporary molecular biological techniques, which have identified specific types of central voltage-operated Ca^{2+} channels, the primary modulating system for cytosolic Ca^{2+} levels in brain neurons. For instance, clinical studies indicate that some Ca^{2+} channel antagonists can improve psychiatric disorders, such as manic depression, panic disorder, and seizures. Moreover, the anxiolytics, diazepam and tandospirone, exhibit Ca^{2+} channel blocking effects, and the antimanic agent, lithium, regulates dysfunctions related to cytosolic Ca^{2+} dynamics. In order to study the complex mechanisms of Chinese herbal medicines acting on the brain, we have focused on the potential relationship between cytosolic Ca^{2+} regulation and expressed pharmacological effects in animal behavioral models.

In experiments we have conducted, *Suan-Zao-Ren-Tang* attenuates emotional disorders in OB rats, and the attenuation is potentiated not only by diazepam, but also by nitrendipine. These effects, however, are not apparent in normal animals administered SZRT. Noteworthy, is that the combined administration of all three drugs together shows greater efficacy as an anxiolytic than the exclusive administrations of any of these three individual drugs. These findings warrant further investigation to understand the precise pharmacological mechanisms by which SZRT can regulate Ca^{2+} channel activity. In addition, more recent results from our laboratory have shown that SZRT reduces extraordinary increases in cytosolic Ca^{2+} levels, induced by 25 mM KCl, in cultured cerebellar granule cells, corroborating the attenuating effects of SZRT on Ca^{2+} channel activity (unpublished data).

To examine another possible pharmacological mechanism, we have also studied the protective effects of Toki-Shakuyaku-San (TSS) on glutamate-induced neuronal death. Its formulation of six different herbal components is effective in preventing neuronal damage resulting from excessive influxes of Ca^{2+} into neuronal cells. The mechanisms of some Chinese herbal medicines affecting the central nervous system may well be explained by the modulation of cytosolic Ca^{2+} dynamics, a common mode of action among modern drugs.

Lastly, Figure 4 summarizes the fundamental elements essential to preclinical research, namely neurochemical techniques and behavioral methods, in the effort to better utilize Chinese and traditional herbal medicines for treating CNS diseases. Empirical pharmacological evaluation of herbal medicines demands not only identification of their purified active agents, but also the dichotomous analysis of the therapeutic effects of both the heterogenous formulations and their individual active components. Agents acting on the CNS should be examined in relation to changes in brain and neuronal substances, such as neurotransmitters, receptors,

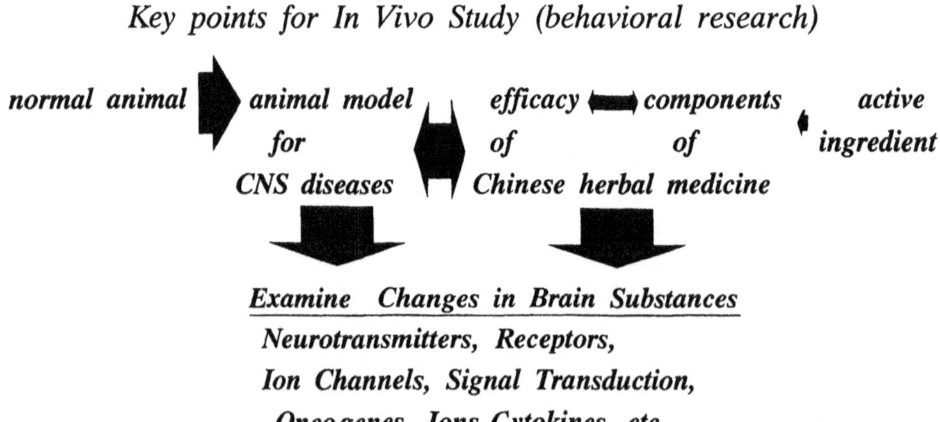

Figure 4 Key elements for the pharmacological study of Chinese herbal medicine.

ion channels, signal transduction, oncogenes, and intra- and extracellular ion concentrations. Moreover, the functional roles of cytokines in the brain must be considered, in light of recent findings concerning neurological disorders which show that the immune system operates within the brain and regulates neuronal dysfunctions. Possible immunological mechanisms of CHM acting on the central nervous system cannot be discounted. Important too is the use of appropriate animal models, such as those employed in our investigations, which overall are more effectual models than normal animals, particularly with respect to the CNS pharmacology of herbal medicines. Available animal models and the development of more suitable models for specific diseases and dysfunctions are essential to behavioral studies, since quantitative pharmacological effects of many CHM are often unexpressed in normal experimental subjects.

The development of more effectual methodologies in neuroscience, together with the ongoing international promotion of brain research to better elucidate brain functions, continues to advance our understanding of the neuropharmacological mechanisms of Chinese herbal medicines; notwithstanding, defining the detailed mechanisms of some herbal medicines will remain challenging due to their specific and complex interactive pharmacological features.

REFERENCES

1. N. Akaike and Y. Watanabe, Novel Ca^{2+} channel blockers with CNS effects. *Clinical Calcium* **4**, 24–31 (1994).
2. Z.-Q. Chen and Z.G. Gao, Clinic application of Tang-Kuei and Peonu Formula. *Mintong Med. J.* **155**, 17–21 (1989).
3. Chinese Herbal Medicines in CNS, In *International Academy for Biomedical and Drug Research*, S.Z. Langer, J. Mendlewicz and G. Racagni, eds., Karger, Basel, pp. 243–280 (1996).
4. D.W. Choi, Calcium-mediated neurotoxicity: relationship to specific channel types and role in ischemic damage. *TiNS.* **11**, 465–469 (1988).
5. M.Y. Chou, Y. Watanabe and T. Shibuya, Neuropsychopharmacological studies of a Ca^{2+} channel blocker on the modulation of brain Ca^{2+} mobilization of spontaneously hypertensive rats under mild stress. *Neurosci. Res.* **12**, 346–355 (1991).
6. R.A. Colvin, J.W. Bennett, S.L. Colvin, *et al.*, Na^{+}/Ca^{2+} exchange activity is increased in Alzheimer's disease brain tissues. *Brain Res.* **543**, 139–147 (1991).
7. M. Crompton and A. Costi, Kinetic evidence for a heart mitochondrial pore activated by Ca^{2+}, inorganic phosphate and oxidative stress, A potential mechanism for mitochondrial dysfunction during cellular Ca^{2+} overload. *Eur. J. Biochem.* **178**, 489–501 (1988).
8. Dictionary of Traditional Chinese Medicine 3, Shang-Hai-Ke-Xue-Ji-Shu-Chu-Ban-She, Shogakukan Inc., Tokyo (1985).
9. S.L. Dubovsky and R.D. Franks, Intracellular calcium ions in affective disorders: A review and a hypothesis. *Biol. Psychiat.* **18**, 781–797 (1983).
10. S.L. Dubovsky, R.D. Franks, S. Allen, *et al.*, Calcium antagonists in mania: a double blind study of verapamil. *Psychiatry Res.* **18**, 309–320 (1986).
11. M.R. Duchen, O.M. McGuinness, L.A. Brown, *et al.*, On the involvement of a cyclosporin A sensitive mitochondrial pore in myocardial reperfusion injury. *Cardiovascular Research* **27**, 1790–1794 (1993).
12. K. Dunlap, J.I. Luebke and T.J. Turner, Exocytotic Ca^{2+} channels in mammalian central neurons. *TiNS.* **18**, 89–98 (1995).
13. R. Etcheberrigaray, E. Ito, C.S. Kim, *et al.*, Soluble β-amyloid induction of Alzheimer's phenotype for human fibroblast K^{+} channels. *Science* **264**, 276–279 (1994).

14. C. Fasolato, B. Innocenti and T. Pozzan, Receptor-activated Ca^{2+} influx: how many mechanisms for how many channels? *TiPS* **15**, 77–83 (1994).
15. M. Fujiwara, Toki-Shakuyaku-San (TJ-23) improves the cognitive disruption caused by central cholinergic dysfunction. *Jpn. J. Neuropshychopharmacol.* **12**, 217–227 (1990).
16. G.E. Gibson and C. Peterson, Review: Calcium and the aging nervous system. *Neurobiology of Aging* **8**, 329–343 (1987).
17. C.J. Grantham, D. Bowman, C.P. Bath, *et al.*, ω-conotoxin MVIIC reversibly inhibits a human N-type calcium channel and calcium influx into chick synaptosomes. *Neuropharmacology* **33**, 255–258 (1994).
18. A. Grove, J.M. Tomich, T. Iwamoto, *et al.*, Molecular design of a dihydropyridine-sensitive calcium channel protein. In *Drugs in Development; Ca^{2+} antagonists in CNS*, A. Scriabine, R.A. Janis and D.J. Triggle, eds., Neva Press, Branford, Connecticut, pp. 21–34 (1993).
19. G. Grynkiewicz, M. Poenie and R.Y. Tsien, A new generation of Ca^{2+} indicators with greatly improved fluorescence properties. *J. Biol. Chem.* **260**, 3440–3450 (1985).
20. N. Hagino, The mode of action of Toki-Shakuyaku-San (TJ-23) on the brain: With special reference to Alzheimer's disease. *Jpn. J. Neuropshychopharmacol.* **12**, 229–234 (1990).
21. H. Komulainen and S.C. Bondy, Increased free intracellular Ca^{2+} by toxic agents: an index of potential neurotoxicity. *TiPS.* **9**, 154–156 (1988).
22. P.W. Landfield, Increased hippocampal Ca^{2+} channel activity in brain aging and dementia. Hormonal and Pharmacologic modulation. *Ann. New York Acad. of Sci.* **747**, 351–364 (1994).
23. G.-Q. Lin, Chinese natural products and new drugs. *Pharmaceutical News* **2**, 10–12 (1995).
24. J.-S. Liu, Y. Watanabe and S. Shibuya, Stimulative and protective effects of Chinese medicinal brain metabolic stimulants, Toki, Senkyu and Tanjin, on glutamine synthetase activity. *J. Trad. Med.* **6**, 220–228 (1995).
25. J. Liu, FK 506 and cyclosporin, molecular probes for studying intracellular signal transduction. *Immunology Today* **14**, 290–295 (1993).
26. R.J. Miller, Multiple calcium channels and neuronal function. *Science* **235**, 46–52 (1987).
27. H. Narita, Kampo-therapy for psychiatric region. *Jpn. J. Neuropsychopharmacol.* **12**, 165–172 (1990).
28. C. Peterson, R.R. Ratan, M.L. Shelanski, *et al.*, Cytosolic free calcium and cell spreading decrease in fibroblasts from aged and Alzheimer donors. *Proc. Natl. Acad. Sci. USA* **83**, 7999–8001 (1986).
29. M.A. Piggott, J.M. Candy and R.H. Perry, [^{3}H]nitrendipine binding in temporal cortex in Alzheimer's and Huntington's disease. *Brain Res.* **565**, 42–47 (1991).
30. A.H. Price, Possible role for calcium in oxidative plant stress. *Free Rad. Res. Commun.* **10**, 345–349 (1990).
31. A. Randall and R.W. Tsien, Pharmacological dissection of multiple types of Ca^{2+} channel currents in rat cerebellar granule neurons. *J. Neurosci.* **15**, 2995–3012 (1995).
32. P.A. Rosenberg, Accumulation of extracellular glutamate and neuronal death in astrocyte-poor cortical cultures exposed to glutamine. *Glia.* **4**, 91–100 (1991).
33. A.P. Sen, P. Boksa and R. Quirion, Brain calcium channel related dihydropyridine and phenylalkylamine binding sites in Alzheimer's, Parkinson's and Huntington's diseases. *Brain Res.* **611**, 216–221.
34. T. Shibuya, Y. Watanabe, H. Matsuda, *et al.*, The pharmacological studies on Suanzaorentang using the EtOH induced sleeping mice. *J. Tokyo Med. Col.* **44**, 1018–1024 (1986).
35. T. Shibuya, H. Matsuda, Y. Watanabe, *et al.*, A pharmacological study on suanzaorentang. 2 The evaluation of behavioral pharmacological studies on the antianxiety and analgesic function. *J. Tokyo Med. Col.* **45**, 176–182.
36. T. Shibuya and Y. Watanabe, Central effects of Ca^{2+} channel blockers: Multiple sites of action. *Folia Pharmacol. Japon.* **100**, 239–247 (1992).
37. B.K. Siesjo, K. Katsura and T. Kristian, The biochemical basis of cerebral ischemia damage. *J. Neurosurg. Anesthesiol.* **7**, 47–52 (1995).
38. T.W. Soong, A. Stea, C.D. Hodson, *et al.*, Structure and functional expression of a member of the low voltage-activated calcium channel family. *Science* **260**, 1133–1136 (1993).
39. E. Sugaya, Pharmacological studies of Kampo medicines: Central Nervous System. In *Recent Advances in the Pharmacology of Kampo (Japanese Herbal) Medicines*, E. Hosoya and Y. Yamamura, eds., Excerpta Medica, Amsterdam, pp. 51–55 (1988).
40. M. Tanaka, Congestion in China – its conception, cause and disease, symptoms and diagnosis, drug and prescription. *Chinese J. Trad. Chin. Med.* **11**, 21–35 (1982).

41. G. Varadi, Y. Mori, G. Mikala, *et al.*, Molecular determinants of Ca^{2+} channel function and drug action. *TiPS.* **16**, 43–49 (1995).
42. H. Watanabe, T. Satoh, K. Matsumoto, *et al.*, Psychopharmacological effects of Shimotsu-To and Shimotsu-To components, Toki, Shakuyaku, Jio and Senkyu, in mice. *J. Trad. Med.* **8**, 102–107 (1991).
43. Y. Watanabe and T. Shibuya, Modulators of cytosolic Ca^{2+} levels related to evaluation of novel psychiatric drugs. *Neuropschychopharmacology* **10**, 426s (1994).
44. Y. Watanabe, X.-Q. Zhang and J.-S. Liu, *et al.*, Protection of glutamate induced neuronal damages in cultured cerebellar granule cells by Chinese herbal medicine, Toki-Shakuyaku-san and its comprised six medical herbs. *J. Trad. Med.* **12**, 93–101 (1995).
45. D.B. Wheeler, A. Randall and R.W. Tsien, Roles of N-types and Q-type Ca^{2+} channels in supporting hippocampal synaptic transmission. *Science* **264**, 107–111 (1994).
46. Y. Yoshimura, Y. Watanabe and T. Shibuya, Inhibitory effects of calcium channel antagonists on motor dysfunction induced by intracerebroventricular administration of paraquat. *Pharmacol. & Toxicol.* **72**, 229–235 (1993).
47. X.-Q. Zhang, Y. Watanabe, M. Ohnishi, *et al.*, Comparative studies on the inhibitory effects of calcium antagonists on cytosolic Ca^{2+} levels increased by high-potassium of glutamate in cultured rat cerebellar granule cells. *Jpn. J. Phamacol.* **62**, 411–414 (1993).

5. PHARMACOLOGICAL STUDY OF THE ANTI-DEMENTIA EFFECT OF SHIMOTSU-TO (SI-WU-TANG)

HIROSHI WATANABE[1], KINZO MATSUMOTO[1] and HARUMI MATSUDA[2]

[1]*Research Institute for Oriental Medicines (Wakan-Yaku), Toyama Medical and Pharmaceutical University, Toyama 930-0194;*
[2]*Toyama Prefectural Hospital, Ishigane, Toyama 930-0966, Japan*

1. INTRODUCTION

Shimotsu-to (Si-Wu-Tang) is a basic prescription of traditional Chinese medicine which consists of four herbs: Japanese angelica root, cnidium rhizome, peony root and rehmannia root. It has been used clinically to treat menstrual problems and autonomic imbalances in women for several centuries in China and Japan (Chen, 1988) (Table 1). Shimotsu-to is also used to treat anemia (Kashiwabuchi, 1992). It was demonstrated that Shimotsu-to improved the decrease in the number of red blood cells, hemoglobin and hematocrit value and also increased the ratio of polychromatic erythroblasts and of orthochromatic erythroblasts in the femoral bone marrow of anemic rats (Kojima *et al.*, 1994, 1995). Shimotsu-to is often applied in combination with other traditional prescriptions for many diseases. Shimotsu-to in combination with Choto-san (Diao-Teng-San) has been shown to be effective for the treatment of various symptoms of cerebrovascular diseases such as cerebral apoplexy, ischemia and infarction, and senile dementia in humans (Table 2) (Narita, 1991). We have been investigating the anxiolytic and anti-depressive effects of Shimotsu-to using animal models of various diseases to clarify the psychopharmacological actions of traditional medicines (Watanabe *et al.*, 1990; 1991a).

Table 1 Components of Shimotsu-to (四物湯) and its indications

Components	Angelica root 3 g, Peony root 3 g, Cnidium rhizome 3 g, Rehmannia root 3 g.
Indications	Feeling of cold, dysmenorrhea, menopause syndrome, dysfunction of autonomic nervous system, infertility, exhaustion after parturition, hypertension, pigmentation in elderly people.
Target group	Relatively weak, pale face, dried skin, trophic disturbance of the skin, reduced abdominal tension, palpitation of the abdominal aorta.
Traditional applications	Yin-disease 1 stage, Yin-excess, Oketsu-syndrome, Ki-deficiency.

Cited from K. Terasawa, *Kampo*, p. 216, K. K. Standard McIntyre, Tokyo (1991).

Table 2 Traditional medicines which have therapeutic effects on dementia in humans

Name of prescription	*Component herbs*
Oren-gedoku-to(4) 黄連解毒湯	Coptis, Scutellaria, Phellodendron, Gardenia.
Choto-san(11) 釣藤散	Uncaria, Citrus, Pinellia, Ophiopogon, Poria, Ginseng, Chrysanthemum, Ledebouriella, Gypsum, Glycyrrhiza, Zingiber. combined with Shigyaku-san (Bupleurum, Paeonia, 四逆散 Glycyrrhiza, Citrus). Shimotsu-to (Cnidium, Angelica, Paeonia, Rehmannia). 四物湯
Kihi-to(12) 帰脾湯	Astragalus, Ginseng, Atractylodes, Poria, Zizyphus, Longan, Angelica, Zingiber, Zizyphus, Polygala, Glycyrrhiza, Aucklandia.
Kami-kihi-to(14) 加味帰脾湯	above 12 herbs + Bupleurum, Gardenia.
Yoku-kan-san-ka-chimpi-hange(9) 抑肝散加陳皮半夏	Pinellia, Citrus, Bupleurum, Uncaria, Poria, Glycyrrhiza, Angelica, Cnidium, Atractylodes.
Ninjin-yoei-to(12) 人参養栄湯	Polygala, Paeonia, Ginseng, Angelica, Citrus, Astragalus, Glycyrrhiza, Cinnamomum, Schizandra, Poria, Rehmannia, Atractylodes.

The numbers in parentheses show the number of herb components of the medicines.

The aim of this chapter is to present data concerning the anti-dementia activity of Shimotsu-to using an animal model of spatial cognitive deficits induced by scopolamine or permanent occlusion of the bilateral common carotid arteries.

2. PSYCHOPHARMACOLOGICAL PROFILE OF SHIMOTSU-TO

Shimotsu-to produces neither effects on general behavior nor spontaneous motor activity in naive mice and rats. However, Shimotsu-to showed anti-stress effects in mice.

An injection of pentobarbital produced a shorter sleeping time in naive mice during a stress situation such as forced shaking stress, long-term isolation or communication box-stress than that in the control animal (Matsumoto *et al.*, 1991a; Nguyen Thi *et al.*, 1995; Ishikawa *et al.*, 1992). A single administration of Shimotsu-to recovered the sleeping time to the control level in mice with repeated forced-shaking in a cold room (Watanabe *et al.*, 1990), whereas it did not affect the pentobarbital-induced sleep of long-term isolated mice (unpublished data). These results indicate that the hypnotic activity of pentobarbital is suppressed by various stressors and is differentially affected by Shimotsu-to.

In contrast, long-term isolated housing of mice produces a fighting behavior (Matsumoto *et al.*, 1991b; Yoshimura and Ogawa, 1984). Shimotsu-to inhibited the fighting behavior of mice following isolated feeding for 4 to 6 weeks. Consequently, Shimotsu-to has anti-stress effects, although the action mechanism is as yet unclear. It appears to have multiple sites of action depending on the kind of stress responses elicited.

Shimotsu-to produced effects neither on the immobility of rats in the forced swimming test which has been used as a screening method for tricyclic antidepressants (Porsolt *et al.*, 1978), nor on the conflict behavior of rats in a simple operant discrimination performance which has been used for the screening of anxiolytic drugs (Geller and Seifter, 1960). These results suggest that Shimotsu-to may not produce antidepressive or anxiolytic effects as evaluated like synthesized drugs. The reference drugs desipramine and diazepam caused sedation and prolonged the pentobarbital sleeping time not only in naive mice but also in stressed animals. Desipramine, a typical tricyclic antidepressant, suppressed the immobility of rats in the forced swimming test and diazepam, an anxiolytic drug, inhibited conflict behavior in a conditioned discrimination test. Thus, the psychopharmacological profile of Shimotsu-to is different from those of desipramine and diazepam (Table 3) (Watanabe *et al.*, 1990). Oren-gedoku-to has been used clinically to treat patients with senile dementia, and was shown to be a useful herbal medicine which is comparable to western medicines (Araki, 1991). The psychopharmacological profile of Oren-gedoku-to is similar to that of Shimotsu-to.

3. EFFECT OF SHIMOTSU-TO ON SPATIAL COGNITIVE DEFICITS INDUCED BY SCOPOLAMINE IN RATS

First, we investigated the effect of Shimotsu-to on the spatial cognitive deficits of rats induced by scopolamine injection using an eight-arm radial maze performance

Table 3 Psychopharmacological profile of Shimotsu-to

	Spontaneous activity	*Forced shaking low temp.*	*Long-term isolation*		*Forced swim.*	*Conditioned behavior*
		Pent.-sleep	*Pent.-sleep*	*Fighting*	*Immobility*	*Conflict*
Shimotsu-to	→	↑	→	↓	→	→
Oren-gedoku-to	→	↑	?	↓	→	→
Hochu-ekki-to	↗	→	?	→	↗	→
Desipramine	↓	↑	?	↑↓	↓	→
Diazepam	↓	↑	↑	↓	→	↓

Temp.: Temperature, Pent.-sleep: pentobarbital-induced sleep, ?: not tested.
↑ : increase, → : no change, ↓ : decrease, ↗: slight increase.

test. In the radial maze performance test, scopolamine (0.075–0.3 mg/kg, intraperitoneally, (i.p.)) dose-dependently induced the spatial cognitive deficit. Since the highest dose of scopolamine completely inhibited the rats' spatial cognition in this experiment, 0.3 mg/kg of scopolamine was administered to animals to induce the disruption of maze performance (Watanabe *et al.*, 1991b). A reference drug, 9-aminotetrahydroacridine (THA; 0.1–1.0 mg/kg, per os (p.o.)), dose-dependently improved the scopolamine-induced spatial cognitive deficit (Akazawa *et al.*, 1996). A single administration of Shimotsu-to (1.0 g dried herbs/kg, p.o.), significantly improved the scopolamine-induced performance deficit (Figure 1). Shimotsu-to improved not only spatial cognition but also motor function in the scopolamine-treated rats. However, this phenomenon may not be due to changes in the motor activity, but might be specific to scoplamine-treated animals, since Shimotsu-to itself does not affect the motor activity in naive animals. The repeated administration of Shimotsu-to (0.5 g/kg, p.o.) for one week also exhibited the reversal of scopolamine-induced impairment in the radial maze performance (Watanabe *et al.*, 1991b). These results clearly demonstrated the beneficial effects of Shimotsu-to on a spatial cognitive deficit.

To clarify which component herbs contribute to the cognition-improving effect of Shimotsu-to, we examined the effect of each componet herb on the scopolamine-

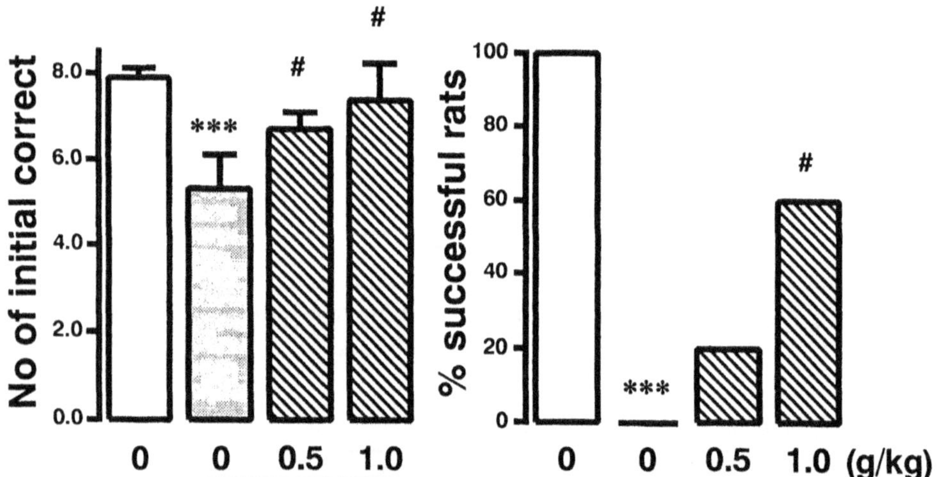

Figure 1 Effect of Shimotsu-to on scopolamine-induced deficit of radial maze performance in rats. The left panel shows the number of correct responses by the rats before committing the first error. The right panel shows the percentage of rats which made no error or only one error among the eight choices. Saline (O) or scopolamine (SCOP, 0.3 mg/kg) was intraperitoneally administered 30 min before testing. Shimotsu-to (0 ; tap water, 0.5 and 1 g/kg) was orally administered 60 min before scopolamine. Each column represents the mean ± S.E.M. (n = 10).

*** $p < 0.001$ compared to O group; $^{\#}$ $p < 0.05$ compared to Shimotsu-to 0 g/kg group (Mann–Whitney U-test for the left panel. and Fisher's probability test for the right panel.).

induced deficit in the radial maze performance. Peony root (0.25 and 1 g/kg, p.o.) and Japanese angelica root (1 g/kg, p.o.) significantly attenuated the scopolamine-induced disruption of the spatial cognition, whereas neither cnidium rhizome (1 g/kg, p.o.) nor rehmannia root (1 g/kg, p.o.) affected it (Ohta *et al.*, 1993a). Hence we examined effects of various fractions separated from the methanol extract of peony root on scopolamine-induced spatial cognitive impairment using the eight-arm radial maze task in rats. The methanol extract of peony root dose-dependently attenuated the scopolamine-induced disruption of the maze performance. The water-soluble fraction of methanol extract, which contains glycosides, also significantly attenuated the scopolamine-induced disruption, whereas the less-polar fraction of methanol extract failed to improve it. Moreover, paeoniflorin (0.1 mg/kg), a major component of this water-soluble fraction, improved the scopolamine-induced impairment dose-dependently at the doses of 0.01–1 mg/kg, p.o. The dose–response relationship of paeoniflorin was bell-shaped, however, since the effect deteriorated at the dose of 3 mg/kg. These results suggest that the aqueous extract of peony root ameliorates spatial working memory deficits in rats, and indicate that such a beneficial effect is attributable mainly to the water-soluble glycocides such as paeoniflorin of peony root extract (Nishi *et al.*, 1994).

The effects of the glycoside constituents of peony root, paeoniflorin, oxypaeoniflorin, benzoylpaeoniflorin and albiflorin, on scopolamine-induced spatial cognitive impairment were also investigated. Paeoniflorin (0.01–0.1 mg/kg, p.o.), oxypaeoniflorin (0.01–0.1 mg/kg, p.o.) and benzoylpaeoniflorin (0.3–3 mg/kg, p.o.), constituents which have a cage-like skeleton with an acetal and hemiketal structure, dose-dependently attenuated the scopolamine-induced impairment of maze performance. The ameliorative effects of these constituents decreased at large doses. Albiflorin (0.01–1 mg/kg, p.o.), which has a lactone ring in the pinane skeleton, had no effect on the maze performance disrupted by scopolamine. These results suggest that not only paeoniflorin but also oxypaeoniflorin and benzoylpaeoniflorin ameliorate spatial cognitive impairment, and that a pinane skeleton including an acetal and hemiketal structure may play an important rôle in the ameliorating effect (Akazawa *et al.*, 1996).

The ameliorating effect of paeoniflorin was blocked by an alpha 1-adrenoceptor antagonist, prazosin, but not by an alpha 2-antagonist, yohimbine (Ohta *et al.*, 1993b). Moreover, a peripherally acting beta-adrenoceptor antagonist, atenolol, antagonized the ameliorating effect of paeoniflorin. These results suggest that alpha 1- and/or beta-adrenergic systems may be involved in the antagonizing effect of paeoniflorin on the scopolamine-induced deficit in radial maze performance in rats (Ohta *et al.*, 1993c, d). Paeoniflorin attenuated a learning impairment in an operant brightness discrimination task which was observed in aged rats (Ohta *et al.*, 1993e, 1994).

As part of the investigation of the action mechanisms, the effects of Shimotsu-to and paeoniflorin, a major component of peony root, on the acetylcholine contents in rat brain tissues were measured. Scopolamine significantly decreased the acetylcholine contents in the hippocampus, cerebral cortex and striatum. Shimotsu-to itself did not affect the acetylcholine content in the naive rat brain, while it antagonized the decrease in the acetylcholine content induced by scopolamine in the rat striatum (unpublished data). Paeoniflorin significantly prevented the scopolamine-induced decrease in the acetylcholine content in the striatum, and also showed a tendency to antagonize the acetylcholine decrease in the hippocampus or cerebral cortex (Ohta *et al.*, 1993a). These data suggest that Shimotsu-to may be a useful prescription as a cognition enhancer in patients with senile dementia, and that Shimotsu-to and paeoniflorin may act via a cholinergic mechanism in the CNS.

4. EFFECT OF SHIMOTSU-TO ON BRAIN LESIONS AND SPATIAL COGNITIVE DEFICIT INDUCED BY PERMANENT OCCLUSION OF THE COMMON CAROTID ARTERIES OF THE RAT

We further studied the effect of Shimotsu-to on brain lesions and a spatial cognitive deficit induced by the permanent occlusion of the bilateral common carotid arteries (2VO) of rats. The 2VO produced a prominent reduction of the blood flow in the cerebral cortex, the hippocampus, the striatum and other brain regions. In the frontal cortex, the blood flow reduced to 30% of the control level immediately after the 2VO and had recovered to 70% of the control level at three weeks after the 2VO. These results suggest that the 2VO creates chronic hypocirculation in the brain.

Various degrees of brain lesion were produced after the 2VO (Ni *et al.*, 1995; Tanaka *et al.*, 1996). A prominent infarction was detected in the cerebral cortex and the striatum 1 month after the 2VO. The ratios of animals with infarction in the cerebral cortex 1 and 4 months after the 2VO were 28.6 and 14.3%, respectively, whereas those in the striatum 1 and 4 months after the 2VO were both 42.9%. Infarction in the hippocampus and hypothalamus was rarely detected 1 month after the 2VO, whereas the rate of infarction was increased 2 months after the 2VO. The number of pyramidal cells in the CA1 region of the hippocampus significantly decreased 1 month after the 2VO, whereas it did not significantly decrease in the CA2 and 3 regions at the same time period. Surprisingly, rarefaction was first found in the corpus callosum in 14.3 and 71.4% of the animals when examined 1 and 4 months after the 2VO, respectively (Figure 2). Thus, the lesion could be detected anywhere in the brain region two months after the 2VO. According to the electron microscopic observation of the brain tissue, the changes in the neuronal cells and glial cells were due to the chronic ischemia (unpublished data).

The acetylcholine content in the striatum was significantly decreased 1 month after the 2VO (by 14.9%), whereas the content in other regions did not change significantly (Ni *et al.*, 1995). It was reported that the 2VO caused an activation of glial cells and that cyclosporin A suppressed the change of glial cells (Wakita *et al.*, 1994, 1995). Taken together, these and the present results clearly show that the infarction and rarefaction induced by the 2VO in the brain were due to the chronic ischemia; this animal model is thus similar to leukoencephalopathy in humans.

The chronic hypocirculation in the brain caused a gradual development of the spatial cognitive deficit in eight-arm radial maze performance. The rats that had not been pretrained could not learn the radial maze performance 1 month after the 2VO. The number of errors in the maze performance did not reach the level of

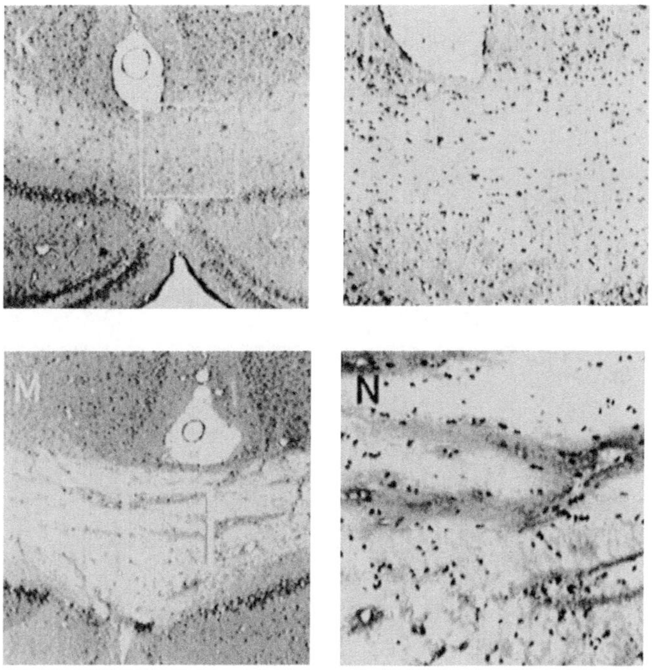

Figure 2 Rarefaction in the corpus callosum 4 months after the 2VO. K and L are from the sham operated rat, and M and N are from the 2VO rat. Magnification: × 4 for K and M, × 20 for L and N.

the sham-operated controls even after repeated training. In contrast, the rats that had been well pretrained exhibited a slight impairment in the retention of radial maze performance 3–7 days after the 2VO, and the deficit gradually recovered to the level of the control group with repeated trials. There was, however, no such recovery 4 months after the 2VO (Ni *et al.*, 1994).

The subchronic administration of Shimotsu-to (0.3 and 1.0 g dried herbs/kg, p.o.) did not improve the spatial cognitive deficit induced by the 2VO (Figure 3). The repeated administration of aminotetrahydroacridine (Tacrine; 0.1 and 0.3 mg/kg, i.p.), a reference drug, from 1 month after the 2VO improved the spatial cognitive deficit in the radial maze task (Murakami *et al.*, 1997).

Shimotsu-to did not affect the induction rate of infarction in the cerebral cortex, striatum or thalamus, or that of rarefaction in the white matter (corpus callosum). However, when the size of the brain lesion was estimated by a computerized imaging system, the development of infarction of the parenchyma and enlargement of the cerebral ventricle were suppressed by Shimotsu-to (0.3 g/kg, p.o.) at two months after the 2VO (Figure 4). The enlargement of the cerebral ventricle was also inhibited by Shimotsu-to (1 g/kg, p.o.) at 4 months after the 2VO (unpublished data). These results indicate that the chronic administration of Shimotsu-to, which has a weak sedative effect on stress-loaded mice as described in Table 3, delayed the development of the infarction and rarefaction following chronic ischemia of the

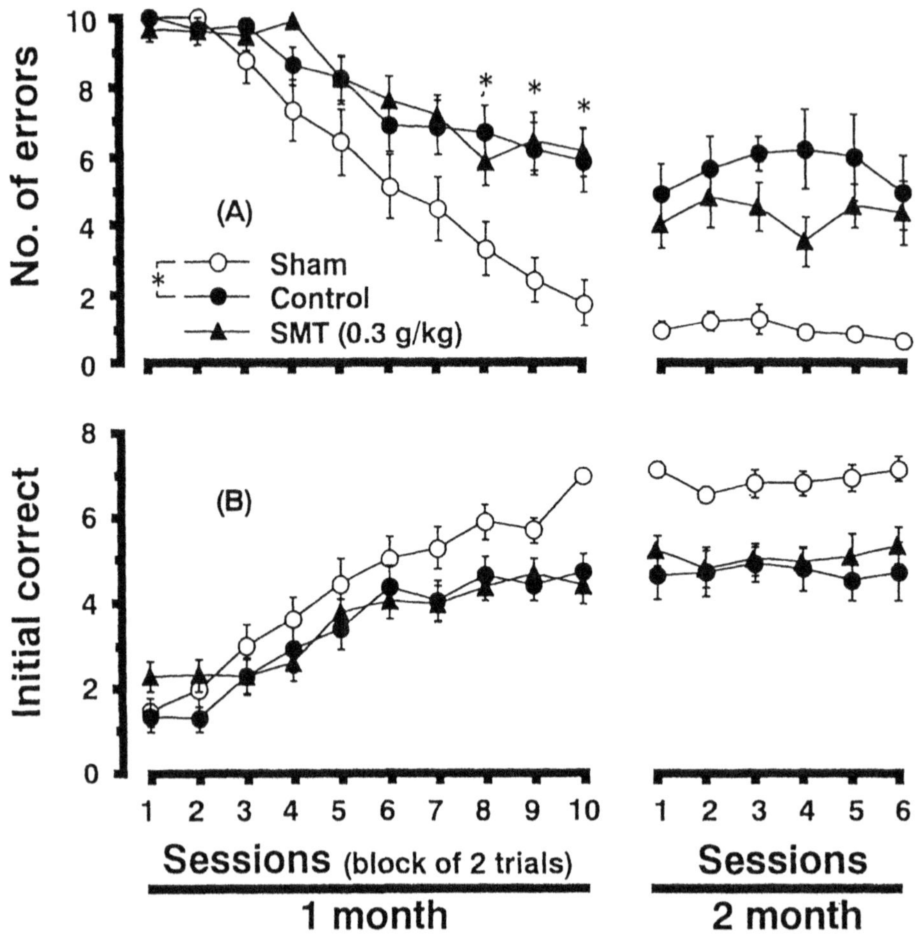

Figure 3 Effect of the subchronic administration of Shimotsu-to (SMT, 0.3 g/kg/day, p.o.) on radial maze learning of 2VO (permanent occulusion of the bilateral common carotid arteries) rats. Shimotsu-to was given from 3 days prior to the 2VO surgery. The learning task started two weeks after the 2VO and continued for two months. Data represent the mean ± S.E.M. (n = 8–9).
* $p < 0.05$ compared to the sham-operated group.

brain. Thus, the present results support the clinical evidence that Shimotsu-to has beneficial effects on various symptoms followed by cerebral vascular diseases.

5. CONCLUSIONS

We established an animal model which is similar to human leukoencephalopathy, produced by chronic brain ischemia. We examined the effect of the subchronic

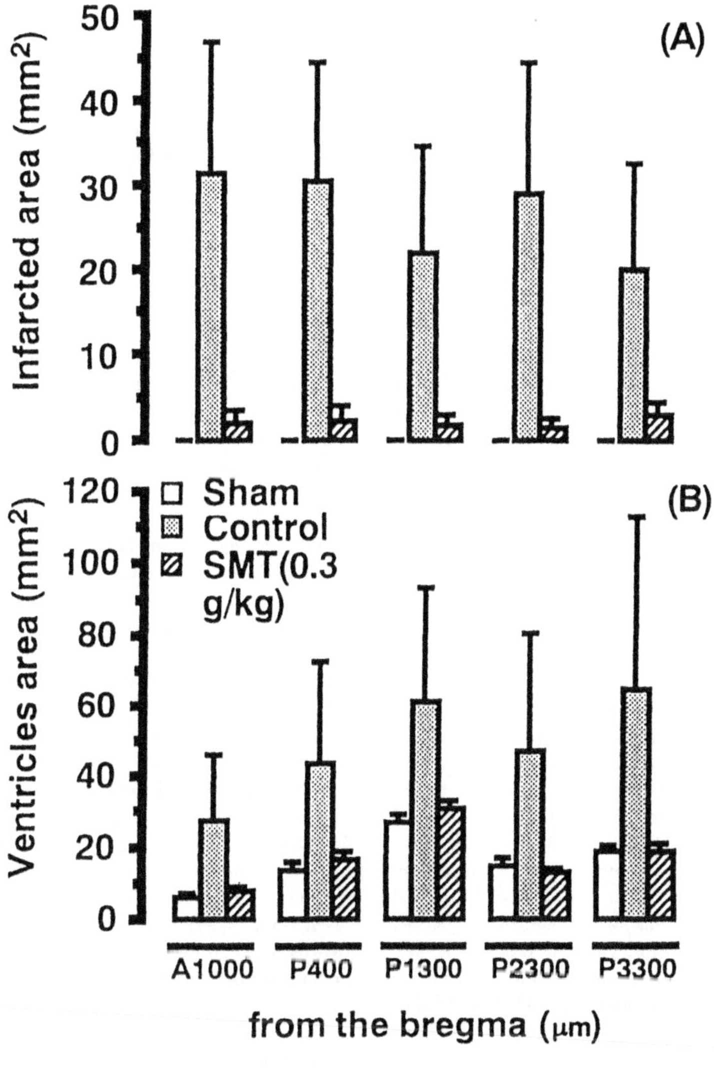

Figure 4 Effect of the subchronic administration of Shimotsu-to (SMT, 0.3 g/kg/day, p.o.) on the cerebral infarction induced by the 2VO. The rat brain was removed after behavioral tests at 2 months after the 2VO and sliced in 0.14 mm coronal sections. After the staining, infarcted area (A) and area of the cerebral ventricle (B) were measured with computerized imaging system. Each column represents the mean ± S.E.M. (n = 8–9).

administration of Shimotsu-to on this model. The present results suggest that a basic prescription of the traditional medicine, Shimotsu-to, prevents the development of brain infarction and rarefaction induced by chronic brain ischemia. Further studies are necessary to clarify the active constituent(s) and action mechanism of Shimotsu-to, with the goal of identifying more effective cognitive enhancers.

ACKNOWLEDGEMENTS

This work was supported in part by a Grant-in-Aid (No. 08457634) from the Ministry of Science, Education, Sports and Culture of Japan. The authors are grateful to Drs. J.-W. Ni and H. Ohta, Mr. Y. Murakami, T. Sakai and Miss E. Tanaka for their collaboration.

REFERENCES

1. K. Akazawa, K. Matsumoto, M. Shimizu, H. Ichiki, M. Maruno and H. Watanabe, Effects of paeoniflorin and paeoniflorin-related glycosides on scopolamine-induced disruption of radial maze performance in rats. *J. Trad. Med.* **13**, 243–247 (1996).
2. G. Araki, Therapeutic effect of Ore-gedoku-to on cerebrovascular diseases followed by senile dementia, In *Syndrome Following Cerebrovascular Disorders in Aged People and Kampo*, G. Araki, ed., Iyaku Journal Inc., Tokyo (1991).
3. S. Chen, select, *Zengguan Taipinghuimin hejijufang*, Enterprise Co., Ltd, Tokyo, p. 150 (1988).
4. I. Geller and J. Seifter, The effects of meprobamate, barbiturates, d-amphetamine and promazine on experimentally induced conflict in the rat. *Psychopharmacol. (Berlin)* **1**, 482–492 (1960).
5. M. Ishikawa, C. Hara, S. Ohdo and N. Ogawa, Plasma corticosterone response of rats with sociopsychological stress in the communication box. *Physiol. & Behav.* **52**, 475–480 (1992).
6. S. Kashiwabuchi, Kampo treatment for anemia induced by iron deficiency. *Kampo Shinryo* (in Japanese) **6**, 40–43 (1992).
7. S. Kojima, N. Ezaki, M. Inoue, K. Yanagisawa and Y. Sugawara, Effects of Shimotsu-to and Kihi-to on the anemic rats. *J. Trad. Med.* (Abstract in English) **11**, 231–235 (1994).
8. S. Kojima, N. Ezaki, M. Inoue, K. Yanagisawa and Y. Sugawara, Effects of Shimotsu-to on the hematopoietic system in the anemic rats. *J. Trad. Med.* (Abstract in English) **12**, 229–234 (1995).
9. K. Matsumoto, T. Satoh, H.-B. Li, H. Ohta and H. Watanabe, Effects of forced shaking stress at low temperature on pentobarbital-induced sleeping in mice. *Gen. Pharmacol.* **22**, 729–733 (1991a).
10. K. Matsumoto, Cai, T. Satoh, H. Ohta and H. Watanabe, Desipramine enhances isolation-induced aggressive behavior in mice. *Pharmacol. Biochem. Behav.* **39**, 167–170 (1991b).
11. Y. Murakami, E. Tanaka, Y. Sakai, K. Matsumoto, H.-B. Li and H. Watanabe, Tacrine improves working memory deficit caused by permanent occlusion of bilateral common carotid arteries in rats. *Jpn. J. Pharmacol.* **75**, 443–446 (1997).
12. H. Narita, Kampo medicines and cerebrovascular disorders. In *Syndrome Following Cerebrovascular Disorders in Aged People and Kampo*, G. Araki, ed., Iyaku Journal Inc., Tokyo (1991).
13. T.H. Nguyen Thi, K. Matsumoto, K. Yamasaki, D. Nguen Minh, N. Nguyen Tho and H. Watanabe, Crude saponin extracted from Vietnamese ginseng and its major constituent Majonoside-R_2 attenuate the psychological stress-and footshock stress-induced antinociception in mice. *Pharmacol. Biochem. Behav.* **52**, 427–432 (1995).
14. J.-W. Ni, H. Ohta, K. Matsumoto and H. Watanabe, Progressive cognitive impairment following chronic cerebral hypoperfusion induced by permanent occlusion of bilateral carotid arteries in rats. *Brain Res.* **653**, 231–236 (1994).
15. J.-W. Ni, K. Matsumoto, H.-B. Li, Y. Murakami and H. Watanabe, Neuronal damage and decrease of central acetylcholine level following permanent occlusion of common carotid arteries in rat. *Brain Res.* **673**, 290–296 (1995).
16. K. Nishi, K. Matsumoto, H. Ohta, M. Shimizu and H. Watanabe, Glycoside fraction of peony root improves thescopolamine-induced disruption of spatial cognition in rats. *J. Trad. Med.* **11**, 118–122 (1994).
17. H. Ohta, J.-W. Ni, K. Matsumoto, H. Watanabe and M. Shimizu, Peony and its major constituent, paeoniflorin, improve radaial maze performance impaired by scopolamine in rats. *Pharmacol. Biochem. Behav.* **45**, 719–723 (1993a).

18. H. Ohta, J.-W. Ni, K. Matsumoto, H. Watanabe and M. Shimizu, Involvement of α1- but not α2-adrenergic systems in the antagonizing effect of paeoniflorin on scopolamine-induced deficit in radial maze performance in rats. *Jpn. J. Pharmacol.* **62**, 199–202 (1993b).
19. H. Ohta, J.-W. Ni, K. Matsumoto and H. Watanabe, The interaction between central cholinergic and peripheral β-adrenergic systems on radial maze performance in rats. *Brain Res.* **662**, 353–356 (1993c).
20. H. Ohta, K. Matsumoto, H. Watanabe and M. Shimizu, Involvement of β-adrenergic systems in the antagonizing effect of paeoniflorin on the scopolamine-induced deficit in radial maze performance in rats. *Japan J. Pharamacol.* **62**, 345–349 (1993d).
21. H. Ohta, K. Matsumoto and H. Watanabe, Impairment in acquisition but not in retention of a simple operant discrimination performance in aged Fischer 344 rats. *Physiol. Behav.* **54**, 443–448 (1993e).
22. H. Ohta, K. Matsumoto, M. Shimizu and H. Watanabe, Paeoniflorin attenuates learning impariment of aged rats in operant brightness discrimination task. *Pharmacol. Biochem. Behav.* **49**, 213–217 (1994).
23. R.D. Porsolt, G. Anton, N. Blavet and M. Jalfre, Behavioural despair in rats: A new model sensitive to antidepressant treatment. *Eur. J. Pharmacol.* **47**, 379–391 (1978).
24. K. Tanaka, N. Ogawa, M. Asanuma, Y. Kondo and M. Nomura, Relationship between cholinergic dysfunction and discrimination learning disabilities in Wistar rats following chronic cerebral hypoperfusion. *Brain Res.* **729**, 55–65 (1996).
25. H. Wakita, H. Tomimoto, I. Akiguchi and J. Kimura, Glial activation and white matter changes in the rat brain induced by chronic cerebral hypoperfusion: an immunohistochemical study. *Acta Neuropathol.* **87**, 484–492 (1994).
26. H. Wakita, H. Tomimoto, I. Akiguchi and J. Kimura, Protective effect of cyclosporin A on white matter changes in the rat brain after chronic cerebral hypoperfusion. *Stroke* **26**, 1415–1422 (1995).
27. H. Watanabe, K. Matsumoto, T. Satoh, H. Ohta and H. Matsuda, Psychopharma-cological studies on Shimotsu-to, Kakkon-to and Oren-gedoku-to using animal models. *J. Med. Pharm. Soc. for Wakan-Yaku* (Abstract in English) **7**, 99–107 (1990).
28. H. Watanabe, T. Satoh, K. Matsumoto, H. Ohta, H. Matsuda and T. Shimada, Psychopharmacological studies on Shimotsu-to and Shimotsu-to components, Toki, Shakuyaku, Jio and Senkyu, in the mouse. *J. Med. Pharm. Soc. for Wakan- Yaku* (Abstract in English) **8**, 102–107 (1991a).
29. H. Watanabe, J.-W. Ni, H. Ohta, X.-H. Ni and K. Matsumoto, A Kampo prescription, Shimotsu-to, improves scopolamine-induced spatial cognition deficits in rats. *Jpn. J. Psycopharmacol.* **11**, 215–222 (1991b).
30. H. Yoshimura and N. Ogawa, Pharmaco-ethological analysis of agonistic behavior between resident and intruder mice: Effects of psychotropic drugs. *Folia Pharmacol. Japon* **84**, 221–228 (1984).

6. EXPERIMENTAL ANALYSIS OF THE ANTI-AGING EFFECT OF *PANAX GINSENG*

NOBUYOSHI NISHIYAMA* and HIROSHI SAITO

Department of Chemical Pharmacology, Faculty of Pharmaceutical Sciences, The University of Tokyo, 7-3-1 Hongo, Bunkyo-ku, Tokyo 113-0033, Japan

INTRODUCTION

Multiple experimental paradigms should be employed in the development of anti-aging drugs. We are currently using three major experimental frameworks:

1. Primary neuronal culture for the determination of neurotrophic effects of the candidate drugs.
2. Electrophysiological experiments including the recording of long-term potentiation as one of the typical examples of synaptic plasticity which underlies learning and memory function.
3. Behavioral experiments which will eventually give the most important and convincing information about the drug under study.

Ginseng (the root of Panax ginseng C.A. Meyer), which has been utilized for more than two thousand years in Chinese medicine and is now popular throughout the world as a natural medicine, is characterized as a drug, Bu-Chi-Yaw, which is used to replenish vital energy and strengthen physical resistance, a concept for which we have no alternative technical terms in modern medicine. The effect of ginseng expected from its chief indications described in old and recent Chinese traditional medical books implicates its beneficial action on the diseases of old age, especially on senile amnesia and/or dementia. Prof. Shibata and his coworkers determined the structure of saponins in ginseng and provided us with crude saponin fractions and individual saponins.

1. BEHAVIORAL AND PHARMACOLOGICAL PROPERTIES OF GINSENG

When the pharmacological study of ginseng was initiated in our laboratory, we adopted general pharmacological screening tests using mice as the first step of investigation.[1–5] Briefly, central nervous system (CNS) depressant action was found in all saponins except ginsenoside Rg_1 (GRg_1). Ginsenoside Rb_1 (GRb_1), in particular, a panaxadiol saponin, had remarkable sedative and tranquilizing actions.

* To whom correspondence should be addressed

Decreases in motor coordination, body tone, grip tone, body temperature, and antipyretic action were observed in GRb_1-treated mice. Spontaneous movement was also apparently decreased by the ginseng saponins. Exploratory movement and conditioned avoidance response examined in a pole-climbing test were remarkably depressed by all ginseng fractions. These results indicate that GRb_1 and similar saponins may have tranquilizing and anti-anxiety effects. Although GRg_1 showed CNS depressant action at high doses, it shortened the response latency in a pole-climbing test and increased motor activity at a lower dose suggesting its biphasic action on the CNS.

Since a decrease in physical strength is one of the characteristic alterations observed in aged people, we produced a fatigue model in mice by oscillating them.[6] Ginseng preparations were administered after the oscillation. Water extract, GRb_1 and GRg_1 showed anti-fatigue activity. Similar results indicating the promotion of recovery from fatigue have also been reported using crude ginseng extract.[7] These results provide experimental evidence for the anti-aging or Bu-Chi-Yaw capacity of ginseng.

In summary, GRb_1 exerts CNS depressant action. Other depressant substances may exist in the water-soluble and lipid-soluble fractions. GRg_1 and lipid-soluble fraction exhibited weak CNS-stimulating activity at low doses. Anti-fatigue activity was observed in both GRb_1 and GRg_1. The multiple pharmacodynamic activities of ginseng originated in its various ingredients. Some ingredients have an action antagonistic towards that of another ingredient; for example, CNS-stimulant and depressant actions.

2. EFFECT OF GRb_1 AND GRg_1 ON LEARNING AND MEMORY

One of the biologically, socially and economically important aspects of aging is senile dementia and/or amnesia. Since ginseng has been prescribed as an anti-amnesiac drug in traditional Chinese medicine, we next studied the effect of ginseng and its components on learning and memory ability in experimental animals. As demonstrated by other laboratories and described in the previous section of this chapter, saponins are thought to be the principal active ingredients in ginseng. We thus investigated the effect of ginseng saponins, especially GRb_1 and GRg_1, on learning and memory ability.

2.1. Effects of Ginseng and its Saponins on Sex and Learning Behaviors in Chronic Stressed Mice

Using the chronic hanging stress method reported by Avakian *et al.*, we succeeded in producing a decrease of sex behavior (licking, mounting and intromission) and an increase in the failure of learning behavior (extinction and retrieval of passive avoidance response) in IV-CS strain adult male mice, and thus developed a new screening method.[8,9] The water extract of ginseng (500 mg/kg, p.o.) and GRg_1 (50 mg/kg, p.o.) showed protective effects against the decrease in both sex and learning behaviors. GRb_1 (50 mg/kg, p.o.) protected the mice from the decrease in sex behavior only on the first day of treatment, but a few days later, it potentiated the decrease in sex behavior for several days. GRb_1 showed the most protective effect against the decrease in learning behavior among the ginseng saponins tested. These results suggest that GRg_1 may have facilitated a recovery from the fatigued state and supported memory retrieval.

2.2. Effects of Ginseng and its Saponins on Experimental Amnesia in Mice

Four tests, i.e., a step-through test, a step-down test, a shuttle-box test and a lever-press test were employed. In the step-down test, GRb_1, GRb_2, GRe and GRg_1 antagonized the electroconvulsive shock-induced inhibition of the retention of memory. Moreover, GRg_1 antagonized the ethanol-induced inhibition of memory retrieval and facilitated the registration of memory. In the step-through test, water extract of ginseng, GRb_1, GRb_2, GRe and GRg_1 antagonized the electroconvulsive shock-induced inhibition of the retrieval of memory. In the shuttle-box and lever-press tests, only GRb_1 had an effect on the acquisition and retrieval of memory; it depressed the retrieval of the conditioned avoidance response in the shuttle-box test. These results indicate that GRb_1 and GRg_1 had beneficial effects on the retrieval of memory and on other learning processes.

Although not reproduced in our mice study with a single oral administration protocol, Benishin *et al.* found that repeated intraperitoneal administrations of GRb_1 (5 mg/kg, 4 days) prevented the memory deficits induced by muscarinic receptor antagonist scopolamine in rats.[10] GRb_1 did not affect the muscarinic receptor binding or acetylcholine esterase activity, but it facilitated the release of acetylcholine from hippocampal slices. They speculated that the increase in acetylcholine release is associated with an increased uptake of choline in nerve endings, and that the ability of GRb_1 to prevent memory deficits may be related to a facilitation of acetylcholine metabolism in the CNS. However, this notion requires further examination, since the muscarinic binding sites were reported to increase significantly in ginseng-treated mice brains.[11] More recently, Yamaguchi *et al.* demonstrated that GRg_1, but not GRb_1, improved the spatial learning impairment induced by scopolamine in rats.[12] The apparent discrepancy in terms of the beneficial effect of GRb_1 against scopolamine-induced learning impairment needs further examination, including investigations of the aspects of the category of learning test, animal species and prescription protocol.

2.3. Effects of Ginseng and its Saponins on Memory and Learning in Naive Rats

Two tests, a pole-climbing test and a Y-maze test, were employed.[13,14] In the pole-climbing test, GRb_1 suppressed the conditioned avoidance response and prolonged the response latency, while GRg_1 had no effect on the conditioned avoidance response. However, GRg_1 (10 mg/kg, i.p.) facilitated the rate of acquisition of a discrimination response in naive young adult rats. To determine whether GRg_1 modulated the acquisition of the conditioned avoidance response via a direct mechanism in the brain, GRg_1 was administered intracerebroventricularly. GRg_1 (10 and 50 μg/animal) did not show the facilitatory effect in terms of the first success of conditioned avoidance response in the learning trial, but it reduced the number of trails required to achieve 10 continuous successes of conditioned avoidance. This result indicates that the target of GRg_1 resides somewhere in the CNS and that this saponin accelerates the acquisition of learning directly in naive animals. In the Y-maze test, GRb_1 (> 10 mg/kg, i.p.) elongated both the period of time between the opening of the start box door and the beginning of animal movement and the time it took for the rat to reach the door of the goal box. GRg_1 (20 mg/kg, i.p.) shortened both of these time parameters. GRg_1 (10 mg/kg, i.p.) also significantly accelerated the rate of acquisition for discrimination reversal learning. These results indicate that GRg_1 had an effect on the acquisition process of learning response.

The favorable effects of ginseng on learning and memory processes were also demonstrated in aged rats.[15] In some cases, the memory-improving effect of ginseng, especially at low doses, was more pronounced in old rats than in young rats. The effect of ginseng on cognition was accompanied by the ginseng-induced increase in the levels of dopamine and noradrenaline and a decrease in the serotonin level in the hippocampus.[15] Watanabe *et al.* reported that chronic feeding with 1% red ginseng powder resulted in a significant increase in D-2 dopamine receptors in aged rats.[16,17] These results indicate that treatment with ginseng may benefit the aged animal by modulating the neurotransmitter signaling in the CNS.

Free radical generation and lipid peroxidation is strongly associated with aging and aging-related disorders.[18–20] Ginseng administered continuously to old rats suppressed the increase in lipid peroxidation products in the blood.[15] Furthermore, GRb_1 antagonized lipid peroxidation, scavenged oxygen free radicals and increased catalase and gluthathion peroxidase activity in rat liver.[11] The reported benefits of ginseng include reductions of serum cholesterol, triglyceride, and free fatty acids.[21] The reduction of these metabolites and their peroxidized products may be related to the anti-aging effect of ginseng.

3. EFFECTS OF GINSENG SAPONINS ON CULTURED NEURONS

To further investigate the effects of ginseng saponins on the nervous system, we used a neuron cell culture technique to determine their effects on neurite extension, neuron survival, and neurotransmitter-synthesizing enzyme activity.

3.1. Effect of Ginseng Saponins on the Neurite Extension of Dorsal Root and Sympathetic Ganglia

Nerve growth factor (NGF) is a polypeptide which promotes neuron survival and neurite outgrowth, and maintains the function of sensory and sympathetic neurons.[22] The stimulation of neural fiber outgrowth from chick embryonic dorsal root or sympathetic ganglia in tissue culture is a biological criterion of NGF activity.[23]

We tested the effects of various ginseng saponins in this tissue culture system.[24,25] GRb_1 and GRg_1 did not promote fiber production by themselves, but the effect of NGF was markedly potentiated by GRb_1. Optimal fiber outgrowth was achieved with 0.83 nM NGF, and in the presence of 30 μM GRb_1 it was achieved with 0.12 nM NGF. These results indicate an about 7-fold potentiation of the original effect of NGF by GRb_1. This potentiation of the NGF effect by GRb_1 was similar in dorsal root and sympathetic ganglia. The ginseng saponins GRd, GF_2 and GM_{6-a} markedly potentiated the neurite extension effect of NGF. They are all 20*S*-protopanaxadiol glycosides and have a few glycosides in their two branches. The removal of glucose or the introduction of a hydroxy group into the side chain of GRd reduced the activity, indicating that the whole structure including the glucose residue is critical for the potentiation of the NGF effect. A neurite extension effect of the lipophilic component in ginseng was also observed in a rat pheochromocytoma cell culture.[26] Determination of the active substance and further examination using a primary neuron will provide important information.

3.2. Effect of GRb_1 on the Neuron Survival of Chick Sympathetic Neurons[27,28]

Chick embryonic sympathetic neurons were cultured for 72 hour in the presence of GRb_1 (30 μM) alone, NGF alone, or GRb_1 (30 μM) plus NGF. Then neurofilament content, one of the indicators of neuron survival, was measured by enzyme-linked immunosorbent assay. NGF enhanced neuron survival markedly, which is consistent with previously documented results. However, GRb_1 had no effect and did not potentiate the effect of NGF. These results indicate that GRb_1 does not enhance neuron survival in spite of its potentiating effect on NGF-induced fiber outgrowth.

3.3. Effects of GRb_1 and GRg_1 on the Survival of Brain Neurons[27]

NGF is a strong stimulator of cell survival of peripheral neural crest-derived neurons, but the situation is almost completely different in central neurons, with a few exceptions. This regional difference may be due to the distribution of the specific receptor for NGF. Similarly, although we did not find any survival-promoting effect of ginseng saponins on peripheral neurons, central neurons derived from NGF-specific areas might respond to ginseng saponins. We tested this notion using several central neurons. Cerebral cortex neurons were cultured first. Since cortical neurons are heterogeneous in neurotransmitter phenotype and cell morphology, they were stained with anti-neurofilament antibody and all surviving neurons were counted. GRb_1 and GRg_1 significantly enhanced the survival of cerebral cortex neurons obtained from chick and rat. NGF had no effect and did not potentiate the effect of these saponins, which may be because cortical neurons do not possess functional NGF receptors. Crude saponins of ginseng root have been demonstrated to promote the neurite extension in cultured rat cerebral cortex neurons.[29] Thus, ginseng saponins exert neurotrophic action on cerebral cortex neurons in culture.

To determine which transmitter phenotype neurons are responsive to ginseng saponins, we examined dopaminergic neurons. The ventral midbrain contains two major dopaminergic nuclei, the substantia nigra and the ventral tegmental area. Ventral midbrain cells prepared from 16-day-old rat embryos were cultured and stained immunohistochemically with anti-tyrosine hydroxylase (TH; the first rate-limiting catecholamine synthesizing enzyme) antibody. The TH-positive cells were counted as surviving dopaminergic neurons. GRb_1 and GRg_1 had no effect on the survival of TH-positive neurons, indicating that these saponins do not support the cell survival of dopaminergic neurons in culture.

A large number of cells in the medial septum are cholinergic neurons which project to the hippocampus. The effect of NGF on the survival of septal area neurons has been established both *in vitro* and *in vivo.* We studied the effect of GRb_1 and GRg_1 on the choline acetyltransferase (acetylcholine synthesizing enzyme) activity as an indicator of cholinergic neuron survival. NGF increased the enzyme activity in the cultured septal neurons from both embryo (16 day) and postnatal (13 day) rats. GRb_1 and GRg_1 had no effect and did not potentiate the effect of NGF in fetal septal cells. Although these ginseng saponins potentiated the NGF-induced neurite extension in peripheral neurons, they did not affect the NGF-mediated neuron survival in central cholinergic neurons. An investigation pursuing this discrepancy may provide new insights into the neurotrophic mechanisms of both ginsenoside and NGF. Neuroprotective effects of red ginseng powder and especially GRb_1 were recently reported in an ischemia-induced hippocampal delayed neuronal death experiment.[30] Utilization of this kind of disease model would be an interesting application for studying the effects of ginseng on neurodegenerative diseases.

In conclusion, GRb_1 and some 20*S*-protopanaxadiol saponins potentiated the NGF-mediated nerve fiber outgrowth in organ cultures of chick dorsal root and sympathetic ganglia, and GRb_1 and GRg_1 significantly supported the neuron survival of both chick and rat cerebral cortex neurons. These results indicate that ginseng saponin might be a novel exogenous neurotrophic factor, and may partly explain the anti-aging action of ginseng.

4. EFFECTS OF GINSENG AND GINSENG SAPONINS ON THE GENERATION OF LONG-TERM POTENTIATION IN RAT HIPPOCAMPUS

Long-term potentiation (LTP) of an evoked response in the hippocampus is one of the manifestations of synaptic plasticity and is considered to be the cellular basis of learning and memory process (for review see 31). Synaptic plasticity in the hippocampus shows a good correlation with the learning performance in rats.[32] Rats with inborn inferior learning performance required a greater threshold to generate LTP in comparison with rats with an inborn superior performance in a shuttle-box test.[33] Ethanol impairs both learning performance and LTP generation.[34,35] From these lines of evidence, it could be speculated that such a drug as ginseng or ginseng saponins which influences learning and memory ability might also affect the LTP generation in the hippocampus.

To test this hypothesis, we first examined the effect of water extract of ginseng on tetanus-induced LTP. LTP cannot be produced in control rats when subthreshold tetanic stimulation (60 Hz, 20 pulses) is applied. However, a single oral ginseng treatment of ginseng water extract (500 mg/kg, 30 mm prior to the tetanus) increased the population spike amplitude evoked by the subthreshold tetanus.[36] LTP is considered to consist of several phases, i.e., an induction phase, an early maintenance phase and a late maintenance phase, which are mediated by different mechanisms.[37] The potentiation which returns to the baseline level within 30 mm is regarded as short-term potentiation (STP), and STP is coupled with the induction of LTP. A ginseng administration might have lowered the threshold or enhanced the stimulating efficacy of a subthreshold tetanus. This STP-promoting effect of ginseng may be correlated with its ameliorative effects on learning performance. However, the ginseng extract at the same dose did not influence the LTP formation induced by a suprathreshold tetanic stimulation, suggesting its adaptive role on synaptic transmission. The search for an active compound in ginseng extract which promotes LTP generation is now in progress. Preliminary results indicate that the non-saponin fraction rather than the saponin fraction enhances LTP generation after oral administration (Kakimoto *et al.*, unpublished observation).

Abe *et al.*[38] tested the effect of ginseng saponins on LTP generation. The intracerebroventricular administration of GRb_1 attenuated the magnitude of LTP induced by suprathreshold tetanus. In contrast, malonylginsenoside Rb_1 ($GRb_{1\text{-}m}$) facilitated the generation of LTP induced by subthreshold tetanus. Since another ginseng saponin, GRg_1, did not affect the induction of LTP at all, the inhibition and facilitation of LTP induction are probably specific actions of GRb_1 and $GRb_{1\text{-}m}$, respectively. A further systematic examination of the responsible chemicals in ginseng and of the optimal route and time-course of administration will provide comprehensive information concerning ginseng's LTP-enhancing effect.

Ginsenoside has been shown to release nitric oxide (NO) from endothelial cells.[39] NO production in the brain is involved in the generation of LTP. The ginseng-induced enhancement of LTP generation may therefore be achieved, at least partly, by enhancing the NO release. Nah *et al.* recently found that a trace saponin ginsenoside, Rf, inhibits high-threshold, voltage-dependent calcium channels through an unidentified receptor linked to pertussis toxin-sensitive G protein.[40] This kind of approach is expected to delineate the specific mechanisms of action of each ginseng saponin.

SUMMARY AND CONCLUSION

The information described in this chapter is summarized in Table 1. GRb_1 exerts a CNS-tranquilizing action, anti-stress action, cognition enhancing action, and neurotrophic action including the modulation of hippocampal synaptic plasticity. GRg_1 and the lipid-soluble fraction exhibited weak CNS-stimulating activity at low doses. An anti-stress action, cognition-enhancing action, and neurotrophic action were also observed in GRg_1. Therefore, the multiple pharmacodynamic activities

Table 1 Pharmacological properties of Ginsenoside Rb_1 and Ginsenoside Rg_1

Ginsenoside Rb_1	*Ginsenoside Rg_1*
Tranquilizing Action Inhibition of CAR (shuttle-box and pole-climbing tests) Prolongation of response latency and running time in PDR performance (Y-maze test) Potentiation of hexobarbital-induced loss of righting reflex Anti-pyretic action	A Weak CNS-Stimulating Action A slight increase in motor activity (Triangular-cage test) Reduction of response latency and running time in PDR performance (Y-maze test)
Anti-Stress Action Protection against acceleration of memory extinction and increase in memory retrieval failure (hanging stress and step-down test) Protection against stress ulcer	Anti-Stress Action Acceleration of recovery from fatigued states (oscillation stress: body temperature, spontaneous and exploratory movements) Protection against decrease in sex behavior and increase in memory retrieval failure (hanging stress and step-down test) Aggravation of stress ulcer
Cognition Enhancing Action Protection against ECS-induced impairment in memory retrieval (step-through test) Protection against ECS-induced impairment in memory consolidation (step-down test)	Cognition Enhancing Action Protection against ECS-induced impairment in memory retrieval (step-through test) Protection against ECS- and ethanol-induced impairment in memory consolidation (step-down test) Acceleration of memory acquisition (step-down and pole-climbing tests) Acceleration of acquisition of CAR and SDR (pole-climbing test), PDR and reversal learning (Y-maze test)
Neurotrophic Action Potentiation of NGF-mediated neurite outgrowth (chick DRG and SymG) Enhancement of neuron survival (chick and rat cerebral cortex neuron) Modulation of hippocampal synaptic plasticity (GRb_1: inhibition, GRb_{1-m}: potentiation)	Neurotrophic Action Enhancement of neuron survival (chick and rat cerebral cortex neuron)

Abbreviations: CAR: conditioned avoidance response, PDR: position discrimination response, ECS: electroconvulsive shock, SDR: sound discrimination response, NGF: nerve growth factor, DRG: dorsal root ganglion, SymG: sympathetic ganglion, GRb_{1-m}: malonyl-ginsenoside Rb_1.

of ginseng originate in its various ingredients. Some ingredients have actions antagonistic to other ingredients; for example, stimulating and depressing the CNS.

REFERENCES

1. K. Takagi, H. Saito and H. Nabata, Pharmacological studies of panax ginseng root: Estimation of pharmacological actions of panax ginseng root. *Jpn. J. Pharmacol.* **22**, 245–259 (1972).
2. K. Takagi, H. Saito and M. Tsuchiya, Pharmacological studies of panax ginseng root: Pharmacological properties of a crude saponin fraction. *Jpn. J. Pharmacol.* **22**, 339–346 (1972).
3. H. Nabata, H. Saito and K. Takagi, Pharmacological studies of neutral saponins (GNS) of panax ginseng root. *Jpn. J. Pharmacol.* **23**, 29–41 (1973).
4. H. Saito, M. Morita and K. Takagi, Pharmacological studies of panax ginseng leaves. *Jpn. J. Pharmacol.* **23**, 43–56 (1973).
5. K. Takagi, H. Saito and M. Tsuchiya, Effect of panax ginseng root on spontaneous movement and exercise in mice. *Jpn. J. Pharmacol.* **24**, 41–48 (1974).
6. H. Saito, Y. Yoshida and K. Takagi, Effect of panax ginseng root on exhaustive exercise in mice. *Jpn. J. Pharmacol.* **24**, 119–127 (1974).
7. I.I. Breckman and I.V. Dardymov, New substances of plant origin which increase nonspecific resistance. *Ann. Rev. Pharmacol.* **9**, 419–430 (1969).
8. H. Saito, N. Nishiyama, H. Fujimori, K. Hinata, T. Kamegaya, Y. Kato and T. Bao, Effects of drugs on sex and learning behaviors and tyrosine hydroxylase activities of adrenal gland and hypothalamic regions in chronic stressed mice. In *Stress, The Role of Catecholamines and Other Neurotransmitters*, F. Usdin, R. Kvetnansky and I.J. Kopin, eds., Gordon and Breach Sciences, New York, pp. 467–480 (1984).
9. N. Nishiyama, T. Kamegaya, A. Iwai, H. Saito, S. Sanada, Y. Ide and J. Shoji, Effects of eleutherococcus senticosus and its components on sex- and learning behaviors and tyrosine hydroxylase activities of adrenal gland and hypothalamic regions in chronic stressed mice. *Syoyakugaku Zasshi* **39**, 238–242 (1985).
10. C.G. Benishin, R. Lee, L.C.H. Wang and H.J. Liu, Effects of ginsenoside Rb1 on central cholinergic metabolism. *Pharmacology* **42**, 223–229 (1991).
11. J.-T. Zhang, Z.-W. Qu, Y. Liu and H.-L. Deng, Preliminary study in the antiamnestic mechanism of ginsenoside Rg1 and Rb1. In Recent Advances in Ginseng Studies, S. Shibata, Y. Ohtsuka and H. Saito, eds., Hirokawa Publishing, Tokyo, pp. 113–123 (1990).
12. Y. Yamaguchi, K. Haruta and H. Kobayashi, Effects of ginsenosides on impaired performance induced in the rat by scopolamine in a radial arm maze. *Psychoneuroendocrinology* **20**, 645–653 (1995).
13. H. Saito, M. Tsuchiya, S. Naka and K. Takagi, Effect of panax ginseng root on conditioned avoidance response in rats. *Jpn. J. Pharmacol.* **27**, 509–516 (1977).
14. H. Saito, M. Tsuchiya, S. Naka and K. Takagi, Effects of panax ginseng root on acquisition of sound discrimination behavior in rats. *Jpn. J. Pharmacol.* **29**, 319–324 (1979).
15. V.D. Petkov, Ginseng as a remedy regulating aging processes in brain. In *Recent Advances in Ginseng Studies*, S. Shibata, Y. Ohtsuka and H. Saito, eds., Hirokawa Publishing, Tokyo, pp. 83–98 (1990).
16. H. Watanabe, H. Ohta and K. Matsumoto, Effect of ginseng on spontaneous motor activity, water-maze learning and central dopaminergic systems in old rats. In *Recent Advances in Ginseng Studies*, S. Shibata, Y. Ohtsuka and H. Saito, eds., Hirokawa Publishing, Tokyo, pp. 73–81 (1990).
17. H. Watanabe, H. Ohta, L. Imamura, W. Asakura, Y. Matoba and K. Matsumoto, Effect of Panax ginseng on age-related changes in the spontaneous motor activity and dopaminergic nervous system in the rat. *Jpn. J. Pharmacol.* **55**, 51–56 (1991).
18. D. Harman, Aging; a theory based on free radical and radiation chemistry. *J. Gerontol.* **11**, 298–300 (1956).
19. M. Yoshikawa and S. Hirai, Lipid peroxide formation in the brain of aging rats. *J. Gerontol.* **22**, 162–165 (1967).
20. P.I. Ceballos, A. Nicole, M. Clement, J.M. Bourre and P.M. Sinet, Age-related changes in antioxidant enzymes and lipid peroxidation in brains of control and transgenic mice overexpressing copper-zinc superoxidase dismutase. *Mut. Res.* **275**, 281–293 (1992).
21. M. Yamamoto, *Yakuyoninjin, '85–its basic and clinical medicine*, A. Kumagai, H. Ooura and H. Okuda, eds., Kyoritsu Publisher, Tokyo, pp. 174–193 (1985).
22. H. Thoenen and Y.A. Barde, Physiology of nerve growth factor. *Physiol. Rev.* **60**, 1284–1335 (1980).

23. E.L. Fenton, Tissue culture assay of nerve growth factor and of the specific antiserum. *Exp. Cell Res.* **59**, 383–392 (1970).
24. H. Saito, K. Suda, M. Schwab and H. Thoenen, Potentiation of the NGF-mediated nerve fiber outgrowth by ginsenoside Rb1 in organ cultures of chicken embryonic dorsal root ganglia. *Jpn. J. Pharmacol.* **27**, 445–451 (1977).
25. Y. Takemoto, T. Ueyama, H. Saito, S. Horio, S. Sanada, J. Shoji, S. Yahara, O. Tanaka and S. Shibata, Potentiation of nerve growth factor-mediated nerve fiber production in organ cultures of chicken embryonic ganglia by ginseng saponins structure-activity relationship. *Chem. Pharm. Bull.* **32**, 3128–3133 (1984).
26. T. Mohri, K. Chiba, M. Yamazaki, M. Shimizu and N. Morita, Activation of PC12 cells by lipophilic components of Panax ginseng. *Planta Medica* **58**, 321–323 (1992).
27. N. Nishiyama, Y. Takemoto, T. Himi, T. Sakai, C. Ohsawa and H. Saito, Effects of ginseng saponins on cultured neuron cells. In *Recent Advances in Ginseng Studies*, S. Shibata, Y. Ohtsuka and H. Saito, eds., Hirokawa Publishing, Tokyo, pp. 125–137 (1990).
28. T. Himi, H. Saito and N. Nishiyama, Effect of ginseng saponins on the survival of cerebral cortex neurons in culture. *Chem. Pharm. Bull.* **37**, 481–484 (1989).
29. A. Sugaya, M. Yuzurihara, T. Tsuda, K. Yasuda, K. Kajiwara and E. Sugaya, Proliferative effect of ginseng saponin on neurite extenuation of primary cultured rat cerebral cortex. *J. Ethnopharmacol.* **22**, 173–181 (1988).
30. T.-C. Wen, H. Yoshimura, S. Matsuda, J.-H. Lim and M. Sakanaka, Ginseng root prevents learning diability and neuronal loss in gerbils with 5-minute forebrain ischemia. *Acta Neuropathol.* **91**, 15–22 (1996).
31. T.V.P. Bliss and G.L. Collingridge, A synaptic model of memory: long-term potentiation in the hippocampus. *Nature* **361**, 31–39 (1993).
32. O.A. Ramirez and H.F. Carrer, Correlation between threshold to induce long-term potentiation in the hippocampus and performance in a shuttle box avoidance response in rats. *Neurosci. Lett.* **104**, 152–156 (1989).
33. O.A. Ramirez, O.A. Orsingher and H.F. Carrer, Differential threshold for long-term potentiation in the hippocampus of rats with inborn high or low learning capacity. *Neurosci. Lett.* **92**, 275–279 (1988).
34. M. Sugiura, Y. Shoyama, H. Saito and N. Nishiyama, Crocin improves the ethanol-induced impairment of learning behaviors of mice in passive avoidance tasks. *Proc. Jpn. Acad. Ser B* **71**, 319–324 (1995).
35. M. Sugiura, Y. Shoyama, H. Saito and K. Abe, Crocin (Crocetin di-gentiobiose ester) prevents the inhibiltory effect of ethanol on long-term potentiation in the dentate gyrus *in vivo*. *J. Pharmacol. Exp. Ther.* **271**, 703–707 (1994).
36. M. Sinriga, H. Saito and N. Nishiyama, Hoelen (Poria Cocos Wolf) and ginseng (Panax Ginseng C.A. Meyer), the ingredients of a Chinese prescription DX-9386, individually promote hippocampal long-term potentiation *in vivo*. *Biol. Pharm. Bull.* **18**, 518–522 (1995).
37. K. Abe, F.-J. Xie and H. Saito, Epidermal growth factor enhances short-term potentiation and facilitates induction of long-term potentiation in rat hippocampal slice. *Brain Res.* **547**, 171–174 (1991).
38. K. Abe, S.I. Cho, I. Kitagawa, N. Nishiyama and H. Saito, Differential effects of ginsenoside Rb1 and malonylginsenoside Rb1 on long-term potentiation in the dentate gyrus of rats. *Brain Res.* **649**, 7–11 (1994).
39. X. Chen and T.J. Lee, Ginsenoside-induced nitric oxide-mediated relaxation of the rabbit corpus cavernosum. *Brit. J. Pharmacol.* **115**, 15–18 (1995).
40. S.Y. Nah, H.J. Park and E.W. McCleskey, A trace component of ginseng that inhibits Ca channels through a pertussis toxin-sensitive G protein. *Proc. Natl. Acad. Sci. USA* **92**, 8739–8743 (1995).

7. EFFECTS OF VIETNAMESE GINSENG ON PSYCHOLOGICAL STRESS–INDUCED CHANGES IN PHARMACOLOGICAL RESPONSES

NGUYEN THI THU HUONG[1,3], NGUYEN THOI NHAM[3], KINZO MATSUMOTO[1], KAZUO YAMASAKI[2] and HIROSHI WATANABE[1]*

[1]*Department of Pharmacology, Research Institute for Oriental Medicines (Wakan-Yaku), Toyama Medical and Pharmaceutical University, 2630 Sugitani, Toyama 930-0194, Japan;* [2]*Department of Biological Active Substances, Institute of Pharmaceutical Sciences, Hiroshima University School of Medicine, 1-2-3 Kasumi, Minami-ku, Hiroshima 734-0037, Japan;* [3]*The Science-Production Centre of Vietnamese Ginseng, Ho Chi Minh University of Medicine and Pharmacy, 41 Dinh Tien Hoang, District 1, Ho Chi Minh City, Vietnam*

1. INTRODUCTION

Most of the species of wild Panax are found in the Himalayas and in China. Yunnan, China and the northern provinces of Vietnam near the China–Vietnam border at about latitude 23°N were previously regarded as the southern limit of the distribution of Asian Panax spp. In 1973, however, a new species of genus Panax L., Araliaceae, was discovered in Ngoc Lay, Dakto District, Gia lai-Kontum Province, Central Vietnam, which is 1800 m above sea-level. This plant is located at latitude 15°N, further south than the predicted limit of the Panax genus. The plant was named "Sam Dot Truc" or "Cu Ngai Rom con", which in Vietnamese means "a ginseng plant with an underground part with internodes like a bamboo tree". It has been used by the Sedang ethnic minority living in the central Vietnam region as a "secret medicinal plant" or a panacea, a life-saving medicinal plant for the treatment of many serious diseases and for enhancing body strength for long journeys across high mountains (Nham *et al.*, 1995).

A preliminary chemical study using thin-layer chromatography (TLC) showed that the plant "Sam Dot Truc" contains saponins quite similar to those found in *Panax ginseng*, and suggested that this plant is a valuable source of ginseng-like medicine (Nham *et al.*, 1989). In 1985, its scientific name was designated as *Panax vietnamensis* Ha et Grushv., Araliaceae (common name: Vietnamese ginseng). Since its discovery, Vietnamese ginseng has been considered one of the most important medicinal plants in Vietnam, and its pharmacognostical, chemical and pharmacological profiles have been thoroughly studied by the collaboration of several institutes. In this chapter, we review the chemical profile and the pharmacological activity of Vietnamese ginseng (crude extract, total saponin) and majonoside-R_2, a dominant component of Vietnamese ginseng saponins, focusing on the effects on

* To whom all correspondence should be addressed

pathophysiological changes caused by stressful stimuli. We also discuss the possible neuronal mechanisms involved in the action of Vietnamese ginseng.

2. CHEMICAL CONSTITUENTS OF VIETNAMESE GINSENG

Chemical studies on the rhizomes and roots of Vietnamese ginseng have identified 14 new dammarane saponins named Vina-ginsenosides-R_1~R_{14}, and 23 saponins which are the same as those previously isolated from other Panax spp. (Duc *et al.*, 1993, 1994a, 1994b). The underground part of Vietnamese ginseng (root and rhizome) contains mainly dammarane saponins such as ginsenoside-Rb_1 and ginsenoside-Rg_1, but it also contains a small amount of oleanolic acid saponins and polyacetylene compounds which have been isolated from ginseng roots and are known to have cytotoxic activity against some cultured cancer cells. However, it is noteworthy that in addition to these compounds, Vietnamese ginseng contains a large amount of ocotillol-type saponins, especially majonoside-R_2. This type of saponin has not been isolated from *Panax ginseng* (Ginseng), *Panax quinquefolium* (American Ginseng) and *Panax notoginseng* (Sanchi Ginseng) (Nham *et al.*, 1995).

For the pharmacological evaluations of Vietnamese ginseng, Vietnamese ginseng crude extract (VG extract), total saponins and majonoside-R_2 have been used. VG extract can be prepared from powdered root and rhizome of this plant (5-year-old plant age) by extracting four times with 96%, 48%, 24%, and 0% v/v ethanol, respectively (a percolation method), and then by evaporating under reduced pressure and lyophilization (yield: 41.2%). The crude saponin fraction can be obtained from this crude extract by the following treatment: i) with ethyl ether to remove lipids, and ii) with water-saturated n-butanol to extract saponins (VG saponin; yield: 13.2%). Majonoside-R_2 (Figure 1) has been purified from the total saponin fraction (yield: 5.29%)(Duc *et al.*, 1993). Quantitative analysis using high performance liquid chromatography (HPLC) revealed that the contents of the major saponosides majonoside-R_2, ginsenoside-Rg_1, -Rb_1 and -Rd were 8.27, 4.18, 1.71 and 1.57 (%, w/w), respectively, in VG extract, and 22.67, 7.38, 3.7 and 3.7 (%, w/w), respectively, in VG saponin.

These pharmacognostical and chemical profiles of Vietnamese ginseng make this plant very interesting from not only a chemotaxonomical but also a pharmacological point of view.

3. PHARMACOLOGICAL PROFILE OF VIETNAMESE GINSENG

Evaluation of the experimental and clinical pharmacological activities of Vietnamese ginseng was begun in Vietnam in the 1980s. Clinical studies have suggested that VG extract exhibits an anti-fatigue action, increases appetite, causes better sleep, and improves muscular force, anemia, arteriosclerosis, mental and sexual asthenia, and chronic granular angina, among other effects (Nham *et al.*, 1989). VG extract and total saponins also exert a variety of activities in experimental

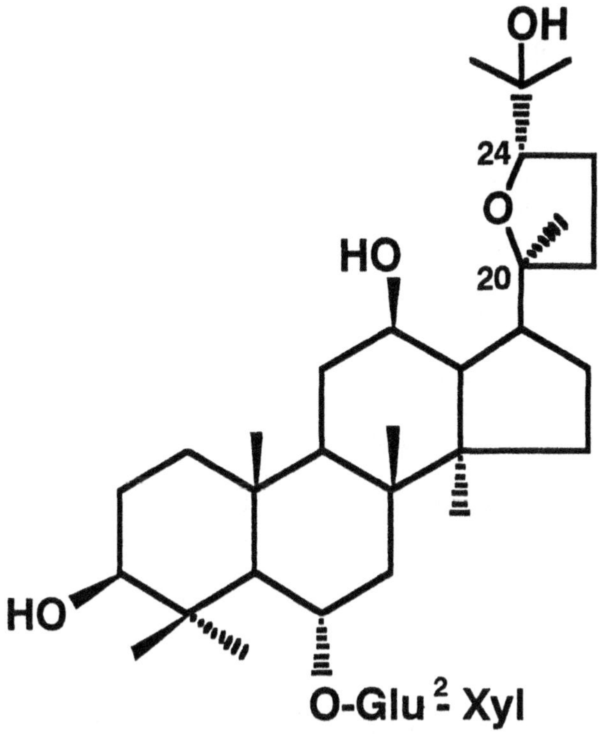

Figure 1 Chemical structure of majonoside-R_2.

animals such as analeptic, anti-fatigue, adaptogenic, gonadotropic, anti-diabetic, and hypertensive actions. Nham *et al.* (1989) have shown that VG extract has a biphasic effect on the central nervous system; i.e., stimulatory and suppressive actions following oral administration of small and large doses, respectively. However, whether these activities of Vietnamese ginseng are due to the same saponin constituents as those found in Ginseng or to the constituents characteristically found in Vietnamese ginseng remains to be clarified.

3.1. Stress Responses and Vietnamese Ginseng Effect

Stress has been widely implicated as a contributing variable in a number of health problems in humans. In laboratory animals, various stressful manipulations cause pathophysiological changes. For instance, footshock, forced swimming stress, and restraint stress produce antinociception, gastric lesions or changes in the hypnotic activity of barbiturates in rodents (Carmody, 1983; Glavin, 1985; Hanada *et al.*, 1985; Willow *et al.*, 1980). It is well known that the physiological responses to stressful stimuli are primarily mediated by the hypothalamic-pituitary-adrenal (HPA) axis. In addition, emotional factors, such as anxiety and fear, also appear to play an important role in these pathophysiological effects of stressful stimuli

(Bodnar, 1986; Ohdo *et al.*, 1989; Snow *et al.*, 1982; Takahashi *et al.*, 1988). In fact, if animals fail to maintain homeostasis following exposure to various physical or psychological stressors, they show distress responses associated with emotional changes.

Medicinal plants capable of reducing stress reactions in the alarm phase or of retarding/preventing the exhaustion phase in the General Adaptation Syndrome described by Selye (1950), and/or providing a certain degree of protection against long-term stress exposure have been defined as 'plant adaptogens' by Brekhman (1980). Ginseng and ginseng-like medicinal plants are thought to be typical adaptogenic plant drugs, since ginseng has been demonstrated to increase the power of resistance against a variety of stressors in animal experiments (Wagner *et al.*, 1994). Vietnamese ginseng also seems to have adaptogenic activity against stressful manipulations. Nham *et al.* (1989) observed that the oral administration of VG extract prolonged the endurance time in forced swimming stress in mice, and suggested that this effect of Vietnamese ginseng may be due to an increase in the biochemical capacity of working skeletal muscle to utilize proportionally more lipid and save carbohydrate source. In addition, they demonstrated that administration of VG extract prolonged the survival time of mice when mice were maintained in housing at 37–42 °C. However, the possible neuronal mechanisms underlying these antistress and adaptogenic activities of Vietnamese ginseng are not yet clear.

3.2. The Effects of Vietnamese Ginseng on Pathophysiological Changes Caused by Psychological Stress or Conditioned Fear Stress

3.2.1. Antinociception caused by psychological stress in mice and conditioned fear stress in rats

Two techniques often employed for exposing experimental animals to psychological stress are the communication box method for mice and the conditioned fear test for rats. The changes in the nociceptive responses of animals exposed to psychological stress were measured by the tail-pinch test in mice (Takagi *et al.*, 1966) and the tail-flick test in rats (D'Amour *et al.*, 1941; Gutstein *et al.*, 1995).

The communication box method was originally described by Ogawa *et al.* (1990), and has been used to investigate behavioral and physiological changes in mice caused by psychological stress. The important feature of this method is that an animal exposed to physical stress such as electric footshock can induce socio-psychological stress in other animals. The communication box is delineated into a "checkerboard" of two types of compartments (A and B) with transparent Plexiglas walls and stainless steel grid floors (Figure 2). One animal is placed in each compartment. A scrambled electric shock is delivered through the grid floor, but the compartment B floors are covered with Plexiglas plates, and although the animals in the B compartments do not receive footshock, they appear to experience psychological stress by watching and hearing the struggle, jumping, and vocalization of the animals shocked in the adjacent compartments. The animals exposed to psychological stress or footshock stress exhibit markedly prolonged latencies to show nociceptive responses than unstressed control animals in the tail-pinch test.

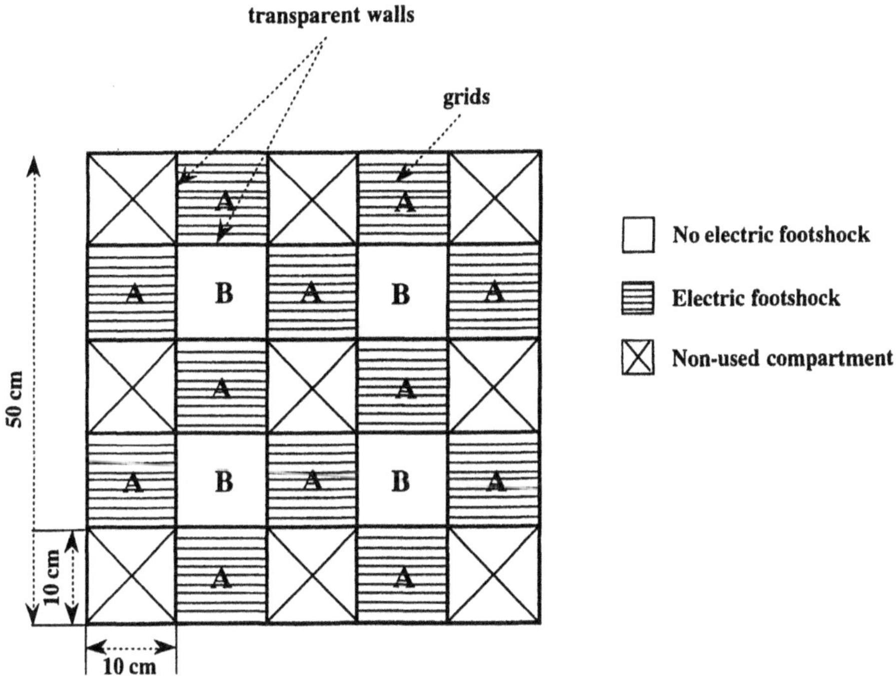

Figure 2 Schematic drawing of the communication box used to expose mice to psychological stress. Mice were placed individually in every compartment (A and B). The animals in the A compartments received footshock through the grid floor (0.5 cm diameter rods 1 cm apart from each other) and those in the B compartments were exposed to psychological stress by watching the behavior or hearing the vocalization of the animals in the A compartments.

The conditioned fear test was originally developed by Fanselow *et al.* (1989). The shock box for conditioned fear stress exposure consists of 4 chambers (20 × 20 × 40 cm each) with transparent Plexiglas walls and a stainless steel grid floor (0.5 cm diameter rods 1 cm apart from each other). Intermittent electric shocks are delivered through the grids by a shock generator. Rats that had been habituated to handling, the injection procedure and the other aspects of the experimental procedure daily for 3 days before the start of experiments were subjected to inescapable electric footshock for 30 min daily for 5 days. Control animals were placed in the apparatus for the same period as the stressed group, without receiving footshock. Twenty-four hours after the last session, each rat was placed in the shock chamber for 15 min without footshock. The animals that had previously received footshock can be said to have developed conditioned fear. The rats exposed to conditioned fear stress also exhibited significantly longer latencies to show nociceptive responses in the tail-flick test compared to unstressed control animals.

Huong *et al.* (1995, 1996a, 1996b) measured the changes in the nociceptive responses of animals exposed to psychological stress, footshock stress, and conditioned fear stress, and tested the effects of Vietnamese ginseng on the antinociception caused by these stressors. As shown in Figure 3, the mice exposed to psychological stress or to footshock stress for 15 min in the communication box showed latencies to produce nociceptive responses in the tail-pinch test longer than those of unstressed control animals. Consistent with another report (Takahashi *et al.*, 1990), these findings indicate that the stressful stimuli cause antinociception in mice. Huong *et al.* (1995, 1996a) also found that the administration of VG extract (100 mg/kg, p.o.) and VG saponin (6.2–25 mg/kg, p.o.) suppressed the anti-nociceptive effect of the psychological stress but had no effect on footshock stress-induced antinociception. Majonoside-R_2 (3.1–12.5 mg/kg, p.o. or i.p.), as well as naloxone (2 mg/kg, i.p.), appeared to attenuate the antinociception induced by psychological stress and footshock stress at half the dose of VG saponin required to produce the effect.

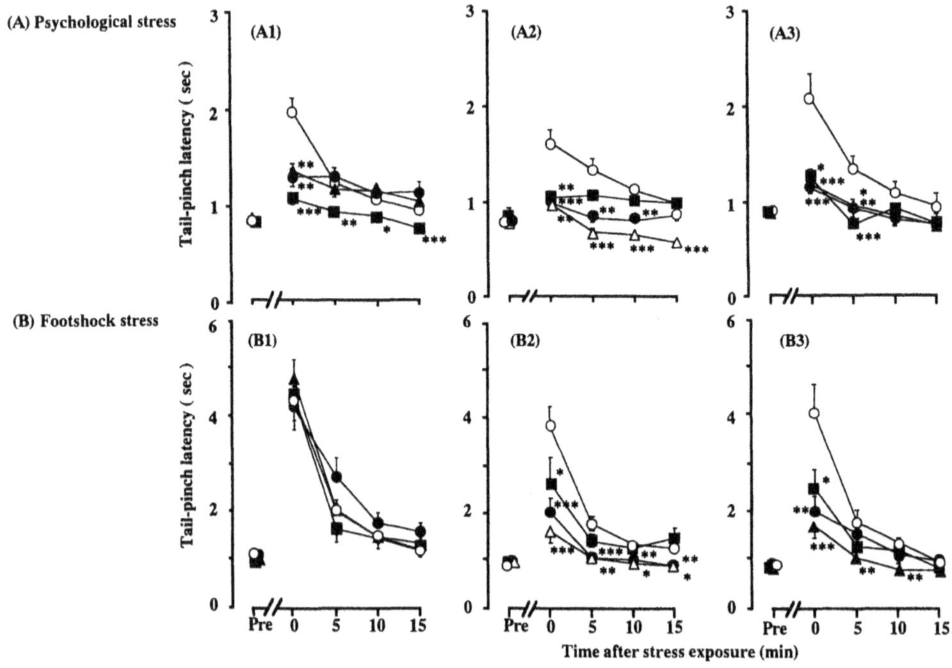

Figure 3 Effect of Vietnamese ginseng saponin (VG saponin), majonoside-R_2 and naloxone on the antinociception induced by psychological stress or footshock stress (1 mA, 1-s duration, 4-s intershock interval). Mice were exposed to psychological stress (A) or footshock stress (B) for 15 min. In A1, B1: Either vehicle (○) or VG saponin (6.2 (■), 12.5 (●) and 25 mg/kg (▲)) was administered p.o. 1 hour prior to stress exposure. In A2 and B2, vehicle (○), naloxone (2 mg/kg, i.p., (△)) and majonoside-R_2 (3 (■), 6.2 (●) mg/kg, p.o.) were given 10, 10 min and 1 hour before stress exposure, respectively. In A3 and B3, vehicle (○) and majonoside-R_2 (3 (■), 6.2 (●) and 12.5 (▲) mg/kg) were administered i.p. 30 min before stress exposure. The latency of nociceptive responses caused by the tail-pinch test was measured every 5 min for 15 min immediately after the termination of stress exposure. Each point is the mean ± S.E.M. of 12–15 mice. $*p < 0.05$, $**p < 0.01$ and $***p < 0.001$ compared with the vehicle-treated group (Dunnett's test).

The repeated exposure of rats to conditioned fear stress for 5 days produced a marked decrease in their nociceptive response in the tail-flick test. VG saponin (25 mg/kg, i.p.) and majonoside-R_2 (6.2 mg/kg, i.p.), as well as naloxone (2 mg/kg, i.p.), also had attenuating effects on the antinociception caused by conditioned fear stress (Figure 4). In contrast, no significant changes in nociceptive responses were observed in the footshock-naive control rats following administration of these agents (Huong *et al.*, 1996b).

The development of adaptation to psychological stress and footshock stress following repeated exposure to the same stressors was reported by Takahashi *et al.* (1992). Huong *et al.* (1995, 1996a) also observed the development of stress adaptation in the tail-pinch test following repeated exposure of mice to the psychological stress or footshock stress for 5 days. Daily oral administrations of VG extract (50 mg/kg),

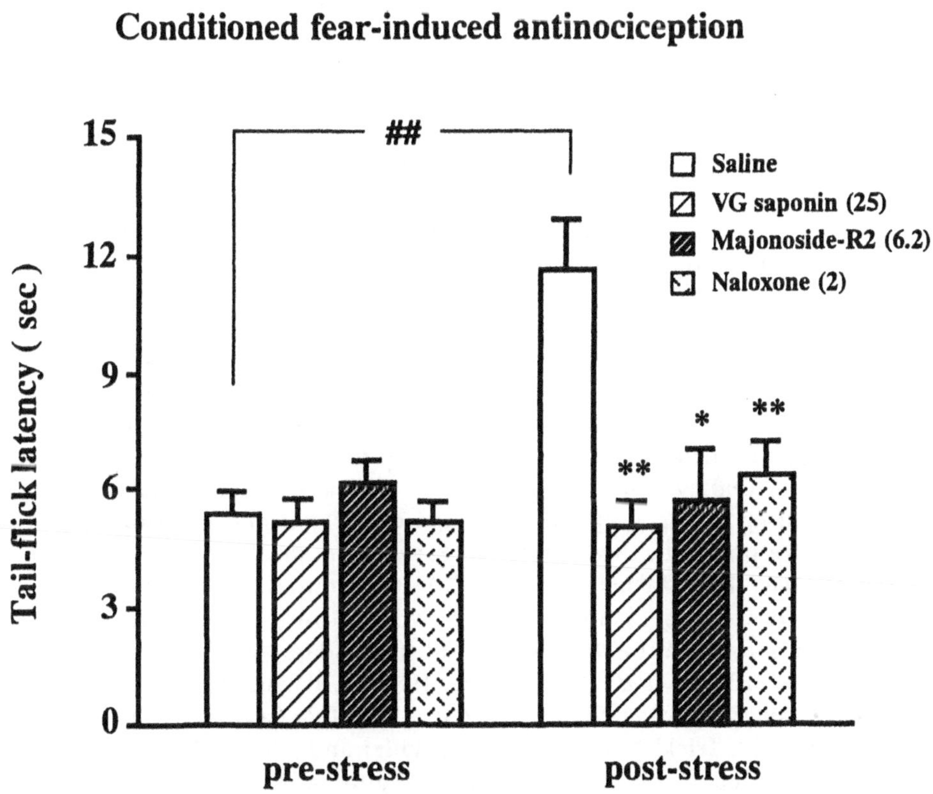

Figure 4 Effects of Vietnamese ginseng saponin (VG saponin), majonoside-R_2 and naloxone on the conditioned fear stress-induced antinociception in rats. Rats were subjected to inescapable electric footshock (2 mA, 1-s duration, 4-s intershock interval) for 30 min for a 5-day period. Twenty-four hours after the last session, the rats were placed in the shock chambers for 15 min without footshock, and then the tail-flick lantency was measured. Test drugs were administered i.p. 15 min before the conditioned fear stress exposure. The numbers in parenthese indicate the dose (mg/kg). Each column represents the mean latency ± S.E.M. (n = 4–7). $*p < 0.05$ and $**p < 0.01$ *vs.* vehicle group (Mann-Whitney test).

VG saponin (12.5 mg/kg) or majonoside-R_2 (6.2 mg/kg) had no effect on the development of adaptation to footshock stress, although VG saponin treatment and majonoside-R_2 treatment showed suppressing effects on the footshock stress-induced antinociception measured on the first, second and third day during the stress exposure period. VG saponin treatment also did not affect the development of adaptation to psychological stress, and the VG extract treatment and majonoside-R_2 treatment both inhibited this development of stress adaptation. At least two explanations are possible for this discrepancy between majonoside-R_2 and VG saponin effects on stress-induced antinociception and stress adaptation. First, other components of VG saponin may counteract the effect of majonoside-R_2. Secondly, taking into account the finding that both VG saponin and majonoside-R_2 had attenuating effects on the conditioned fear stress-induced antinociception measured by the tail-flick test in rats, it is possible that VG saponin and majonoside-R_2 exhibit different pharmacological activities depending on the type of stressors applied (physical or emotional stressor), the species difference of experimental animals or the nociceptive tests used.

The effects of Vietnamese ginseng on stress-induced antinociception and stress adaptation seem to be quite different from those of the standardized *Panax ginseng* extract G115 (trademark for the standardized ginseng extract containing 4% ginsenosides), since it was reported that, although the acute administration of G115 had no effect on footshock stress- or psychological stress-induced antinociception, the repeated administration of this extract suppressed the development of adaptation to psychological stress but not to footshock stress (Takahashi *et al.*, 1992). Since the ginseng saponin fraction does not contain the new saponin compounds (Vina-ginsenosides) or ocotillol-type saponins isolated from Vietnamese ginseng, especially majonoside-R_2, which accounts for over 50% of the VG saponins (Duc *et al.*, 1993), the different action profiles of VG saponins and ginseng saponins may be due to the difference in chemical composition between these two saponin extracts.

It has been reported that the opioid antagonist naloxone reverses the antinociception caused by footshock stress, psychological stress and conditioned fear stress, indicating the involvement of an opioid mechanism in the suppression of nociceptive responses by these stressors (Fanselow *et al.*, 1989; Izumi *et al.*, 1983; Przewlocka *et al.*, 1990; Takahashi *et al.*, 1990). It is known that both monoaminergic and GABA receptor mechanisms are involved not only in stress-induced antinociception but also in the development of stress adaptation following repeated stress exposure (Biggio *et al.*, 1990; Bodnar, 1986; Corda and Biggio, 1986; Losada, 1989; Oluyomi and Hart, 1990; Rovati *et al.*, 1990; Takeda *et al.*, 1994; Tokuyama *et al.*, 1992, 1993; Tricklebank *et al.*, 1984; Willer and Ernst, 1986). Huong *et al.* (1996c, 1997) recently found that Vietnamese ginseng and its major component majonoside-R_2 attenuated mu- and selective kappa-opioid receptor agonist-induced antinociception by acting at spinal and supraspinal levels. Supraspinal $GABA_A$ receptor systems also appeared to play a role in the effect of majonoside-R_2 on opioid-induced antinociception. Their findings raise the possibility that the attenuating effect of majonoside-R_2 on the stress-induced antinociception is at least partly due to a modulation of opioid systems or $GABA_A$ systems or both in the brain. Taking into account the evidence that the spinal $alpha_2$ adrenoceptors are involved in conditioned fear stress-induced antinociception (Rochford *et al.*, 1992), the involvement of the descending noradrenergic system in the effect of majonoside-R_2 cannot be excluded.

3.2.2. *Psychological stress-induced decrease in pentobarbital sleep in mice*

Psychological stimuli such as noise, fear, and anxiety increase the arousal level and cause insomnia in humans (Martin, 1983; Hamon, 1994). Most such stress stimuli not only induce antinociception but also affect the activity of hypnotic drugs such as barbiturates. The changes in the effects of hypnotic drugs caused by stressful stimuli seem to be dependent on the type of stressors and/or the strength of the stimuli. For instance, cold swim stress and footshock stress prolong pentobarbital-induced sleep in mice, while social isolation stress and repeated forced shaking stress shorten pentobarbital-induced sleeping time in mice (Hanada *et al.*, 1985; Matsumoto *et al.*, 1991, 1996). Huong *et al.* (1996d) reported that mice previously exposed to psychological stress exhibited shorter durations of pentobarbital sleep than those of unstressed control animals, and that VG extract (50 mg/kg, p.o.), VG saponin (25 mg/kg, p.o.), and majonoside-R_2 (3.1–12.5 mg/kg, p.o. or i.p.) each restored the hypnotic activity of pentobarbital decreased by psychological stress to the level of unstressed control mice, without affecting pentobarbital sleep in unstressed animals (Table 1). Diazepam (0.1 mg/kg, i.p.), a reference drug, prolongs pentobarbital sleep in both unstressed and psychologically stressed animals. Interestingly, the effects of majonoside-R_2 and diazepam on the psychological stress-induced decrease in pentobarbital sleep can be antagonized by flumazenil (1 mg/kg, i.v.), a selective benzodiazepine antagonist (Figure 5). Although the exact mechanism underlying this apparent antagonistic interaction between majonoside-R_2 and flumazenil in the psychological stress-induced decrease in pentobarbital sleep remains unclear, it can be suggested that the $GABA_A$-benzodiazepine receptor complex may participate in the action of majonoside-R_2. The clarification of whether majonoside-R_2 interacts directly with GABA receptors in the brain requires further investigation such as *in vitro* binding experiments.

In contrast to Vietnamese ginseng and majonoside-R_2, *Panax ginseng* extract (50 and 100 mg/kg, p.o.) and naloxone (0.5–5 mg/kg, i.p.) each had no effect on pentobarbital sleep in both unstressed and psychologically stressed groups. This finding demonstrates another difference in the pharmacological profiles between Vietnamese ginseng and *Panax ginseng*. Opioid systems also appear to be involved in the pathological states induced by stressful stimuli since naloxone, an opioid receptor antagonist, modulates the stress responses of animals (Carmody, 1983; Dai and Chan, 1983; Ohdo *et al.*, 1989). In stress-induced antinociception, both VG saponin and majonoside-R_2, as well as naloxone, suppressed psychological stress-induced antinociception in mice and rats, suggesting the possible involvement of opioid receptor mechanisms in the effects of VG saponin and majonoside-R_2. Taken together, the findings described above suggest that functional changes in opioid systems caused by psychological stress exposure may participate in the effect of Vietnamese ginseng on pentobarbital sleep in psychologically stressed mice. However, this seems unlikely, because naloxone produced no significant effect on pentobarbital sleep in psychologically stressed mice.

Table 1 Effects of VG extract, VG saponin and its major constituent majonoside-R_2 on the psychological stress-induced decrease in pentobarbital-induced sleep in mice

Drugs	*Dose (mg/kg)*	*Sleeping time (min)* *Unstressed*	*Stressed*	*Interaction between stress and drug*
Exp. I				
Vehicle		73.0 ± 1.9	56.2 ± 2.3$^{##}$	
VG extract	25	65.4 ± 3.0	56.5 ± 3.1	$F(3.169) = 5.696, p < 0.01$
	50	70.3 ± 4.5	79.5 ± 5.3**	
	100	77.7 ± 6.3	72.0 ± 4.3	
PG extract	50	76.7 ± 3.6	65.2 ± 4.3	$F(2,136) = 4,693, p < 0.05$
	100	64.3 ± 4.5	65.8 ± 3.5	
Exp. II				
Vehicle		70.2 ± 1.6	56.9 ± 1.9$^{##}$	
VG saponin	12.5	68.9 ± 2.6	65.0 ± 2.8	$F(2,118) = 4.064, p < 0.05$
	25	75.5 ± 4.0	71.3 ± 4.7**	
Majonoside-R_2	6.2	73.4 ± 2.0	80.0 ± 6.3**	$F(2,118) = 4.105, p < 0.05$
	12.5	76.4 ± 4.0	69.9 ± 4.0*	
Exp. III				
Vehicle		73.1 ± 2.9	58.3 ± 2.2$^{#}$	
Majonoside-R_2	3.1	71.6 ± 3.7	80.7 ± 4.2**	$F(3,101) = 5.523, p < 0.01$
	6.2	66.9 ± 2.8	72.2 ± 1.5*	
	12.5	74.5 ± 4.3	74.6 ± 4.1*	

Mice were divided into 2 groups, and the stressed group was exposed to psychological stress for 30 min. Test drugs were administered p.o. 1 hour before stress exposure (except for Exp. III, in which majonoside-R_2 was injected i.p. 30 min before stress exposure). Each datum represents the mean ± S.E.M. of 12–15 mice. $^{#}p < 0.05$ and $^{##}p < 0.01$ compared with respective unstressed group. $^{*}p < 0.05$ and $^{**}p < 0.01$ compared with respective vehicle treatment (Tukey's test).

3.2.3. *Psychological stress-induced gastric lesions in mice*

Stress ulceration of the stomach is a frequent life-threatening condition in humans. The relationship between stress, ulcers and gastric secretion has been extensively studied using a variety of stress lesion models in which animals are subjected to various kinds of physical stimuli (Glavin *et al.*, 1985, 1991). It is well known that psychological stress such as fear and anxiety exhibits a marked depressant effect on the stomach, and such stress has been considered one of the causes of gastric ulceration (Ichimaru and Gomita, 1987; Nomura *et al.*, 1994). The recent advances in the use of anxiolytic and antidepressant drugs in the management of stress-related gastric mucosal injury further support the contention that a brain–gut axis, which may have nervous, peptidergic and classic monoaminergic components, modulates the intricate and complicated pattern of communication between the brain and the stomach (Glavin *et al.*, 1991). Using the communication box method described by Nomura *et al.* (1994), Huong *et al.* (1996d) evaluated the effect of

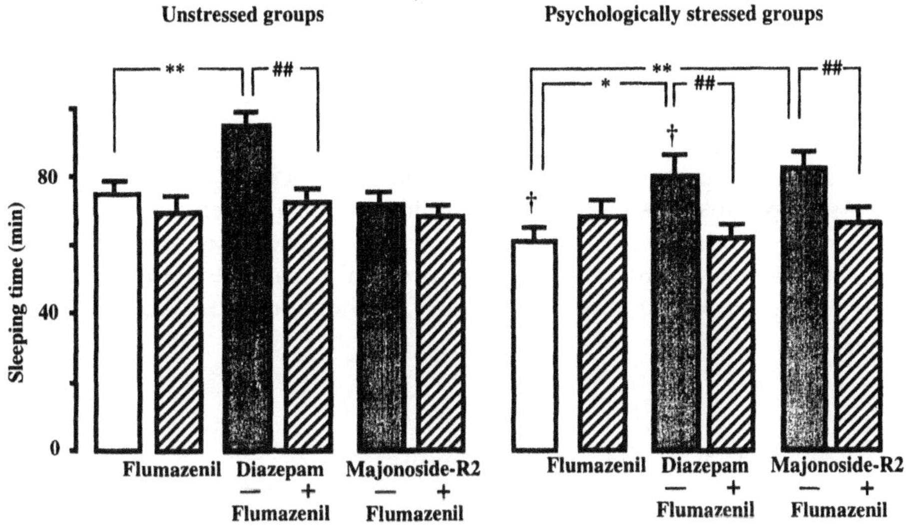

Figure 5 Effects of majonoside-R_2 and diazepam on the psychological stress-induced decrease in pentobarbital sleep, and the antagonistic action of flumazenil on the effects of majonoside-R_2 and diazepam. Majonoside-R_2 (3.1 mg/kg) or diazepam (0.1 mg/kg) was administered i.p. 30 min before stress exposure. Flumazenil (1 mg/kg) was injected i.v. just before stress exposure. The unstressed animals were treated with test drugs in the same way as the stressed mice. Each value is the mean ± S.E.M. of 12 mice. $^{*}p < 0.05$ and $^{**}p < 0.01$ compared with the respective vehicle control groups, $^{\#\#}p < 0.01$ compared with the corresponding non-flumazenil groups (majonoside-R_2 or diazepam alone) and † $p < 0.05$ compared with the corresponding unstressed groups (Tukey's test).

Vietnamese ginseng on stress-induced gastric ulceration in mice. They fasted mice for 24 hrs and then randomly divided the animals into 2 groups: one group was subjected to psychological stress in a communication box for 16 hours, and the other mice were placed in compartments individually, without being exposed to psychological stress or footshock, for the same period as the psychologically stressed group. The animals were killed by decapitation and their stomachs were rapidly removed and immersed in saline containing 1% formaldehyde. The gastric lesion severity was scored according to the criteria reported by Tsukamoto *et al.* (1981) (score: 0 = no pathology; 1 = mucosal edema; 2 = petechia; 3 = gross mucosal edema; 4 = severe erosion; 5 = perforated ulcer). Gastric lesion scores of more than 2 were defined as the lesion incidence. The 16-hour psychological stress exposure significantly increased gastric lesion incidence (76.9 and 17.5% in stressed and unstressed mice, respectively) and gastric lesion severity (3.5 ± 0.3 and 1.4 ± 0.2 in stressed and unstressed mice, respectively). Pretreatment with VG extract (25 and 50 mg/kg, p.o.) or with majonoside-R_2 (6.2 and 12.5 mg/kg, p.o.) reduced the incidence and severity of gastric lesions in psychologically stressed mice without affecting these parameters in unstressed control animals. The reference drugs diazepam (10 mg/kg, p.o.) and naloxone (5 mg/kg, i.p.) also appeared to show protective actions against psychological stress-induced gastric lesions, while *Panax ginseng* extract (50 mg/kg) exhibited no significant protective effect against gastric lesions.

Although there are conflicting reports on the roles of central opioid systems in stress-induced gastric ulceration (Dai and Chan, 1983; Glavin, 1985), the above findings of Huong *et al.* (1996d) suggest the possible involvement of opioid and $GABA_A$ systems in psychological stress-induced gastric lesions in mice. Taking into account the possible role of $GABA_A$ systems in mediating the effects of Vietnamese ginseng and majonoside-R_2 on pentobarbital sleep in psychologically stressed mice, it can be speculated that $GABA_A$ systems are also partially implicated in the protective effect of majonoside-R_2 against psychological stress-induced gastric lesions.

4. GENERAL DISCUSSION AND SUMMARY

As described above, Vietnamese ginseng and its major component majonoside-R_2 attenuate psychological stress-induced antinociception, restore the stress-induced decrease in hypnotic activity of pentobarbital to the normal level, and show protective effects against stress-induced gastric lesions, whereas *Panax ginseng* extract had no effect on these pathophysiological changes. The differences in chemical compositions between Vietnamese ginseng and *Panax ginseng* may explain their different action profiles. The observation that the characteristic ocotillol-type saponin majonoside-R_2, as well as VG extract and VG saponin, attenuates the pathophysiological changes caused by psychological stress supports this contention, and gives further evidence that majonoside-R_2 plays an important role in the effects of Vietnamese ginseng.

At least three mechanisms may account for the antistress effect of Vietnamese ginseng and majonoside-R_2. First, Vietnamese ginseng may be able to interact directly or indirectly with opioid systems in the brain. Although to date no information is available on the bioavailability or metabolism of majonoside-R_2 following systemic administration, the findings obtained by Huong *et al.* (1997) indicate that majonoside-R_2 itself modulates the opioid receptors at the spinal and supraspinal levels.

Secondly, it is possible that majonoside-R_2 exerts its pharmacological activity by enhancing the function of $GABA_A$ receptors in the brain. Huong *et al.* (1997) have implicated the $GABA_A$-benzodiazepine receptor complex at the supraspinal level in the modulating effect of majonoside-R_2 on opioid-induced antinociception. Recent evidence indicates that neuroactive steroids capable of modulating the $GABA_A$ receptor complex are involved in the decrease of hypnotic activity of pentobarbital following long-term social isolation in mice (Matsumoto *et al.*, 1996). Thus, since majonoside-R_2 has a steroid-like structure, it is possible that majonoside-R_2 itself or its metabolites may act like neuroactive steroids and modulate $GABA_A$ receptor function.

In summary, Vietnamese ginseng attenuated the physical stress (footshock)- and emotional stress (psychological stress and conditioned fear stress)-induced antinociception in the tail-pinch test in mice and tail-flick test in rats. Vietnamese ginseng also had a protective effect on pathophysiological changes (sleeping time, gastric ulcer) following psychological stress exposure in mice. Cumulative findings also strongly indicate that majonoside-R_2 plays an important role in the effect of Vietnamese ginseng. The effects of Vietnamese ginseng and majonoside-R_2 may be mediated by opioid and $GABA_A$ receptor mechanisms. Nevertheless, further investigation is necessary to clarify the exact mechanisms underlying the protective effects of Vietnamese ginseng and majonoside-R_2 on the pathophysiological changes caused by psychological stress.

5. CONCLUSION

Vietnamese ginseng exerts antistress activity and prevents pathophysiological changes caused by unfavorable environmental and psychological stressors. Its major component, majonoside-R_2, contributes to the effects of Vietnamese ginseng. The pharmacological profile of Vietnamese ginseng is quite different from that of *Panax ginseng*. Pharmacological evidence suggests the adaptogenic activity of Vietnamese ginseng.

ACKNOWLEDGMENTS

The authors gratefully acknowledge Dr. Shoji Shibata, an emeritus professor of Tokyo University, and Dr. Osamu Tanaka, an emeritus professor of Hiroshima University, for their encouragement. We also thank Dr. Nguyen Minh Duc and Mr. Nguyen Minh Cang for their helpful assistance.

REFERENCES

1. G. Biggio, A. Concas, M.G. Corda, O. Giorgi, E. Sanna and M. Serra, GABAergic and dopaminergic transmission in the rat cerebral cortex: effect of stress, anxiolytic and anxiogenic drugs. *Pharmac. Ther.* **48**, 121–142 (1990).
2. R.J. Bodnar, Neuropharmacological and neuroendocrine substrates of stress-induced analgesia. *Ann. N. Y. Acad. Sci.* **467**, 345–360 (1986).
3. I.I. Brekhman, Man and Biologically active Substances, *The Effect of Drugs, Diet and Pollution on Health*, Pergamon Press Ltd., Oxford (1980).
4. J. Carmody, Effects of electric footshock on barbiturate sensitivity, nociception and body temperature in mice. *Eur. J. Pharmacol.* **89**, 119–123 (1983).
5. M.G. Corda and G. Biggio, Stress and GABAergic transmission: biochemical and behavioral studies. In *GABAergic Transmission and Anxiety*, G. Biggio and E. Costa, eds., New York, Raven Press, pp. 121–136 (1986).
6. S. Dai and M.Y. Chan, Effects of naloxone on serum corticosterone and gastric lesions in stressed rats. *Pharmacology* **27**, 180–184 (1983).
7. F.E. D'Amour and D.L. Smith, A method for determining loss of pain sensation. *J. Pharmacol. Exp. Ther.* **72**, 74–79 (1941).
8. N.M. Duc, R. Kasai, K. Ohtani, A. Ito, N.T. Nham, K. Yamasaki and O. Tanaka, Saponins from Vietnamese ginseng, *Panax vietnamensis* Ha et Grushv. Collected in Central Vietnam. I. *Chem. Pharm. Bull.* **41**, 2010–2014 (1993).

9. N.M. Duc, R. Kasai, K. Ohtani, A. Ito, N.T. Nham, K. Yamasaki and O. Tanaka, Saponins from Vietnamese ginseng, *Panax vietnamensis* Ha et Grushv. collected in central Vietnam. II. *Chem. Pharm. Bull.* **42**, 115–122 (1994a).
10. N.M. Duc, R. Kasai, K. Ohtani, A. Ito, N.T. Nham, K. Yamasaki and O. Tanaka, Saponins from Vietnamese ginseng, *Panax vietnamensis* Ha et Grushv. collected in central Vietnam. III. *Chem. Pharm. Bull.* **42**, 634–640 (1994b).
11. M.S. Fanselow, D.J. Calcagnetti and F. J. Helmstetter, Role of mu and kappa opioid receptors in conditional fear-induced analgesia: The antagonistic actions of nor-binaltorphimine and the cyclic somatostatin octapeptide, $Cys^2Tyr^3Orn^5Pen^7$-Amide. *J. Pharmacol. Exp. Ther.* **250**, 825–830 (1989).
12. G.B. Glavin, Effects of morphine and naloxone on restraint-stress ulcers in rats. *Pharmacology* **31**, 57–60 (1985).
13. G.B. Glavin, R. Murison, J.B. Overmier, W.P. Pare, H.K. Bakke, P.G. Henke and D.E. Hernandez, The neurobiology of stress ulcers. *Brain Res. Rev.* **16**, 301–343 (1991).
14. H.B. Gutstein, K.A. Trujillo and H. Akil, Does chronic nociceptive stimulation alter the development of morphine tolerance? *Brain Res.* **680**, 173–179 (1995).
15. M. Hamon, Neuropharmacology of anxiety: perspectives and prospects. *Trends Pharmacol. Sci.* **15**, 36–39 (1994).
16. S. Hanada, Y. Deguchi and H. Kaneto, Diversity of underlying mechanisms in the production of analgesic and pentobarbital-hypnosis prolonging effects of various analgesic drugs and stresses. *Jpn. J. Pharmacol.* **39**, 117–119 (1985).
17. D.E. Hernandez, Neuroendocrine mechanisms of stress ulceration: focus on thyrotropin-releasing hormone (TRH). *Life Sci.* **39**, 279–296 (1986).
18. N.T.T. Huong, K. Matsumoto, K. Yamasaki, N.M. Duc, N.T. Nham and H. Watanabe, Vietnamese ginseng crude saponin and its major component majonoside-R_2 attenuate the psychological stress- and footshock stress-induced antinociception in mice. *Pharmacol. Biochem. Behav.* **52**, 427–432 (1995).
19. N.T.T. Huong, K. Matsumoto, N.M. Duc, K. Yamasaki, N.T. Nham and H. Watanabe, Vietnamese ginseng extract and its saponin constituent majonoside-R_2 attenuate psychological stress-induced antinociception. In Preclinical and Clinical Strategies for the Treatment of Neurodegerative, Cerebrovascular and Mental Disorders. *Int. Acad. Biomed. Drug Res.* Vol. 11, T. Shibuya, ed., Karger, Basel, pp. 265–269, (1996a).
20. N.T.T. Huong, K. Matsumoto, K. Yamasaki, N.M. Duc, N.T. Nham and H. Watanabe, Effects of Vietnamese ginseng on opioid agonist- and conditioned fear stress-induced antinociception. *Phytomed.* **3**, 33–39 (1996b).
21. N.T.T. Huong, K. Matsumoto, K. Yamasaki, N.M. Duc, N.T. Nham and H. Watanabe, The possible involvement of $GABA_A$ systems in the antinarcotic effect of majonoside-R_2, a major constituent of Vietnamese ginseng. *Jpn. J. Pharmacol.* **71**, 345–349 (1996c).
22. N.T.T. Huong, K. Matsumoto, K. Yamasaki, N.M. Duc, N.T. Nham and H. Watanabe, Effects of majonoside-R_2 on pentobarbital sleep and gastric lesion in psychologically stressed mice. *Pharmacol. Biochem. Behav.* **53**, 957–963 (1996d).
23. N.T.T. Huong, K. Matsumoto, K. Yamasaki, N.M. Duc, N.T. Nham and H. Watanabe, Majonoside-R_2, a major constituent of Vietnamese ginseng, attenuates opioid-induced antinociception. *Pharmacol. Biochem. Behav.* **57**, 285–291 (1997).
24. Y. Ichimaru and Y. Gomita, A new screening method for anti-ulcer agents: Psychological stress produced by intraspecies emotional communication. *Pharmacology* **34**, 176–180 (1987).
25. K. Iimori, M. Tanaka, Y. Kohno, Y. Ida, R. Nakazawa, Y. Hoaki, A. Tsuda and N. Nagasaki, Psychological stress enhances noradrenaline turnover in specific brain regions in rats. *Pharmacol. Biochem. Behav.* **16**, 637–640 (1982).
26. R.J. Inglefield and K.C. Kellogg, Hypothalamic $GABA_A$ receptor blockade modulates cerebral cortical systems sensitive to acute stressors. *Psychopharmacology (Berlin)* **116**, 339–345 (1994).
27. M. Ishikawa, S. Ohdo, H. Watanabe, C. Hara and N. Ogawa, Alteration in circadian rhythm of plasma corticosterone in rats following sociopsychological stress induced by communication box. *Physiol. Behav.* **57**, 41–47 (1995).
28. R. Izumi, M. Takahashi and H. Kaneto, Involvement of different mechanisms, opioid and non-opioid forms, in the analgesia induced by footshock and immobilized-water immersion stress. *Jpn. J. Pharmacol.* **33**, 1104–1106 (1983).

29. H.S. Kim, K.W. Oh, H.M. Rheu and S.H. Kim, Antagonism of U-50,488H-induced antinociception by ginseng total saponins is dependent on serotonergic mechanisms. *Pharmacol. Biochem. Behav.* **42**, 587–593 (1992).
30. T. Kimura, P.A. Saunders, H.S. Kim, H.M. Rheu, K.W. Oh and I.K. Ho, Interactions of ginsenosides with ligand-bindings of $GABA_A$ and $GABA_B$ receptors. *Gen. Pharmacol.* **25**, 193–199 (1994).
31. R.L. Kleiman-Wexler, K.S. Ephgrave and C.G. Adair, Naloxone and restraint stress: effects on gastric mucosal injury and gastric function. *Pharmacotherapy* **12**, 61–67 (1992).
32. M.E.O. Losada, Acute stress and GABAergic function in the rat brain. *Br. J. Pharmacol.* **96**, 507–512 (1989).
33. J.B. Martin, The sleep-wake cycle and disorders of sleep. In *Harrison's Principles of International Medicine*, 10th edn., R.G. Petersdorf, R.D. Adams, E. Braunwald, K.J. Isselbacher, J.B. Martinand and J.D. Wilson, eds., McGraw-Hill Book Company, New York, pp. 118–124 (1983).
34. K. Matsumoto, T. Satoh, H.B. Li, H. Ohta and H. Watanabe, Effects of forced shaking stress at low temperature on pentobarbital-induced sleeping in mice. *Gen. Pharmacol.* **22**, 729–733 (1991).
35. K. Matsumoto, K. Ojima and H. Watanabe, Neurosteroidal modulation of social isolation-induced decrease in pentobarbital sleep in mice. *Brain Res.* **708**, 1–6 (1996).
36. N.T. Nham, Study on *Panax vietnamensis* Ha et Grushv. Araliaceae. Botany – Tissue culture – Chemistry – Biological properties. *Herba Polonica* **35**, Suppl. II (1989).
37. N.T. Nham, P.V. De, T.C. Luan, N.M. Duc, S. Shibata, O. Tanaka and R. Kasai, Pharmacognostical and chemical studies on Vietnamese ginseng *Panax vietnamensis* Ha et Grushv. Araliaceae. *J. Jpn. Bot.* **70**, 1–10 (1995).
38. K. Nomura, N. Maeda, T. Yoshino and I. Yamaguchi, A mechanism of 5-HT3 receptor mediation is involved etiologically in the psychological stress lesion in the stomach of the mouse. *J. Pharmacol. Exp. Ther.* **271**, 100–106 (1994).
39. N. Ogawa, C. Hara and M. Ishikawa, Characteristic of socio-psychological stress induced by the communication box method in mice and rats. In *Environmental stress*, O. Mannine, ed., ACES Publishing Ltd., Tampere, pp. 417–427 (1990).
40. S. Ohdo, H. Yoshimura and N. Ogawa, Alteration in hypnotic effect of pentobarbital following repeated agonistic confrontations in mice. *Psychopharmacology (Berlin)* **97**, 30–34 (1989).
41. O.A. Oluyomi and L.S. Hart, α-adrenoceptor involvement in swim stress-induced antinociception in mouse. *J. Pharm. Pharmacol.* **42**, 778–784 (1990).
42. B. Przewlocka, A. Sumova and W. Lason, The influence of conditioned fear-induced stress on opioid systems in the rat. *Pharmacol. Biochem. Behav.* **37**, 661–666 (1990).
43. S. Puri, A. Ray, A.K. Chakravarti and P.A. Sen, A differential dopamine receptor involvement during stress ulcer formation in rats. *Pharmacol. Biochem. Behav.* **47**, 749–752 (1994).
44. J. Rochford, B. Dube and P. Dawes, Spinal cord alpha-2 noradrenergic receptors mediate conditioned analgesia. *Psychopharmacology* **106**, 235–238 (1992).
45. L.C. Rovati, P. Sacerdote, P. Fumagalli, M. Bianchi, P. Mantegazza and A.E. Panerai, Benzodiazepines and their antagonists interfere with opioid-dependent stress-induced analgesia. *Pharmacol. Biochem. Behav.* **36**, 123–126 (1990).
46. H. Selye, *Stress*, Montreal, Acta Medical Publisher (1950).
47. T. Shibasaki, T. Imaki, M. Hotta, N. Ling and H. Demura, Psychological stress increases arousal through brain corticotropin-releasing hormone without significant increase in adrenocorticotropin and catecholamine secretion. *Brain Res.* **618**, 71–75 (1993).
48. A.E. Snow, S.M. Tucker and W.L. Dewey, The role of neurotransmitters in stress-induced antinociception. *Pharmacol. Biochem. Behav.* **16**, 47–50 (1982).
49. H. Takagi, T. Inukai and M. Nakama, A modification of Haffner's method for testing analgesics. *Jpn. J. Pharmacol.* **16**, 287–294 (1966).
50. M. Takahashi, T. Senda, S. Tokuyama and H. Kaneto, Further evidence for the implication of a κ-opioid receptor mechanism in the production of psychological stress-induced analgesia. *Jpn. J. Pharmacol.* **53**, 487–494 (1990).
51. M. Takahashi, S. Tokuyama and H. Kaneto, Distinctive implication of emotional factors in various types of stress-induced analgesia. *Jpn. J. Pharmacol.* **46**, 418–420 (1988).
52. M. Takahashi, S. Tokuyama and H. Kaneto, Antistress effect of ginseng on the inhibition of the development of morphine tolerance in stressed mice. *Jpn. J. Pharmacol.* **59**, 399–404 (1992).

53. H. Takeda, K. Yanagawa, T. Shibuya and T. Matsumiya, The relationship between stress adaptation and brain serotonin dynamics. *Can. J. Physiol. Pharmacol.* **72**, 453 (1994).
54. S. Tokuyama, M. Takahashi and H. Kaneto, Participation of GABAergic systems in the production of antinociception by various stresses in mice. *Jpn. J. Pharmacol.* **60**, 105–110 (1992).
55. S. Tokuyama, M. Takahashi and H. Kaneto, Involvement of serotonergic receptor subtypes in the production of antinociception by psychological stress in mice. *Jpn. J. Pharmacol.* **61**, 237–242 (1993).
56. D.M. Tricklebank, H.P. Hutson and G. Curzon, Involvement of dopamine in the antinociceptive response to footshock. *Psychopharmacology (Berlin)* **82**, 185–188 (1984).
57. Y. Tsukamoto, S. Nakazawa, K. Segawa, J. Taminaga and C. Chujoh, Gastric mucosal damage of rats in hypoxemia. In *Advances in experimental ulcers*, S. Umehara and H. Ito, eds., The 4th International Conference for Experimental Ulcer (ICEU), pp. 129–135 (1981).
58. P.F. VonVoigtlander, R.A. Lewis and G.L. Neff, Kappa opioid analgesia is dependent on serotonergic mechanisms. *J. Pharmacol. Exp. Ther.* **231**, 270–274 (1984).
59. H. Wagner, H. Norr and H. Winterhoff, Plant Adaptogens. *Phytomed.* **1**, 63–76 (1994).
60. L.R. Watkins and D.J. Mayer, Organization of endogenous opiate and nonopiate pain control systems. *Science* **216**, 1185–1192 (1982).
61. C.J. Willer and M. Ernst, Diazepam reduces stress-induced analgesia in humans. *Brain Res.* **362**, 398–402 (1986).
62. M. Willow, J. Carmody and P. Carroll, The effect of swimming in mice on pain perception and sleeping time in response to hypnotic drugs. *Life Sci.* **26**, 219–224 (1980).
63. Y. Yoneda, K. Kanmori, S. Ida and K. Kuriyama, Stress-induced alterations in metabolism of γ-aminobutyric acid in rat brain. *J. Neurochem.* **40**, 350–356 (1983).

8. NEURODEGENERATIVE AND NEUROPROTECTIVE ACTIONS OF GLUTAMATE-RELATED COMPOUNDS OF NATURAL ORIGIN

HARUHIKO SHINOZAKI

Department of Pharmacology, The Tokyo Metropolitan Institute of Medical Science, 3-18-22, Honkomagome, Bunkyo-ku, Tokyo 113-0021, Japan

1. INTRODUCTION

In 1952, L-glutamic acid was first reported to induce epileptic seizures in mammals, and subsequently L-glutamate and L-aspartate were demonstrated to cause excitation of mammalian central neurons and were therefore named 'excitatory amino acids'. At that time, the excitation was believed to result from a non-selective pharmacological action of these amino acids which did not involve synaptic transmission. Gradually, these excitatory amino acids were found to be major candidates for an excitatory neurotransmitter at the invertebrate neuromuscular junction (NMJ) and in the mammalian central nervous system (CNS). At present, it is widely believed that L-glutamic acid is an excitatory neurotransmitter at many synapses in vertebrates and invertebrates. In order to identify such an excitatory amino acid as a neurotransmitter, it is desirable to have highly selective and potent agonists and antagonists for their receptors. In its early stages, the study of glutamate neurotransmission was based primarily on the investigation of the structure–activity relationships of agonists for glutamate receptors. N-Methyl-D-aspartic acid (NMDA) was one of the first compounds to undergo such investigation, and has had a marked influence on progress in the study of the physiological functions of glutamate receptors (Cotman and Iversen, 1987). Powerful heterocyclic excitatory amino acids, such as kainic acid, ibotenic acid, quisqualic acid and domoic acid (see below), which contain the common chemical moiety of glutamate, were also found to be potent agonists for glutamate receptors. The introduction to neuroscience of these potent agonists of natural origin has allowed new opportunities for elucidating the physiological functions of L-glutamic acid as an excitatory neurotransmitter (Shinozaki, 1978, 1991; McGeer *et al.*, 1978). In particular, the discovery of the potent excitatory actions of kainic acid (Shinozaki and Konishi, 1970; Shinozaki, 1978; Takemoto, 1978), quisqualic acid (Figure 1) (Shinozaki and Shibuya, 1974) and NMDA has greatly contributed to the pharmacological classification of the subtypes of glutamate receptors in the mammalian CNS (Monaghan *et al.*, 1989).

The powerful excitatory actions of kainic acid on the mammalian central neurons gave rise to the excitotoxic concept that L-glutamate destroys neurons by excessive activation of excitatory receptors (Olney *et al.*, 1974; Olney, 1978), and L-glutamate has been implicated in the pathogenesis of neuronal damage in the mammalian CNS (Choi and Rothman, 1990; Olney, 1990). Thus, the pharmacological study of

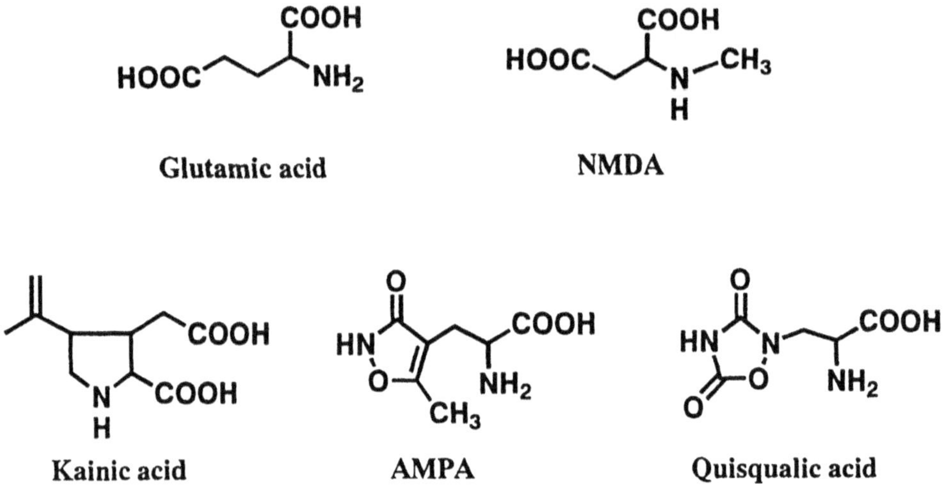

Figure 1 Chemical structures of glutamic acid, NMDA, kainic acid, quisqualic acid and AMPA.

excitatory amino acids has progressed rapidly from the identification of putative transmitters to the elucidation of the physiological functions of each receptor subtype in development, learning, memory, neuropathology and neuroetiology. In particular, from the point of view of the causal relation of excitatory amino acids to neuronal degenerative disorders, the relationship between excitatory amino acids and neuron damage is being examined with keen interest (Olney, 1990; Choi and Rothman, 1990; Shinozaki, 1994). In this chapter, neural degeneration and neuroprotective actions of certain glutamate receptor agonists extracted from medicinal herbs are described. Representative excitatory amino acids of natural origin are summarized in Table 1.

Table 1 Excitatory amino acids of natural origin

Compounds	*Origin*
Kainic acid	*Digenea simplex*
Domoic acid	*Chondria armata*
Acromelic acid	*Clitocybe acromelalga*
Quisqualic acid	*Quisqualic indica L.*
Ibotenic acid	*Amanita strobiliformis, Amanita muscaria*
Tricholomic acid	*Tricholoma muscarium*
Stizolobic acid	*Stizolobium hassjoo*
2-(carboxycyclopropyl)glycine (CCG)	*Blighia sapida, Aesculus parviflora*
β-N-oxalylamino-L-alanine (BOAA)	*Lathyrus sativus*
β-N-methylamino-L-alanine (BMAA)	*Cycas circinalis*
Willardiine	*Acasia willardiana*

2. GLUTAMATE RECEPTOR SUBTYPES

Glutamate receptors have been functionally divided into two main groups: ionotropic (iGluRs) and metabotropic glutamate receptors (mGluRs) (Monaghan *et al.*, 1989). At an early stage, iGluRs were subdivided into three types which demonstrated particular affinity to NMDA, kainic acid and quisqualic acid, respectively. Subsequently, electrophysiological and biochemical analyses of responses to quisqualic acid and ibotenic acid led to the discovery of a new and highly selective agonist for quisqualate receptors, α-amino-3-hydroxy-5-methylisoxazole-4-propanoic acid (AMPA), and quisqualate receptors are now referred to as AMPA receptors due to their particularly high and selective affinity for this compound.

Xenopus oocytes provide a system for the expression of exogenous mRNA that has allowed detailed study of the structure and function of many receptors, including iGluRs. Hollmann *et al.* (1989) first isolated a complementary DNA (cDNA) clone by screening a rat brain cDNA library expressed in *Xenopus* oocytes for kainate-gated ion channels. The cDNA encodes a single protein which, on expression in oocytes, forms a functional ion channel possessing many of the electrophysiological and pharmacological properties of iGluR subtypes of the glutamate receptor family in the mammalian CNS. Since then, a number of similar cDNAs encoding different subunits and isoforms have been cloned. A family of genes encoding subunits of iGluRs has been cloned, and the gene product has been characterized after expression in *Xenopus* oocytes. Human glutamate receptor cDNAs were successively isolated and sequenced, and the chromosomal localization of its gene was determined.

The NMDA family of glutamate receptors is divided into two subunits, NMDAR1 and NMDAR2. The latter is further subdivided into 4 subunits, NMDAR2A – NMDAR2D. The physiological functions of NMDA-type receptors have been well documented, and this can be mainly attributed to the discovery of selective antagonists which facilitated the pharmacological studies. The NMDA receptor-activated channel is highly permeable to Ca^{2+} and is the major route by which glutamate induces a toxic Ca^{2+} influx.

The non-NMDA family of glutamate receptors comprises a growing number of structurally related subunits [GluR1 to GluR4 (GluRA to GluRD); GluR5 to GluR7; KA-1 and KA-2]. GluR1 to GluR4 (GluRA to GluRD) appear to constitute the major AMPA receptor subtypes, and it is known that GluR2 limits Ca^{2+} permeability, but the functional and pharmacological characteristics of the other subunits are not yet fully resolved. AMPA and quisqualic acid effectively stimulate AMPA-type receptors which appear to underlie glutamate-induced fast excitatory postsynaptic potentials (EPSPs). Kainate-type receptors can be generated *in vitro* from subunits GluR5, GluR6 and GluR7 and KA-1 or KA-2, yet the physiological role of kainate receptors is at present unknown. However, their excessive activation *in vivo* leads to characteristic and selective neuron damage in the mammalian CNS. Kainate-type agonists are more markedly potent than other types of excitatory amino acids on a molar basis in causing depolarization and neuron damage.

mGluRs are a novel family of recently cloned receptors, and are coupled to multiple second messenger systems that include increases in phosphoinositide (PI) hydrolysis, decreases or increases in cAMP formation, and changes in ion channel function (Schoepp and Conn, 1993). At present, at least eight mGluR cDNAs have been cloned and subdivided into three groups, Group I (mGluR1 and 5), Group II (mGluR2 and 3) and Group III (mGluR4, 6, 7 and 8), on the basis of their linking second messenger system and affinities for some agonists. The clarification of the physiological roles of mGluRs has steadily progressed, but much less is known of the pharmacology and physiology of mGluRs compared to iGluRs (Nakanishi, 1992; Nakanishi and Masu, 1994). This is due in no small part to the lack of potent and highly selective agonists and antagonists for mGluRs.

Table 2 presents a summary of the classification of the subunits of glutamate receptors.

3. KAINIC ACID

In 1953, kainic acid (Figure 1) was first isolated as an effective ascaricidal component of a seaweed, *Digenea simplex.* The seaweed is one of the red algae, *Rhodomelacea,* and has long been used in Japan as a pharmacopoeia for killing intestinal worms. The common name of *Digenea simplex* in Japanese is "*kainin-so*", thus, this excellent ascaricidal component was named 'kainic acid'. Kainate molecules contain the chemical moiety of glutamic acid, enclosed in a heterocyclic ring. It was thought that the ascaricidal actions of kainic acid might result from marked muscle contraction in ascarides through excessive activation of excitatory receptors at the presumed glutamatergic NMJ, and we first demonstrated that kainic acid has

Table 2 Classification of glutamate receptor subunits

Ionotropic Receptors

Receptor Subunits		*Receptor Subtypes*
Rat	*Mouse*	
GluR1 – GluR4 or GluRA – GluRD	$\alpha 1$-$\alpha 4$	AMPA receptor
GluR5 – GluR7	$\beta 1$-$\beta 3$	low affinity kainate receptor
KA1, KA2	$\gamma 1$, $\gamma 2$	high affinity kainate receptor
$\delta 1$, $\delta 2$	$\delta 1$, $\delta 2$	unknown
NMDAR1	$\xi 1$	NMDA receptor
NMDAR2A – NMDAR2D or NR2A – NR2D	$\varepsilon 1$-$\varepsilon 4$	NMDA receptor

Metabotropic Receptors

Group	*Subtype*	*Agonist*
Group I	mGluR1, mGluR5	quisqualic acid
Group II	mGluR2, mGluR3	DCG-IV, L-CCG-I
Group III	mGluR4, mGluR6 – mGluR8	L-AP4

marked excitatory action on rat cortical neurons which respond to L-glutamate (Shinozaki and Konishi, 1970).

The iontophoretic administration of kainic acid to rat cortical neurons induces marked spike discharges which are effectively blocked by GABA (Shinozaki and Konishi, 1970). L-α-allo-kainic acid also causes spike discharges with a weaker activity than kainic acid. In terms of the frequency of spike discharges produced by iontophoretic currents of the same strength, kainic acid seems to be more potent than L-glutamate. Spike discharges induced by kainic acid persist for a much longer time, even after cessation of the iontophoretic application than those evoked by L-glutamate. Such a prolonged generation of spike discharges is one of the characteristics of the responses to kainic acid, probably reflecting a slow dissociation of kainic acid from receptors.

In the isolated spinal cord preparation of newborn rats, kainic acid and other excitatory amino acids induce depolarization of spinal motoneurons in a concentration-dependent manner. Kainic acid is much more potent than L-glutamate in inducing depolarization on a molar basis, but is slightly less potent than quisqualic acid. Selective NMDA blockers, such as 3-((+)-2-carboxypiperazin-4-yl)propyl-1-phosphonic acid (CPP), D-(–)-2-amino-5-phosphonopentanoic acid (D-AP5), (+)-5-methyl-10,11-dihydro-5H-dibenzo[a,d]cyclohepten-5,10-imine maleate (MK-801) and Mg^{2+}, did not affect the kainate-induced depolarization. 6-Cyano-7-nitroquinoxaline-2,3-dione (CNQX), 6,7-dinitroquinoxaline-2,3-dione (DNQX) and 2,3-dihydroxy-6-nitro-7-sulphamoylbenzo(F)quinoxaline (NBQX) (Sheardown *et al.*, 1990) each effectively blocked the kainate-induced depolarization, so that at high concentrations the depolarization was completely abolished, suggesting that the inhibition is in a competitive manner (Fletcher *et al.*, 1988). However, these quinoxalinediones are not always selective for responses to kainic acid because they depress responses to AMPA, quisqualic acid and NMDA as well as those to kainic acid. The NMDA-blocking action of these quinoxalinediones is considered to be due to their antagonism of the coagonist activity of glycine at the NMDA receptor-channel complexes.

Systemic administration of kainic acid to the rat causes characteristic neurological behaviors. The most pronounced behavioral changes in the rat seen after a systemic injection of kainic acid are immobility, increased incidence of automatism, 'wet dog shakes (WDS)', severe 'rearing, praying, salivating' limbic motor seizures, severe catalepsy, and tonic-clonic convulsions. Kainic acid also induces selective neuron damage in the hippocampus, pyriform cortex, amygdala, several thalamic nuclei and lateral septum (Ben-Ari, 1985; Nadler *et al.*, 1978; Nadler, 1981).

Epileptic seizures induced by kainic acid have been used as a model of temporal lobe epilepsy, and kainic acid has made a considerable contribution to the advances in the study of epilepsy (Ben-Ari, 1985). Recurrent seizures leading to status epilepticus were induced in rats by a subcutaneous injection of kainic acid. Spontaneous epileptic seizures were sometimes seen a considerable number of days after the systemic administration of kainic acid, and disturbances of memory were reported in a water maze test, but marked behavioral changes were not observed despite the severe degeneration of hippocampal neurons. Kainic acid impaired but did not preclude the acquisition of the water-maze task in rats. Kainic acid also caused deficits in the task of learning to approach a visible platform in a water maze. After a 4-day recovery period, the kainic acid-treated animals showed a syndrome of increased activity in an open field, hyper-reactivity in response to handling, and deficits in the acquisition of both passive avoidance and spatial learning tasks.

Kainic acid is a powerful neurotoxin that produces selective neuronal lesions in the mammalian CNS. The excitotoxic actions of kainic acid have been well documented, and kainic acid has been a useful tool for the study of the mechanism of neuron damage. In recent years, considerable interest has focused on the neurotoxic properties of excitatory amino acids and their possible relevance for the study of human neurodegenerative disorders. The recent development of selective agonists and antagonists for excitatory amino acid receptors has also aided in the elucidation of excitotoxic mechanisms. Pharmacological blockade of excitotoxicity may constitute a novel therapeutic strategy for the treatment of neurological disease states such as ischemic neuron damage and epilepsy. A particularly significant feature of lesions made by excitatory amino acids is their sparing of fibers that pass to or through the zone of neuronal degeneration. The local administration of kainic acid in the rat causes damage of almost all neurons in the injected region, but axons of passage or of termination are spared. Accordingly, excitotoxic amino acids, especially kainic acid, are being employed as tools to destroy postsynaptic elements in the brain while preserving presynaptic structures. In addition, kainic acid can be used as a tool in neurobiology to produce an animal model of neuro-degenerative disorders such as Huntington's disease and epilepsy (Coyle, 1987).

There are interesting differences in the pathological manifestations following the intracerebral as compared to the systemic administration of kainic acid. The intracerebral or intraventricular injection of kainic acid produces neuronal loss selectively in the CA3/CA4 cell fields of the hippocampus, which correlates well with the high density of kainate receptors in this area (Foster *et al.*, 1981; Monaghan and Cotman, 1982). In contrast, systemic kainic acid causes seizures, extensive brain edema and neuronal damage in the hippocampus CA1 region where there is a low density of kainate receptors (Foster *et al.*, 1981; Monaghan and Cotman, 1982). It is of great interest that the distribution of the lesions induced by kainic acid does not always correlate with the regions of high density of kainate receptors. It was reported that the extent of neuron damage resulting from the systemic administration of kainic acid is dependent on the extent of seizure activity (O'Shaughnessy and Gerber, 1986) and that the toxicity of kainic acid is uniquely dependent on neuronally released glutamate (McGeer and Zhu, 1990). Friedman *et al.* (1994) showed that, in the hippocampal CA3/CA4 area, the expression of GluR2 and GluR3 was markedly decreased at 12 and 24 hours after an intraperitoneal administration of kainic acid. GluR2 is the non-NMDA receptor subunit that limits Ca^{2+} permeability. This finding raises the possibility that an increased formation of Ca^{2+}-permeable non-NMDA receptors in the CA3/CA4 area may enhance glutamate pathogenicity.

Glutamate neurotoxicity is mediated through the activation of glutamate receptors and entails sustained depolarization of postsynaptic dendrosomal membranes, increased membrane permeability and impaired ion homeostasis, probably by a toxic influx of extracellular Ca^{2+} (Choi, 1987, 1988). Only NMDA receptors were thought to open a membrane channel that is highly permeable to Ca^{2+} (MacDermott *et al.*, 1986), but some types of neurons sensitive to AMPA or kainic acid have also recently proved to have high Ca^{2+}-permeability, which negatively depends on the expression of the GluR2 subunit. The NMDA receptor-activated channel is the major route by which glutamate induces a toxic Ca^{2+} influx, although additional Ca^{2+} entry probably occurs through voltage-activated Ca^{2+} channels, the Na^{+}/Ca^{2+} exchanger, and non-specific membrane leakage (Choi, 1988). Although the mechanism of glutamate neurotoxicity has not yet been fully elucidated, glutamate neurotoxicity has several specific characteristics consistent with an important role in the pathogenesis of ischemic brain damage (Choi and Rothman, 1990), and the CA1 damage caused by systemic kainic acid closely resembles that produced by cerebral ischemia. NMDA receptors clearly participate in certain forms of ischemic neuron damage (Choi and Rothman, 1990); however, it seems likely that the destruction of CA3/CA4 pyramidal cells induced by intrahippocampal kainic acid is mediated primarily by non-NMDA receptors, and that the loss of these cells induced by kainic acid is associated with the neurobehavioral effect of an intra-hippocampal administration of kainic acid, because the kainate-induced CA3/CA4 damage is not affected by MK-801 pretreatment, although high concentrations of MK-801 reduce the amount of kainate-induced cell loss in the dentate gyrus granule cells and, to some extent, the CA1 pyramidal cell damage in the hippocampus. Furthermore, ischemic neuron damage was effectively protected by NBQX, which preferentially blocks responses to quisqualic acid and AMPA (Sheardown *et al.*, 1990). Sheardown *et al.* (1990) have hypothesized that glutamate release during ischemia may trigger a post ischemic release of glutamate or another endogenous substance, which acts at non-NMDA receptors. This hypothesis is supported by the observation that destruction of the CA3 area of the hippocampus which forms glutamatergic synapses in CA1 areas protects CA1 neurons from the degeneration induced by temporal ischemia of the whole brain.

4. QUISQUALIC ACID

Quisqualic acid is an active anthelmintic constituent of *Quisqualis indica L.*, which has been in medicinal use in China and Japan (Takemoto, 1978). We considered that the chemical structure of quisqualic acid (Figure 1) closely resembles that of glutamic acid, and examined the effect of quisqualic acid at the crayfish NMJ, where glutamate is believed to be an excitatory transmitter. We found that quisqualic acid is about 1000 times more potent than L-glutamic acid (Shinozaki and Shibuya, 1974). Since that finding, quisqualic acid has been shown to be one of the most potent agonists for glutamate receptors in various preparations, and has been very useful in the pharmacological classification of glutamate receptors. Quisqualic acid is a contributory compound for promoting the development of AMPA. In the course of the study of glutamate receptors, the measurement of conductance using patch clamp methods and ligand binding experiments has revealed that quisqualic acid is not a pure iGluR agonist but also activates mGluRs. AMPA, derived from a chemical modification of ibotenic acid, has since received widespread use as a pure agonist for an iGluR subtype (AMPA receptor) in place of quisqualate. It is known that ibotenic acid, a mother compound of AMPA, has a high affinity for both NMDA receptors and mGluRs; however, AMPA directly opens non-NMDA-type ion channels, which are related to the production of fast EPSPs in the mammalian CNS. AMPA has no efficacy as an mGluR agonist.

Although quisqualic acid shows complex responses in the mammalian CNS, it caused simple excitation at the crayfish NMJ as a pure agonist. In contrast, AMPA, a pure agonist in the mammalian CNS, did not show any depolarizing activity at the crayfish NMJ, and NMDA, one of the representative agonists for iGluRs, also did not cause a depolarization of the crayfish muscle. At the crayfish NMJ, glutamate and quisqualate seemed to activate a common receptor because they caused cross-desensitization of receptors, and kainic acid markedly augmented depolarizing responses to glutamate, while low concentrations of kainoids (kainate analogues such as kainate, domoate and acromelate; see below) inhibited the depolarizing responses to quisqualate in a competitive manner. Thus, there is a considerable difference between the mammalian CNS and the invertebrate NMJ in their pharmacological responses to excitatory amino acids.

5. DOMOIC ACID

Domoic acid, an excitatory amino acid structurally related to kainic acid, is also one of the potent excitotoxic kainoids of natural origin. Domoic acid was first isolated from a seaweed, *Chondria armata*, the same family of the red algae as *Digenea simplex* from which kainic acid was isolated, and successively some derivatives of domoic acid such as domoilactones and isodomoic acids were also isolated from the seaweed. The common name of *Chondria armata* in Japanese is '*Domoi*' or '*Hanayanagi*', and its excellent ascaricidal component, therefore, was named domoic acid. The seaweed has also been noted to have insecticidal properties. Domoic acid is a tricarboxylic amino acid that is structurally related to glutamic acid, and contains a chemical moiety of glutamate similar to that of kainic acid (Figure 2). Domoic acid is a potent and selective agonist for kainate receptors, and is more active than kainic acid in producing neurotoxic effects through the activation of kainate receptors.

Electroencephalographic, behavioral and neuropathological experiments using systemic or intracerebroventricular administrations of domoic acid were conducted in mice, rats and monkeys. The systemic administration of domoic acid to mice caused characteristic neurological behaviors quite similar to kainic acid, and it was almost impossible to differentiate pharmacologically and histopathologically between the effects of kainic acid and domoic acid. Intraperitoneal injections of either domoic acid or kainic acid produced a series of characteristic behavioral changes including sedation, rigidity, stereotypy (scratching, head nodding), balance loss, and discrete or generalized convulsions and death; however, involuntary scratching of the shoulders with the hind legs is particularly prominent after the administration of domoic acid, but is observed only infrequently after kainic acid administration. Domoic acid was a much more potent excitotoxin in the CA3 region than in the CA1 region of the rat hippocampal slice, whereas no such regional difference has been observed with quisqualate or NMDA (Debonnel *et al.*, 1989a).

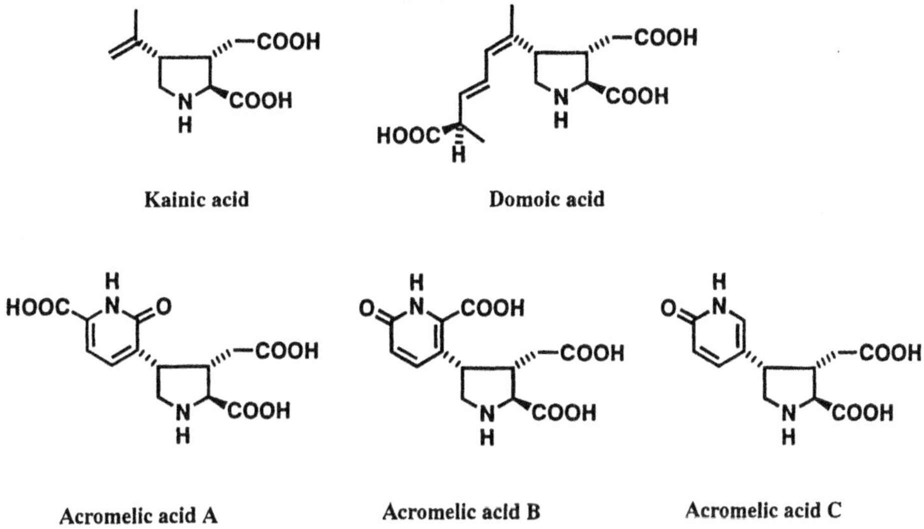

Figure 2 Chemical structures of kainoids such as kainic acid, domoic acid and acromelic acid A, B and C.

The differential regional depolarizing response of hippocampal CA1 and CA3 pyramidal neurons to kainic acid is attributable to the extremely high density of kainate receptors in the CA3 region.

In 1987, an intoxication caused by the consumption of cultured mussels harvested from eastern Prince Edward Island produced neurological problems such as headache, confusion, and loss of memory, which was severe in certain cases. The patients had a rapid onset of confusion, disorientation, and memory loss occurring within 24 hours of eating mussels from the island. Severe neurologic impairment and some deaths occurred, and pathological examinations of patients intoxicated with the mussels revealed extensive damage of the hippocampus, as well as spotty damage of thalamic and forebrain regions (Perl *et al.*, 1990; Teitelbaum *et al.*, 1990), which closely resembled pathological changes seen after systemic administration of kainic acid in the rat. Some of the survivors apparently sustained permanent brain damage and have continued to show profound memory impairment. Domoic acid and its isomers were later identified as the toxin that poisoned the people who had consumed these contaminated blue mussels (*Mytilus edulis*) (Iverson *et al.*, 1989; Grimmelt *et al.*, 1990). Microinjections of domoic acid into the hippocampal formation of rats produces degeneration of CA3 and CA1 pyramidal cells and dentate gyrus granule cells. Loss of neurons in the thalamic reticular nucleus may underlie some types of the attentional deficits observed in humans following domoic acid poisoning (Friedberg and Ross, 1993). A domoic acid-containing mussel extract was more neurotoxic for cultured neurons than was purified domoic acid, and this increase in neurotoxicity was selectively due to domoic acid potentiation of the excitotoxic effect of L-glutamic acid and L-aspartic acid, which are present in high concentrations in mussel tissue. Depolarizing responses to L-glutamate are significantly augmented by low concentrations of kainic acid in rat cortical neurons

(Shinozaki and Konishi, 1970), and the reason why the muscle itself was not damaged by domoic acid would be due to low sensitivity to domoic acid of the mussels. At the NMJ of the invertebrates such as the crayfish, glutamate functions as an excitatory neurotransmitter. Unlike the mammalian CNS, kainic acid does not potently excite the muscle membrane. However, domoic acid was about 10 times more potent than kainic acid at the crayfish NMJ (Shinozaki and Shibuya, 1976; Shinozaki *et al.*, 1986).

A disproportionately high percentage of severely affected individuals were elderly. As Olney (1990) described, the apparently heightened vulnerability of elderly individuals to the neuropsychological consequences of domoic acid poisoning is of particular interest, as this suggests that the kainate receptor is a potentially sensitive exogenous mediator of excitotoxic neuropathology in old age. When kainic acid was administered subcutaneously in various doses to young, middle-aged, and old rats, the middle-aged and old rats were significantly more sensitive than the young rats to the neurotoxic actions of kainic acid (Wozniak *et al.*, 1991). Low doses of kainic acid, which were nontoxic to young rats, frequently triggered status epilepticus, associated brain damage, and precipitous death in old rats. Middle-aged rats were more sensitive than young rats, but less sensitive than old rats to kainic acid neurotoxicity. Thus, considerable interest has focused on the neurotoxic properties of domoic acid and kainic acid (Debonnel *et al.*, 1989a, 1989b; Tryphonas and Iverson, 1990; Huettner, 1990; Patneau and Mayer, 1990).

6. ACROMELIC ACID

Acromelic acids A and B (Figure 2), which possess a constitutional moiety of kainic acid, have been isolated from the mushroom *Clitocybe acromelalga*, one of the most poisonous mushrooms found in Japan (Konno *et al.*, 1983, 1988). Another kainoid, acromelic acid C, has also been isolated from the mushroom (Fushiya *et al.*, 1990). Acromelic acid C is not an isomer of acromelic acid A or B, but has a structure of decarboxylated acromelic acid B. In the present chapter, acromelic acid refers to acromelic acid A, unless otherwise specified. Ingestion by humans of the mushroom *Clitocybe acromelalga* causes a sharp pain, and about a week after ingestion a marked reddish edema (erythromelalga) appears in the hand and foot which lasts for about a month. Fever, gastrointestinal abnormalities, renal disturbances and signs of abnormal CNS function are not observed following intoxication with this mushroom, distinguishing acromelic acid intoxication from that with either kainic acid or domoic acid. Other compounds such as clitidine, clithioneine and 4-aminoquinolinic acid have thus far been isolated from this mushroom, but it is not yet known whether these compounds contribute to the symptoms of poisoning. According to a Japanese medical journal, axonal damage may have occurred in the peripheral nerves of a patient poisoned by the mushroom. Yet rats did not develop reddish edema of the hindlimbs after a systemic injection of acromelic acid, suggesting that acromelic acid is not a causal compound of the erythromelalga in the intoxication.

In the isolated spinal cord of newborn rats, acromelic acid caused a more marked depolarization of motoneurons than either kainic acid or domoic acid, and has proven to be one of the most potent excitatory amino acids in both vertebrates and invertebrates (Ishida and Shinozaki, 1988a, 1991). When depolarizing responses of motoneurons to these kainoids were extracellularly recorded from L_3–L_5 ventral roots of newborn rats, their threshold concentrations for excitation were almost identical (about 1 μM), but acromelic acid demonstrated the highest depolarizing activities among these three kainoids, followed in descending order by domoic acid and kainic acid (Ishida and Shinozaki, 1988a; Shinozaki *et al.*, 1991), and there were also some differences in electrophysiological features of their depolarizing responses, including the slope of the dose–response curves and the time-course of their action (Ishida and Shinozaki, 1988a). In the newborn rat spinal preparation, acromelic acid induced the depolarizing responses which reversed more rapidly than those induced by other kainoids. This short time-course is one of the characteristics of depolarizing responses to acromelic acid in the isolated newborn rat spinal cord. Several factors such as receptor desensitization, equilibrium constants of agonist-receptor complexes, and differences in rates of uptake of agonists would affect the time-course and the peak amplitude of depolarizing responses but have not yet been sufficiently analyzed. Mg^{2+} and selective NMDA blockers such as CPP, D-AP5 and MK-801 do not affect the acromelate-induced depolarization in the newborn rat spinal cord, but the quinoxalinediones, such as CNQX, DNQX and NBQX, effectively depress depolarizing responses to acromelic acid in a concentration-dependent manner, suggesting that acromelic acid is a more potent agonist for non-NMDA-type receptors than kainic acid or domoic acid. Cyclothiazide, which is known to inhibit the development of the desensitization of AMPA receptors rather than kainate receptors, significantly augmented acromelate- and AMPA-induced depolarization in the spinal motoneuron of newborn rats, but did not increase domoate- and kainate-induced depolarization as much. At present, a highly selective antagonist for kainate receptors is not available; therefore, it is difficult to determine pharmacologically whether acromelic acid acts selectively on kainate receptors.

The dorsal root C-fiber may be used for differentiating kainic acid-type agonists from other non-NMDA-type agonists, where domoic acid and kainic acid demonstrate a particularly high depolarizing activity, while AMPA is much less active than kainic acid, and NMDA does not cause depolarization of the dorsal root even at high concentrations (Agrawal and Evans, 1986). In the dorsal root of newborn rats, acromelic acid causes depolarization in a dose-dependent manner, and its depolarizing activity is lower than that of domoic acid, but slightly higher than that of kainic acid. The rank order of their depolarizing activity in the dorsal roots is reversed compared with that of the motoneuron. As in the motoneuron, the depolarization of dorsal roots induced by acromelic acid is effectively blocked by CNQX, but not by Mg^{2+} ions or selective NMDA blockers.

Desensitization to various pharmacological agents has provided the means for analyzing the role of particular types of receptors in drug effects. In particular, crossed desensitization between two excitatory compounds has been considered an effective way to determine their identity or to examine their physiological and pharmacological properties. A marked development of receptor desensitization is one of the characteristic features in depolarizing responses in the dorsal root of newborn rats. When kainic acid was applied to the dorsal root for a prolonged time, initial depolarization was not maintained but decreased to a sustained plateau level, demonstrating the development of desensitization of receptors. Under these conditions, acromelic acid and domoic acid did not cause any further depolarization of the dorsal root. From these results and the lack of effect of NMDA receptor ligands on acromelate responses mentioned above, it seems likely that acromelic acid is an agonist for kainate receptors.

Although kainic acid and domoic acid are not particularly active on the glutamatergic NMJ in invertebrates, acromelic acids A and B cause a large depolarization of the crayfish opener muscle fiber at markedly low concentrations (Shinozaki *et al.*, 1986). Acromelic acids A and B are of almost equal potency at the crayfish NMJ, and their pharmacological properties are quite similar to those of kainic acid (Shinozaki; 1988). The order of depolarizing activity in terms of its threshold concentration is as follows: acromelic acid (100) > domoic acid (10) > kainic acid (1). Acromelic acid is among the most potent of the excitatory amino acids tested to date on insect neurons (Sattelle *et al.*, 1994).

From the structural similarity of acromelic acid to kainic acid and domoic acid, it seems reasonable to propose that acromelic acid demonstrates kainate-like excitotoxic actions in the mammalian CNS. However, the systemic administration of acromelic acid to the rat causes abnormal neurological behaviors distinct from that of kainic acid or domoic acid, and there is a striking difference in the regional distribution of neuron damage (Figure 3) (Shinozaki *et al.*, 1989, 1990; Kwak *et al.*, 1990). The

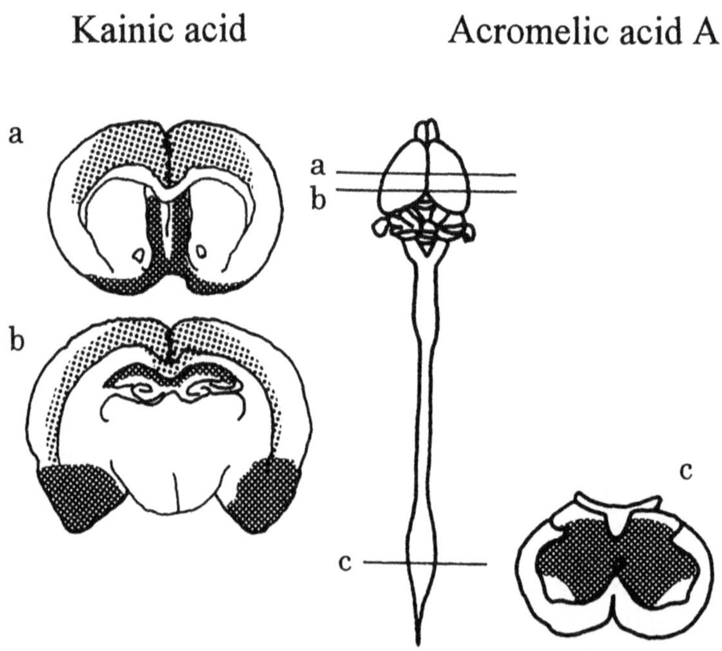

Figure 3 A striking difference in the regional distribution of neuron damage induced by acromelic acid and kainic acid. Thin dots: area of slightly degenerating neurons, thick dots: severely degenerated. a, b, and c: the level of the transection.

most pronounced behavioral changes in the rat seen after a single systemic injection of acromelic acid were the consecutive appearance of tonic extension of the hindlimbs, transient flaccid paralysis, and persistent spastic paraplegia (Shinozaki *et al.*, 1989, 1991; Kwak *et al.*, 1990, 1992). The tonic extension of the hindlimbs very closely resembled strychnine-induced convulsions, although the fore part of the rat body did not show any convulsive action. Before this stage, motor discoordination became much more marked in the hindlimbs. Both hindlimb heels were gradually raised, and the rats fell forward as they stood. They were apparently seized with a cramp in both hindlimbs, and this occurred repeatedly. Afterwards, the rats developed marked generalized tonic-clonic convulsions, which were the only common behavioral sign seen after systemic administrations of either acromelic acid or kainic acid, and almost all rats died at this stage. The surviving rats showed severe seizures, whose duration was markedly shorter than that evoked by kainic acid. About 2 hours after the injection of acromelic acid, the tonic extension of the hindlimbs and tail usually, but not inevitably, took the place of the severe acromelate-induced seizures, followed by a transient flaccid paralysis of the lower part of the body. On the day following a single injection of acromelic acid, the surviving rats that had shown the transient flaccid paralysis developed a severe spastic paraplegia, which was significantly reinforced by sensory stimuli such as touching and poking. The incidence and duration of the transient flaccid paralysis on the day of acromelate injection, which seemed to be a premonitory symptom towards the development of persistent spastic paraplegia, must play key roles in the development of ensuing spastic paraplegia, because the surviving rats that had experienced the heavier transient flaccid paralysis of the hindlimbs developed the more severe spastic paraplegia. The spastic paraplegia persisted until the rats were sacrificed for histological analysis (at least three months after the single application of acromelic acid). The chain of abnormal behavioral signs induced by acromelic acid, beginning with the tonic extension of the hindlimbs, was quite distinct from that induced by kainic acid. Kainic acid never caused spastic paraplegia nor tonic extension of the hindlimb of the rats even at lethal doses, but, instead, caused severe limbic motor seizures. There were no apparent limbic seizures in rats receiving acromelic acid, and only a few incidences of WDS even at lethal doses. The behavioral signs suggested that the spinal cord was exquisitely sensitive to acromelic acid but the convulsions probably reflected action at the brain rather than at the lower spinal cord level. Both rats and mice demonstrate quite similar behavior after the systemic administration of acromelic acid, but the incidence of the spastic paraplegia in the mouse is significantly lower than in the rats.

Histological examination of the paraplegic rats which received systemic acromelic acid revealed that the pathological changes were confined almost completely to the spinal cord and, in particular, to the lumbar and sacral segments, as predicted from the behavioral signs. In these segments, there were several pyknotic cells and a marked decrease in the numbers of small neurons in the dorsal and middle portion of the gray matter. There was marked gliosis in these areas which were diffusely distributed, and glial fibrillary proliferation was also observed in the ventrolateral gray matter. These pathological observations indicate that the selective loss of spinal interneurons, which may exert inhibitory actions on the spinal arc, is probably responsible for the development of persistent spastic paraplegia. No apparent pathological changes were found in hippocampal neurons, in striking contrast to the vulnerability of hippocampal pyramidal cells to kainic acid (McGeer *et al.*, 1978; Nadler *et al.*, 1978; Nadler, 1981; Ben-Ari, 1985), although a mild glial reaction has been detected in the CA4 area of the hippocampus by immunohistochemical staining for glial fibrillary acidic protein (GFAP) (Kwak *et al.*, 1990). The lack of detectable and definitive histological changes in the brain may be due to the transient nature of the convulsions.

The absence of pathological change in the large anterior horn cells and the myelinated spinal tracts indicated that both the primary and secondary pyramidal tracts and the ascending sensory tracts were not affected. Severe rigidospastic paraparesis with extensive damage of spinal interneurons has been known to be induced in the dog, cat, rabbit, and rat after spinal ischemia by the ligation of the thoracic aorta. The distribution of neuron damage induced by acromelic acid closely resembles that of neuron damage by this spinal ischemia. In contrast, as mentioned above, systemic administration of kainic acid to the rat causes extensive neuron damage in the CA1 region of the hippocampus, which closely resembles that induced by brain ischemia. The similarity of neuron damage induced after ischemia to those induced by kainoids suggests the possibility of a common mechanism underlying the neuron damage after ischemia and kainoid neurotoxicity, although the precise mechanism of delayed neuronal death after brain ischemia (Kirino, 1982) or glutamate neurotoxicity is still not known.

At present there is no clear-cut explanation for the regional difference in neuronal degeneration caused by acromelic acid and that caused by kainic acid, but the idea that there are two or more kinds of subtypes of kainate receptors is very attractive. Acromelic acid did not bind to high-affinity [^{3}H]AMPA and [^{3}H]kainic acid binding sites. Smith and Mallhinney (1992) reported that in the absence of chaotropic ions, acromelic acid is the most potent displacer of [^{3}H]AMPA binding yet described, and they suggested that acromelic acid distinguishes two kainic acid binding sites in rat brain synaptic plasma membranes. Indeed, whole-cell voltage-clamp recording data from cultured rat hippocampal neurons revealed the presence of two kainate responses which differed in their rectification properties and their permeability to Ca^{2+} (Iino *et al.*, 1990). Recent data on cloned glutamate receptor molecules suggest a multiplicity of kainate receptors in the rat brain. Kainate receptors or AMPA receptors are heterooligomeric proteins, and genes encoding each subunit express their transcripts differently in brain areas (Boulter *et al.*, 1990; Keinanen *et al.*, 1990; Sommer *et al.*, 1990).

The selective marked loss of interneurons in the spinal cord without an extensive change in large motoneurons in the ventral horns or the corticospinal tract is one of the characteristic pathological features of human 'stiff-man' syndrome, a progressive neurological disease characterized by slowly ascending rigidospasticity of the axial and limb musculature with painful muscle spasm. Although there is no direct evidence that acromelate induces stiff-man syndrome, the close similarity between the clinical and neuropathological pictures of this disorder and those of the rat with spastic paraplegia indicates that further study of the physiological role of the kainate receptors in neurodegenerative disorders is deserved. In fact, an etiological role of excitatory amino acids has been proposed in several neurological diseases affecting predominantly the spinal cord, including amyotrophic lateral sclerosis (Spencer *et al.*,

1987; Young and Fagg, 1990) and neurolathyrism (Rao *et al.*, 1964; Bridges *et al.*, 1989). Elucidation of the mechanism of excitotoxic neuron damage may provide a valuable clue to the selective neuronal death in neurodegenerative disorders (Stewart *et al.*, 1990) (see below).

7. IBOTENIC ACID

Ibotenic acid (Figure 4), isolated from *Amanita muscaria*, is an analogue of glutamate which is partially restricted in conformational flexibilities. Ibotenic acid was found to be a powerful excitant of cat spinal interneurons (Johnston *et al.*, 1968) and more effective than glutamate in producing spike discharges in rat cortical neurons (Shinozaki and Konishi, 1970). When ibotenic acid was iontophoretically applied to the rat cortical neurons, it showed a potent excitatory action but was less potent than kainic acid. The excitation induced by ibotenic acid almost disappeared in the presence of a selective NMDA blocker, and ibotenic acid therefore appears to activate NMDA receptors, but in the mammalian CNS it also interacts with AMPA and kainate receptors and with mGluRs. Since ibotenic acid has potent excitotoxic actions, it has been used to induce selective neuronal degeneration. However, there appear to be differences in the neurons affected by ibotenic acid and kainic acid (Olney, 1990).

Before the appearance of ibotenic acid in the neuroscience research field, all known excitatory amino acids were a straight-chain type of compounds such as glutamic acid and NMDA; ibotenic acid was the first excitotoxic compound consisting of a heterocyclic ring. This ring is of great use in determining the active conformation of glutamate at a given receptor. Muscimol, obtained from the decarboxylation of ibotenic acid, more potently activates GABA receptors as a selective $GABA_A$ agonist than GABA. A decarboxylated derivative of kainic acid, 3-carboxymethyl-4-isopropenylpyrrolidine, did not show any GABA-like action at all, although it possesses a chemical moiety of GABA; this result is probably due to a conformational difference between these derivatives.

Ibotenic acid is chemically unstable, but activates both NMDA receptors and mGluRs. In an attempt to develop chemically stable compounds with high potency

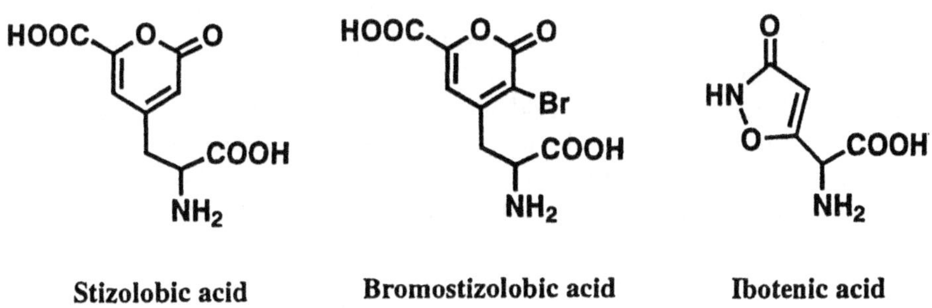

Figure 4 Chemical structures of ibotenic acid, stizolobic acid and its bromine derivative.

and specificity, a number of analogues have been synthesized. Among them, AMPA was found to be a highly selective and potent agonist for a non-NMDA iGluR subtype (Krogsgaard-Larsen *et al.*, 1980), that was later named as AMPA receptors. AMPA receptors and kainate receptors, however, are difficult to distinguish pharmacologically due to a lack of antagonists which selectively discriminate between them, and these two subtypes of excitatory amino acid receptors are often collectively referred to as non-NMDA receptors. The recently developed quinoxalinedione antagonists, CNQX, DNQX and NBQX, act selectively at non-NMDA receptors, but they do not completely distinguish AMPA receptors from kainate receptors (Honore *et al.*, 1988; Sheardown *et al.*, 1990).

Since many of the aforementioned compounds were first isolated due to their effects on invertebrate preparations, a phylogenetic examination is also important (Shinozaki, 1988; Shinozaki and Ishida, 1988b). At the crayfish NMJ, ibotenic acid presynaptically inhibited excitatory junctional potentials. This inhibitory action was blocked by picrotoxin. However, ibotenic acid did not affect glutamate-induced depolarization of the opener muscle. As mentioned above, NMDA did not show any depolarizing activity at the invertebrate NMJ, and it is thus possibile that ibotenic acid activates mGluRs at the invertebrate NMJ. Furthermore, glutamate activated both types of extra-junctional glutamate receptors on the locust muscle to cause depolarization and hyperpolarization (Cull-Candy, 1976); however, ibotenic acid had no effect on locust excitatory neuromuscular synapses, but reacted with extra-junctional receptors to increase selective Cl^- permeability, leading to an increase in membrane conductance. This experimental result stimulated research as to whether glutamate activates Cl^- channels in the mammalian CNS, and some studies have demonstrated glutamate-induced hyperpolarization or biphasic (excitatory-inhibitory) effects on the basis of activating Cl^- channels.

8. STIZOLOBIC ACID

Stizolobic acid (Figure 4) was found together with stizolobinic acid in the sap exuded from the cut surface of the epicotyl tips of etiolated seedlings of *Stizolobium hassjoo* (Hattori and Komamine, 1959). Stizolobinic acid possesses a chemical moiety of α-aminopimelic acid forming a pyron ring in its molecule. α-Aminopimelic acid has a slight depolarizing activity in the cat spinal cord but is considerably less potent than L-glutamate or L-aspartate. The conformation of the molecules of stizolobic acid or stizolobinic acid is to some extent flexible, but rather more restricted than that of α-aminopimelic acid.

At the crayfish NMJ, stizolobic acid and its halogen substituents such as 3-chloro-, 5-fluoro- and 5-bromostizolobic acid depressed the depolarizing responses to quisqualate and L-glutamate in a competitive manner without affecting the resting membrane potential and depolarizing responses to kainoids and GABA (Shinozaki and Ishida, 1988). These halogen substituents and stizolobic acid were almost equipotent at the crayfish NMJ. The amplitudes of the excitatory junctional potentials were depressed by stizolobic acid in a dose-dependent manner.

In contrast to the competitive blocking action of stizolobic acid on quisqualate responses at the crayfish NMJ, stizolobic acid caused marked depolarization of spinal motoneurons of newborn rats and it increased the peak amplitudes of depolarizing responses to quisqualate, kainate, L-glutamate and NMDA in an additive manner, suggesting that stizolobic acid acts as an L-glutamate agonist in the mammalian CNS (Ishida and Shinozaki, 1988b). Most cortical neurons of the rat responding to L-glutamate were also excited by the iontophoretic administration of stizolobic acid. The depolarization induced by stizolobic acid was not affected by selective NMDA blockers but was effectively depressed by CNQX and by NBQX in the isolated spinal cord of newborn rats. The amplitude of the depolarization was about twice as large as that of L-glutamate in the newborn rat spinal motoneuron at the same concentration. A bromine derivative of stizolobic acid, 3-bromostizolobic acid, possessed a potent depolarizing activity in the isolated rat spinal cord, being about 5 times more potent than stizolobic acid. 3-Chloro-, 5-fluoro- and 5-bromostizolobic acid were also more potent than stizolobic acid when their depolarizing activities were compared at the same concentration, and 3- and 5-bromostizolobic acids were almost equipotent.

In contrast, 3- or 5-benzene-, and 5-naphthalenestizolobic acid depressed the depolarizing responses to NMDA-type agonists in a dose-dependent manner in the isolated spinal cord of newborn rats. They were less potent in reducing the NMDA response than CPP, but were quite insensitive to depolarizing responses to quisqualate and kainate in the isolated spinal cord of newborn rats. Among these derivatives of stizolobic acid, 5-naphthalenestizolobic acid is particularly potent and it depresses spinal reflexes almost completely in relatively high concentrations. An increase in π-electron density on C5-substitution may contribute to the increase in the inhibitory action on depolarizing responses to NMDA. It is of great interest that there is an almost linear relationship between the inhibitory activity of these derivatives on quisqualate responses at the crayfish NMJ and the depolarizing activity in the rat spinal cord. In addition, the derivatives of stizolobic acid showing inhibitory actions on the depolarizing responses to NMDA in the rat spinal cord demonstrated weaker inhibitory actions than stizolobic acid at the crayfish NMJ.

9. CCG

2-(Carboxycyclopropyl)glycine (CCG), isolated from young fruits of the plants *Aesculus parviflora* and *Blighia sapida* (Fowden *et al.*, 1969), is a conformationally restricted glutamate analogue, in which the cyclopropyl group fixes the glutamate chain of CCG in either an extended or a folded form. CCG possesses three asymmetric carbon atoms in its molecules, and thus has eight stereoisomers (Figure 5B). The carbon chain of glutamate is able to conform almost completely with that of CCG; therefore, CCG could provide useful information about the relationship between the conformation of glutamate molecules and the activation of subtypes of glutamate receptors, and therefore aid the development of new selective agonists for each subtype of glutamate receptors (Shinozaki *et al.*, 1989a, 1989b; Ohfune and Shinozaki, 1993). The eight CCG stereoisomers demonstrated a large variety

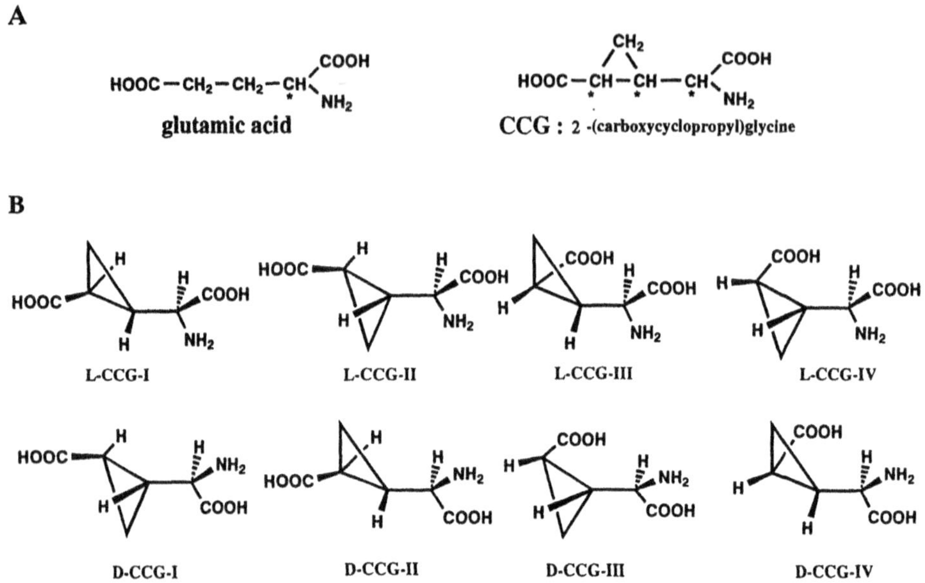

Figure 5 A: Chemical structures of glutamate and 2-(carboxycyclopropyl)glycine (CCG). There is close structural similarity between them. * marks represent asymmetrical carbon atoms. B: Stereochemical display of eight stereoisomers of CCG. *trans*-L-CCG and *cis*-L-CCG, which are natural CCGs, correspond to L-CCG-I and L-CCG-III, respectively.

of depolarizing activities and pharmacological properties (Shinozaki *et al.*, 1989a). The (2*S*,1'*R*,2'*S*)-isomer of CCG (L-CCG-IV) and all D-CCGs caused an NMDA-like depolarization of the spinal motoneuron of newborn rats that was effectively depressed by selective antagonists for NMDA receptors. The rank order of depolarizing activities in the spinal motoneuron of newborn rats was D-CCG-II > L-CCG-IV > NMDA > D-CCG-IV > L-glutamic acid > D-CCG-III > D-CCG-I. However, other data have revealed regional differences in the rank order. L-CCG-IV produced a more significant increase in intracellular Ca^{2+} concentration than NMDA in cultured neurons of the rat hippocampus (Kudo *et al.*, 1991). The (2*S*,1'*S*,2'*R*)-isomer of CCG (L-CCG-III), another folded form of L-CCG, markedly augmented the depolarizing responses to L-glutamate, D- and L-aspartate in the isolated spinal cord of newborn rats, but not to kainate, quisqualate, NMDA and D-glutamate. The potentiation of the L-glutamate response was affected by some factors such as inhibition of uptake of agonists from the neuronal environment by neurons or glias, an increase in transmitter releases from the presynaptic terminal, and changes in physiological properties of postsynaptic receptors. In the case of L-CCG-III, it is reasonable to posit that L-CCG-III blocks a Na^+-dependent uptake of excitatory amino acids from the vicinity of their receptors by the surrounding neurons and glias (Robinson *et al.*, 1993; Nakamura *et al.*, 1993).

The conformation of the (2*S*,1'*S*,2'*S*)-isomer of CCG (L-CCG-I) is an extended form, and it was more potent in inducing depolarization than L-glutamate in the isolated spinal cord of newborn rats, being almost equipotent to (1*S*,3*R*)-1-aminocyclopentane-1,3-dicarboxylic acid [(1*S*,3*R*)-ACPD]. However, the depolarization induced by L-CCG-I was not blocked by some known antagonists for glutamate receptors. From the pharmacology of L-CCG-I-induced depolarization (Shinozaki *et al.*, 1989b), the depolarizing activity in low temperature, the oscillatory responses in *Xenopus* oocytes injected with rat brain mRNA (Ishida *et al.*, 1990), and the stimulation of PI formation (Nakagawa *et al.*, 1990), L-CCG-I was suggested to be a potent mGluR agonist. L-CCG-I preferentially activates cloned mGluR2 and mGluR3 (Group II) that are negatively coupled to adenylyl cyclase, and also stimulates cloned mGluR1 (Group I) and mGluR4 (Group III) (Hayashi *et al.*, 1992). Thus, the conformationally extended form of the glutamate skeleton seems essential for an effective activation of mGluRs (Ohfune and Shinozaki, 1993). Therefore, we have obtained various CCG derivatives whose glutamate skeleton takes a conformationally extended form, including (2*S*,1'*R*,2'*R*,3'*R*)-2-(2,3-dicarboxycyclo-propyl)glycine (DCG-IV) (Ishida *et al.*, 1993) and *cis*- and *trans*-2-(2-carboxy-3-methoxymethylcyclopropyl)glycine (*cis*- and *trans*-MCG-I) (Ishida *et al.*, 1994, 1995) (Figure 6).

DCG-IV possesses the glutamate skeleton of both L-CCG-I and L-CCG-IV (extended and folded form, respectively), and is a highly potent and selective agonist for mGluR2 and mGluR3 (Group II mGluRs) with no substantial activity at mGluR1 or mGluR4 (Group I or Group III mGluRs, respectively) (Genazzani *et al.*, 1993; Hayashi *et al.*, 1993). As far as is known, mGluR agonists preferentially reduce the amplitude of monosynaptic excitation in the spinal cord of newborn rats, probably due to a depression of transmitter release (Ishida *et al.*, 1993), while polysynaptic

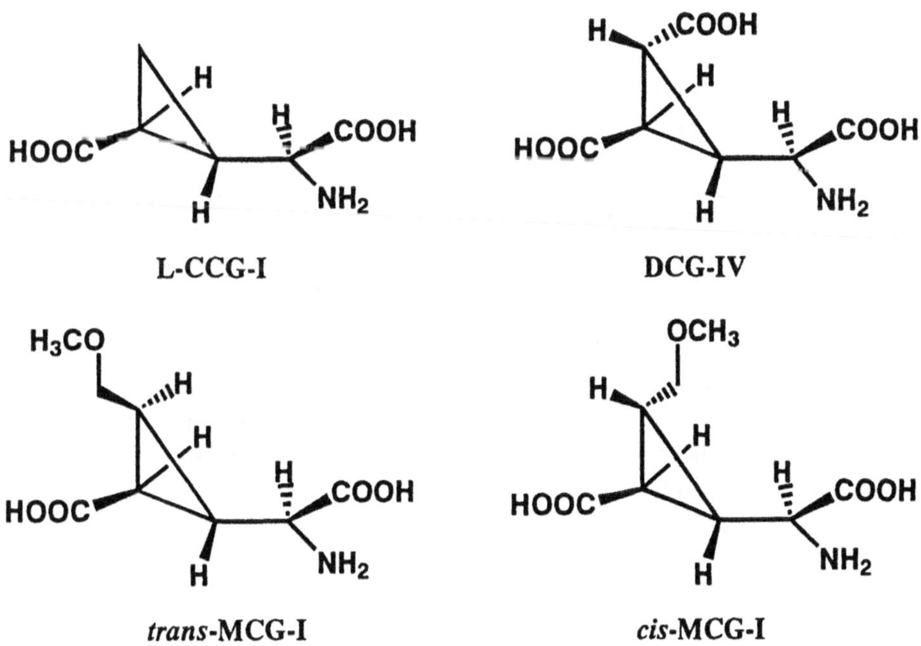

Figure 6 New agonists for mGluRs derived from CCG. All agonists take an extended form.

excitation was reduced to some extent only in higher concentrations. DCG-IV markedly reduced monosynaptic excitation in the isolated spinal cord of newborn rats much more effectively (about 10 times) than L-CCG-I and (1*S*,3*R*)-ACPD on a molar basis. The higher Ca^{2+} concentrations shifted the dose–response curve for the inhibition of monosynaptic excitation to the higher concentration side of DCG-IV, suggesting a suppression of presynaptic Ca^{2+} entry by DCG-IV. DCG-IV selectively depressed field EPSPs evoked by stimulation of mossy fibers in the rat hippocampal CA3 (Kamiya *et al.*, 1996), and reduced the frequency of miniature inhibitory postsynaptic currents (IPSCs) in CA3 pyramidal cells of the rat hippocampal slices (Poncer *et al.*, 1995). In the substantia nigra of the newborn rat, excitatory post synaptic currents (EPSCs) were reduced about 10 times more effectively by DCG-IV than IPSCs on a molar basis. DCG-IV presynaptically reduced the amplitudes of EPSCs preferentially at low concentrations (0.1–0.3 μM), whereas at higher concentrations (1–3 μM) DCG-IV reduced both IPSCs and EPSCs. DCG-IV selectively suppressed the IPSCs recorded from Purkinje cells of rats without significantly affecting the EPSCs produced by the stimulation of parallel fibers.

In the presence of 0.5 μM tetrodotoxin, (1*S*,3*R*)-ACPD, L-CCG-I and DCG-IV caused the depolarization of newborn rat motoneurons in a dose-dependent manner, although their concentration ranges and pharmacological properties were quite different. The depolarizations induced by (1*S*,3*R*)-ACPD and L-CCG-I seemed to be related to the stimulation of PI hydrolysis (one of the characteristic properties of Group I), and was to some extent decreased by relatively high concentrations of a non-selective antagonist for mGluRs, (*S*)-α-methyl-4-carboxyphenylglycine (MCPG). DCG-IV is a potent agonist for Group II mGluRs, but also caused NMDA-like depolarization at concentrations above 5 μM in spinal motoneurons of newborn rats (Ishida *et al.*, 1993), the minimum effective dose for the NMDA-like depolarization being about 300 times higher than that for activation of mGluRs in the spinal cord of newborn rats. Thus, there is a problem of receptor selectivity when higher concentrations (greater than 5 μM) are used. In contrast, *cis*- and *trans*-MCG-I did not cause any depolarization, even at concentrations as high as 1 mM. *trans*- and *cis*-MCG-I effectively inhibited forskolin-stimulated cyclic AMP formation in a concentration-dependent manner in the cultured spinal neurons of rats. They effectively depressed monosynaptic excitation in the spinal reflex of newborn rats with IC_{50} values of 0.3 and 3 μM, respectively, which was sensitive to MCPG. As mentioned above, they did not cause any depolarization of spinal motoneurons in the newborn rat; however, after treatment with quisqualate, *cis*-MCG-I caused a depolarization in a concentration-dependent manner with a threshold concentration of about 1 μM (quisqualate priming). The depolarizing activity developed after quisqualate treatment gradually decreased but lasted for more than 2 hours. The depolarization induced by *cis*-MCG-I seemed pharmacologically similar to that of phosphonate-containing analogues of glutamate such as L-2-amino-4-phosphonobutanoic acid (L-AP4) or L-2-amino-6-phosphonohexanoic acid (L-AP6) under the 'quisqualate priming'.

Except for DCG-IV, none of the newly developed CCG-related mGluR agonists, at a dose of 1 μmol/rat, caused neuron damage or marked excitation in the rat. High intraventricular doses of DCG-IV (5–50 nmol/rat) caused selective neuron damage limited to the cingulate cortex and the hippocampal subiculum. Since, as mentioned above, DCG-IV activates NMDA receptors as well at relatively high concentrations, the protective effect of a competitive antagonist for NMDA receptors was examined on the neurotoxicity of DCG-IV. CPP completely blocked the neurotoxicity of NMDA (100 nmol) at a dose of 0.1 nmol, but to decrease the neurotoxicity of DCG-IV (50 nmol) to a considerable extent, 30 times higher doses of CPP were necessary. The synergistic activation of mGluRs and NMDA receptors is suggested as a possible mechanism underlying the selective neuronal damage induced by DCG-IV, although it remains to be determined whether Group II mGluRs also participate directly in the development of glutamate neurotoxicity.

The intraventricular injection of quite low doses of DCG-IV (less than 1 nmole) to rats caused marked sedation, and retarded the recovery from halothane anesthesia (Miyamoto *et al.*, 1994). When DCG-IV was infused to the lateral ventricle of the rat for at least 15 hours via an osmotic pump, limbic seizures induced by intraventricular kainic acid were considerably alleviated in a dose-dependent manner. In the presence of DCG-IV (24–80 pmol/h/rat), kainate-induced limbic seizures sometimes occurred but were not maintained, and the frequency of limbic seizures was considerably lower. Histological examination showed that the neuron damage caused by kainic acid in the hippocampal CA3 area and the amygdala was considerably decreased by DCG-IV (Figure 7), although DCG-IV did not affect the kainate-induced damage of the hippocampal CA1 area as much. Thus, DCG-IV protects some kinds of neuron damage against the intraventricular administration of kainic acid (Miyamoto *et al.*, 1996). DCG-IV also protects cultured neurons against the neuronal degeneration induced by excitotoxic concentrations of NMDA or kainate (Bruno *et al.*, 1994, 1995). The activation of mGluRs by *trans*-ACPD is known to attenuate NMDA neurotoxicity in cortical cultures. Our newly developed mGluR agonists will provide useful probes for elucidating the relationship between neuron damage and the physiological function of mGluRs.

10. OTHER NEUROTOXINS

In addition to the above-described excitatory amino acids, naturally occurring compounds such as quinolinic acids, β-N-methylamino-L-alanine (BMAA), β-N-oxalylamino-L-alanine (BOAA) and willardiine (Figure 8) also have both neuro-excitatory and neurotoxic properties. Quinolinic acid is a weak NMDA receptor agonist but, since its release is not evoked by physiological stimuli, it is not considered to have a neurotransmitter function, although this substance has been proposed to play a role in chronic neurodegenerative diseases such as Huntington's disease and AIDS-related dementia (Stone and Burton, 1988; Meldrum, 1992). BMAA is a constituent of the seeds of false sago palm, *Cycas circinalis*, which were formerly used by the Chamorro population for food and medicinal purposes in Guam, one of the Mariana Islands, where amyotrophic lateral screrosis–Parkinsonism–Dementia complex (ALS-PD) occurred at an abnormally high incidence. BMAA behaves as a mixed agonist at iGluRs and Group I mGluRs. Spencer *et al.* (1987) have offered a hypothesis that BMAA is primarily responsible for the disease ALS-PD. BMAA caused neurodegenerative changes in primates similar to those of ALS-PD, although

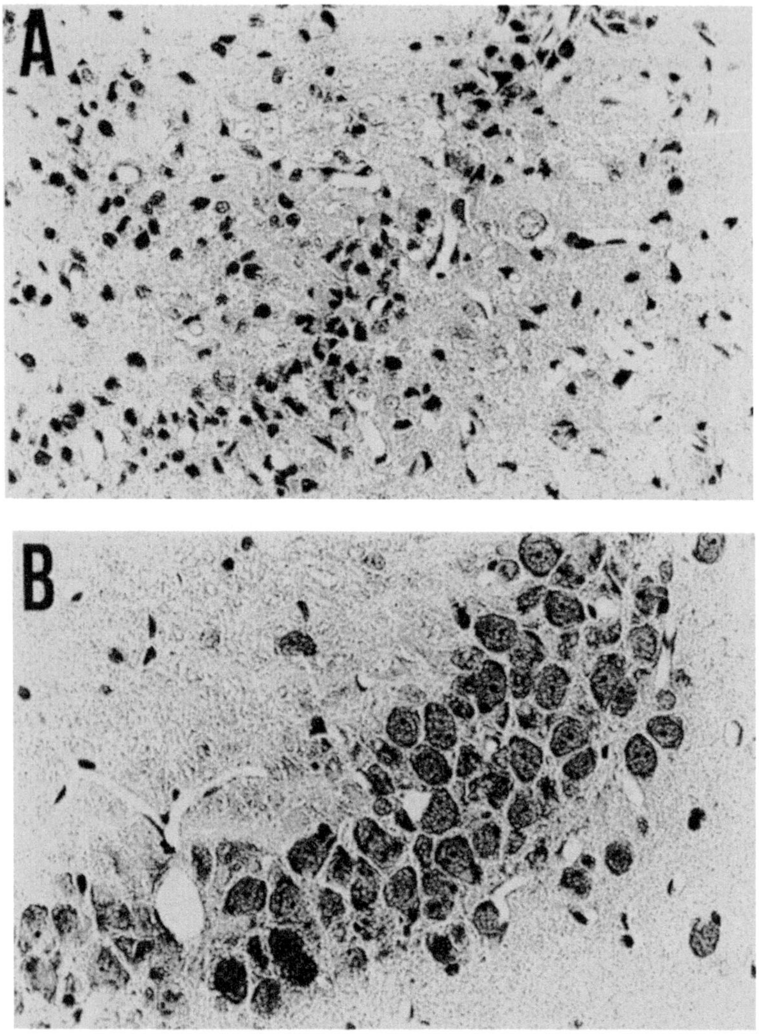

Figure 7 Neuron damage induced by kainic acid in the rat hippocampus CA3 area and protective action of DCG-IV. Kainic acid was intraventricularly administered at a dose of 2 nmoles, and DCG-IV (24 pmoles/hr) was continuously infused into the ventricurum for 15 hours before the administration of kainic acid. A: Control, kainic acid only. B: DCG-IV plus kainic acid.

BMAA is not an acidic amino acid, and BMAA itself does not excite neurons. In mouse cortical cell cultures, the neurodegenerative and excitatory properties of BMAA could only be observed in bicarbonate-containing buffers (Weiss and Choi, 1988). The activation of mGluRs might also contribute to the pathophysiology of the slow neuronal degeneration observed in Macaque monkeys treated with BMAA. It is known that BOAA, closely related to BMAA, isolated from *Lathyrus sativus* (Rao

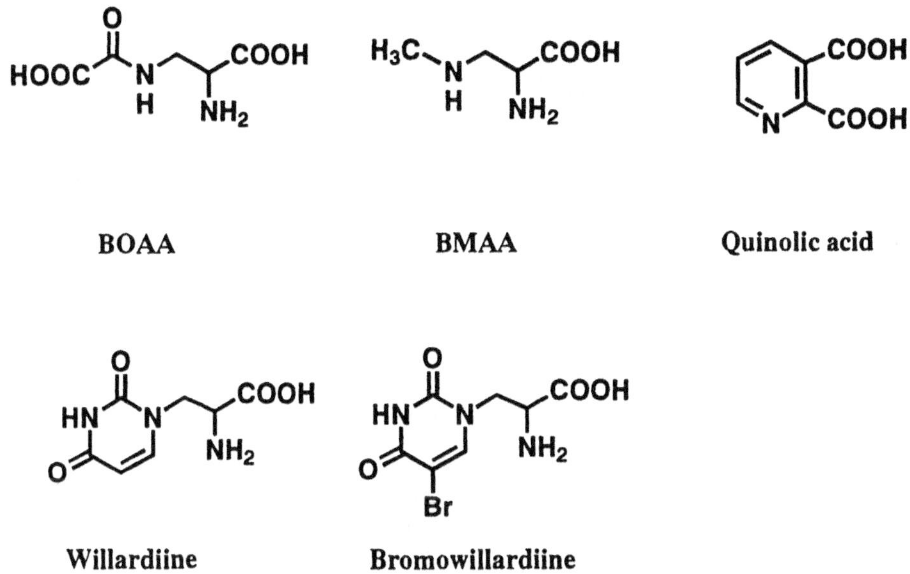

Figure 8 Chemical structures of BOAA, BMAA, quinolinic acid, willardiine and bromowillardiine.

et al., 1964), causes a marked excitation of central neurons and induces neurolathyrism in experimental animals, characterized by irreversible spastic paralysis of the legs. BOAA preferentially acts on AMPA receptors but also shows some affinity to kainate binding sites and weak interaction with NMDA receptors.

Willardiine, from *Acasia willardiana*, which is structurally close to quisqualic acid, also demonstrates powerful depolarizing activity, which is potentiated by the substitution of Br at the C-5 position. Willardiine is known to activate AMPA receptors. Willardiine did not have any potent depolarizing activity in the C-fibers, but Br-willardiine caused depolarization with an about 3 times higher activity than kainic acid (Agrawal and Evans, 1986).

11. SUMMARY AND CONCLUSIONS

The pharmacological features of potentially valuable amino acids of natural origin such as kainic acid, domoic acid, acromelic acid, quisqualic acid, ibotenic acid and CCG, which are all related structurally to glutamate and aspartate, have been described with special reference to neuronal degeneration. As is common in the pharmacological study of neurotransmitters, several agonists and antagonists of natural origin have played a key role as valuable pharmacological tools for research on receptors for neurotransmitters (Shinozaki, 1980, 1988). Highly selective and potent agonists and antagonists for glutamate receptors are still required for further elucidation of neurodegenerative and neuroprotective mechanisms. At present there are, unfortunately, no antagonists of plant origin for excitatory amino acid receptors. At the crayfish NMJ, theanine (Shinozaki and Ishida, 1978), matrin (Ishida and Shinozaki, 1984) and tuberostemonine (Shinozaki and Ishida, 1985) inhibit depolarizing responses to glutamate or quisqualate, probably due to open channel block. The usefulness of these compounds in the pharmacological research of the mammalian CNS is not yet known, but it seems most likely that several compounds which affect the glutamatergic system in the mammalian CNS remain to be identified.

REFERENCES

1. S.G. Agrawal and R.H. Evans, The primary afferent depolarizing action of kainate in the rat. *Br. J. Pharmacol.* **87**, 345–355 (1986).
2. Y. Ben-Ari, Limbic seizure and brain damage produced by kainic acid: mechanisms and relevance to human temporal lobe epilepsy. *Neurosci.* **14**, 375–403 (1985).
3. J. Boulter, M. Hollmann, A. O'Shea-Greenfield, M. Hartley, E. Deneris, C. Maron and S. Heinemann, Molecular cloning and functional expression of glutamate receptor subunit genes. *Science* **249**, 1033–1037 (1990).
4. R.J. Bridges, D.R. Stevens, J.S. Kahle, P.B. Nunn, M. Kadri and C.W. Cotman, Structure-function studies on N-oxalyl-diamino-dicarboxylic acids and excitatory amino acid receptors: evidence that β-L-ODAP is a selective non-NMDA agonist. *J. Neurosci.* **9**, 2073–2079 (1989).
5. V. Bruno, A. Copani, G. Battaglia, R. Raffaele, H. Shinozaki and F. Nicoletti, Protective effect of the metabotropic glutamate receptor agonist DCG-IV, against excitotoxic neuronal death. *Eur. J. Pharmacol.*, **256**, 109–112 (1994).
6. V. Bruno, G. Battaglia, A. Copani, R.G. Giffard, G. Raciti, R. Raffaele, H. Shinozaki and F. Nicoletti, Activation of class II or III metabotropic glutamate receptors protects cultured cortical neurons against excitotoxic degeneration. *Eur. J. Neurosci.* **7**, 1906–1913 (1995).
7. D.W. Choi, Ionic dependence of glutamate neurotoxicity in cortical cell culture. *J. Neurosci.* **7**, 369–379 (1987).
8. D.W. Choi, Calcium-mediated neurotoxicity: Relationship to specific channel types and role in ischemic damage. *Trends in Neurosci.* **11**, 465–469 (1988).
9. D.W. Choi and S.M. Rothman, The role of glutamate neurotoxicity in the hypoxic-ischemic neuronal death. *Ann. Rev. Neurosci.* **13**, 171–182 (1990).
10. C.W. Cotman and L.L. Iversen, Excitatory amino acids in the brain – focus on NMDA receptors. *Trends in Neurosci.* **10**, 263–265 (1987).
11. J.T. Coyle, Kainic acid: insights into excitatory mechanisms causing selective neuronal degeneration. *Ciba Found. Symp.* **126**, 186–203 (1987).
12. S.G. Cull-Candy, Two types of extrajunctional L-glutamate receptors in locust muscle fibres. *J. Physiol.* **255**, 449–464 (1976).
13. G. Debonnel, L. Beauchesne and C. de Montigny, Domoic acid, the alleged "mussel toxin", might produce its neurotoxic effect through kainate receptor activation: an electrophysiological study in the rat dorsal hippocampus. *Can. J. Physiol. Pharmacol.* **67**, 29–33 (1989a).
14. G. Debonnel, M. Weiss and C. de Montigny, Reduced neuroexcitatory effect of domoic acid following mossy fiber denervation of the rat dorsal hippocampus: further evidence that toxicity of domoic acid involves kainate receptor activation. *Can. J. Physiol. Pharmacol.* **67**, 904–908, (1989b).
15. E.J. Fletcher, G. Martin, J.A. Aram, D. Lodge and T. Honoré, Quinoxalinediones selectively block quisqualate and kainate receptors and synaptic events in rat neocortex and hippocampus and frog spinal cord *in vitro*. *Br. J. Pharmacol.* **95**, 585–597 (1988).
16. A.C. Foster, E.E. Mena, D.T. Monaghan and C.W. Cotman, Synaptic localization of kainic acid binding sites. *Nature* **289**, 73–75 (1981).
17. L. Fowden, A. Smith, D.S. Millington and R.C. Shepard, Cyclopropane amino acids from *Aeculus* and *Blighia*. *Phytochem.* **8**, 437–443 (1969).

18. E.B. Friedberg and D.T. Ross, Degeneration of rat thalamic reticular neurons following intrathalamic domoic acid injection. *Neurosci. Lett.* **151**, 115–119 (1993).
19. L.K. Friedman, D.E. Pellegrini-Giampietro, E.F. Sperber, M.V. Bennett, S.L. Moshe and R.S. Zukin, Kainate-induced status epilepticus alters glutamate and $GABA_A$ receptor gene expression in adult rat hippocampus: an *in situ* hybridization study. *J. Neurosci.* **14**, 2697–707 (1994).
20. S. Fushiya, S. Sato, T. Kanazawa, G. Kusano and S. Nozoe, Acromelic acid C, a new toxic constituent of *Clitocybe acromelalga. Tetra. Lett.* **31**, 3901–3904 (1990).
21. A.A. Genazzani, G. Casabona, M.R. L'Episcopo, D.F. Condorelli, P. Dell'Albani, H. Shinozaki and F. Nicoletti, Characterization of metabotropic glutamate receptors negatively linked to adenylyl cyclase in brain slices. *Brain Res.* **622**, 132–138 (1993).
22. B. Grimmelt, M.S. Nijjar, J. Brown, N. Macnair, S. Wagner, G.R. Johnson and J.F. Amend, Relationship between domoic acid levels in the blue mussel (Mytilus edulis) and toxicity in mice. *Toxicon* **28**, 501–508 (1990).
23. S. Hattori and A. Komamine, Stizolobic acid: a new amino acid in Stizolobium Hassjoo. *Nature* **183**, 1115–1117 (1959).
24. Y. Hayashi, Y. Tanabe, I. Aramori, M. Masu, K. Shimamoto, Y. Ohfune and S. Nakanishi, Agonist analysis of 2-(carboxycyclopropyl)glycine isomers for cloned metabotropic glutamate receptor subtypes expressed in Chinese hamster ovary cells. *Br. J. Pharmacol.* **107**, 539–543 (1992).
25. Y. Hayashi, A. Momiyama, T. Takahashi, H. Ohishi, R. Ogawa-Meguro, R. Shigemoto, N. Mizuno and S. Nakanishi, Role of a metabotropic glutamate receptor in synaptic modulation in the accessory olfactory bulb. *Nature* **366**, 687–690 (1993).
26. M. Hollmann, A. O'Shea-Greenfield, S.W. Rogers and S. Heinemann, Cloning by functional expression of a member of the glutamate receptor family. *Nature* **342**, 643–648 (1989).
27. T. Honoré, S.N. Davies, J. Drejer, E.J. Fletcher, P. Jacobsen, D. Lodge and F.E. Nielsen, Quinoxalinediones: Potent competitive non-NMDA glutamate receptor antagonists. *Science* **241**, 701–703 (1988).
28. J.E. Huettner, Glutamate receptor channels in rat DRG neurons: activation by kainate and quisqualate and blockade of desensitization by Con A. *Neuron* **5**, 255–266 (1990).
29. M. Iino, S. Ozawa and K. Tsuzuki, Permeation of calcium through excitatory amino acid receptor channels in cultured rat hippocampal neurones. *J. Physiol.* **424**, 151–165 (1990).
30. M. Ishida, H. Akagi, K. Shimamoto, Y. Ohfune and H. Shinozaki, A potent metabotropic glutamate receptor agonist: electrophysiological actions of a conformationally restricted glutamate analogue in the rat spinal cord and Xenopus oocytes. *Brain Res.* **537**, 311–314 (1990).
31. M. Ishida, Y. Ohfune, Y. Shimada, K. Shimamoto and H. Shinozaki, Changes in preference for receptor subtypes of configurational variants of a glutamate analog: conversion from the NMDA-type to the non-NMDA type. *Brain Res.* **550**, 152–156 (1991).
32. M. Ishida, T. Saitoh, K. Shimamoto, Y. Ohfune and H. Shinozaki, A novel metabotropic glutamate receptor agonist: marked depression of monosynaptic excitation in the newborn rat isolated spinal cord. *Br. J. Pharmacol.* **109**, 1169–1177 (1993).
33. M. Ishida, T. Saitoh, Y. Nakamura, K. Kataoka and H. Shinozaki, A novel metabotropic glutamate receptor agonist: (2*S*,1'*S*,2'*R*,3'*R*)-2-(2-carboxy-3-methoxymethylcyclopropyl)glycine (*cis*-MCG-I). *Eur. J. Pharmacol.* **268**, 267–270 (1994).
34. M. Ishida, T. Saitoh, K. Tsuji, Y. Nakamura, K. Kataoka and H. Shinozaki, Novel agonists for metabotropic glutamate receptors: *trans*- and *cis*-2-(2-carboxy-3-methoxymethylcyclopropyl)glycine (*trans*- and *cis*-MCG-I). *Neuropharmacol.* **34**, 821–827 (1995).
35. M. Ishida and H. Shinozaki, Glutamate inhibitory action of matrine at the crayfish neuromuscular junction. *Brit. J. Pharmacol.* **82**, 523–531 (1984).
36. M. Ishida and H. Shinozaki, Acromelic acid is a much more potent excitant than kainic acid or domoic acid in the isolated rat spinal cord. *Brain Res.* **474**, 386–389 (1988a).
37. M. Ishida and H. Shinozaki, Excitatory action of a plant extract, stizolobic acid, in the isolated spinal cord of the rat. *Brain Res.* **473**, 193–197 (1988b).
38. M. Ishida and H. Shinozaki, Novel kainate derivatives: Potent depolarizing actions on spinal motoneurones and dorsal root fibres in newborn rats. *Br. J. Pharmacol.* **104**, 873–878 (1991).
39. F. Iverson, J. Truelove, E. Nera, L. Tryphonas, J. Campbell and E. Lok, Domoic acid poisoning and mussel-associated intoxication: preliminary investigations into the response of mice and rats to toxic mussel extract. *Food Chem. Toxicol.* **27**, 377–384 (1989).

40. G.A.R. Johnston, D.R. Curtis, W.C. De Groat and A.W. Duggan, Central actions of ibotenic acid and muscimol. *Biochem. Pharmacol.* **17**, 2488–2489 (1968).
41. H. Kamiya, H. Shinozaki and C. Yamamoto, Activation of metabotropic glutamate receptor type 2/3 suppresses transmission at rat hippocampal mossy fiber synapse. *J. Physiol.* **493**, 447–455 (1996).
42. K. Keinänen, W. Wisden, B. Sommer, P. Werner, A. Herb, T.A. Verdoorn, B. Sakmann and P.H. Seeburg, A family of AMPA-selective glutamate receptors. *Science* **249**, 556–560 (1990).
43. T. Kirino, Delayed neuronal death in the gerbil hippocampus following ischemia. *Brain Res.* **239**, 57–69 (1982).
44. K. Konno, H. Shirahama and T. Matsumoto, Isolation and structure of acromelic acid A and B, new kainoids of *Clitocybe acromelalga*. *Tetra. Lett.* **24**, 939–942 (1983).
45. K. Konno, K. Hashimoto, Y. Ohfune, H. Shirahama and T. Matsumoto, Acromelic acids A and B. Potent neuroexcitatory amino acids isolated from *Clitocybe acromelalga*. *J. Amer. Chem. Soc.* **110**, 4807–4815 (1988).
46. P. Krogsgaard-Larsen, T. Honore, J.J. Hansen, D.R. Curtis and D. Lodge, New class of glutamate agonist structurally related to ibotenic acid. *Nature* **284**, 64–66 (1980).
47. Y. Kudo, K. Akita, M. Ishida and H. Shinozaki, A significant increase in intracellular Ca^{2+} concentration induced by (2*S*,3*R*,4*S*)-2-(carboxycyclopropyl)glycine, a new potent agonist, in cultured rat hippocampal neurons. *Brain Res.* **567**, 342–345 (1991).
48. S. Kwak, H. Aizawa, M. Ishida, Y. Gotoh and H. Shinozaki, Small neurons in the lower spinal cord are selectively damaged in the spastic rat by acromelic acid. In *Amino Acids: Chemistry, Biology and Medicine*, G. Lubec and G.A. Rosenthal, eds., ESCOM, Leiden, pp. 318–326 (1990).
49. S. Kwak, H. Aizawa, M. Ishida and H. Shinozaki, Acromelic acid, a novel kainate analogue, induces long-lasting paraparesis with selective degeneration of interneurons in the rat spinal cord. *Exp. Neurol.* **116**, 145–155 (1992).
50. S. Kwak, M. Miyamoto, M. Ishida and H. Shinozaki, Neurotoxicity of (2*S*,1'*R*,2'*R*,3'*R*)-2-(2,3-dicarboxycyclopropyl)glycine, a potent agonist for metabotropic glutamate receptors, in the rat. *Neurosci.* **73**, 687–695 (1996).
51. A.B. MacDermott, M.L. Mayer, G.L. Westbrook, S.J. Smith and J.L. Barker, NMDA-receptor activation increases cytoplasmic calcium concentration in cultured spinal cord neurones. *Nature* **321**, 519–522 (1986).
52. E.G. McGeer, J.W. Olney and P.L. McGeer, eds., *Kainic Acid as a Tool in Neurobiology*, Raven Press, New York (1978).
53. E.G. McGeer and S.G. Zhu, Lamotrigine protects against kainate but not ibotenate lesions in rat striatum. *Neurosci. Lett.* **112**, 348–351 (1990).
54. B.S. Meldrum, Excitatory amino acid receptors and disease. *Current Opinion Neurol. Neurosurg.* **5**, 508–513 (1992).
55. M. Miyamoto, M. Ishida, S. Kwak and H. Shinozaki, Agonists for metabotropic glutamate receptors in the rat delay recovery from halothane anesthesia. *Eur. J. Pharmacol.* **260**, 99–102 (1994).
56. M. Miyamoto, M. Ishida and H. Shinozaki, Anticonvulsive and neuroprotective actions of a potent agonist (DCG-IV) for Group II metabotropic glutamate receptors against intraventricular kainate in the rat. *Neurosci.* **77**, 131–140 (1997).
57. D.T. Monaghan, R.J. Bridges and C.W. Cotman, The excitatory amino acid receptors: Their classes, pharmacology, and distinct properties in the function of the central nervous system. *Ann. Rev. Pharmacol. Toxicol.* **29**, 365–402 (1989).
58. D.T. Monaghan and C.W. Cotman, Distribution of ^{3}H-kainic acid binding sites in rat CNS as determined by autoradiography. *Brain Res.* **252**, 91–100 (1982).
59. J.V. Nadler, Minireview. Kainic acid as a tool for the study of temporal lobe epilepsy. *Life Sci.* **29**, 2031–2042 (1981).
60. J.V. Nadler, B.W. Perry and C.W. Cotman, Intraventricular kainic acid preferentially destroys hippocampal pyramidal cells. *Nature* **271**, 676–677 (1978).
61. Y. Nakagawa, K. Saitoh, T. Ishihara, M. Ishida and H. Shinozaki, (2*S*,3*S*,4*S*)-α-(Carboxycyclopropyl)-glycine is a novel agonist of metabotropic glutamate receptors. *Eur. J. Pharmacol.* **184**, 205–206 (1990).
62. Y. Nakamura, K. Kataoka, M. Ishida and H. Shinozaki, (2*S*,3*R*,4*R*)-2-(caroboxycyclopropyl)glycine, a potent and competitive inhibitor of both glial and neuronal uptake of glutmate. *Neuropharmacol.* **32**, 833–837 (1993).

63. S. Nakanishi, Molecular diversity of glutamate receptors and implications for brain function. *Science* **258**, 597–603 (1992).
64. S. Nakanishi and M. Masu, Molecular diversity and functions of glutamate receptors. *Annu. Rev. Biophys. Biomol. Struct.* **23**, 319–348 (1994).
65. Y. Ohfune and H. Shinozaki, L-2-(Carboxycyclopropyl)glycines: conformationally constrained L-glutamate analogues. In *Drug Design for Neuroscience*, A. Kozikowski, ed., Raven Press, New York, pp. 261–283 (1993).
66. J.W. Olney, Neurotoxicity of excitatory amino acids. In *Kainic Acid as a Tool in Neurobiology*, E.G. McGeer, J.W. Olney and P.L. McGeer, eds., Raven Press, New York, pp. 95–121 (1978).
67. J.W. Olney, Excitotoxic amino acids and neuropsychiatric disorders. *Ann. Rev. Pharmacol. Toxicol.* **30**, 47–71 (1990).
68. J.W. Olney, V. Rhee and O.L. Ho, Kainic acid: A powerful neurotoxic analogue of glutamate. *Brain Res.* **77**, 507–512 (1974).
69. D. O'Shaughnessy and G.J. Gerber, Damage induced by systemic kainic acid in rats is dependent upon seizure activity – a behavioral and morphological study. *Neurotoxicol.* **7**, 187–202 (1986).
70. D.K. Patneau and M.L. Mayer, Structure-activity relationships for amino acid transmitter candidates acting at N-methyl-D-aspartate and quisqualate receptors. *J. Neurosci.* **10**, 2385–2399 (1990).
71. T.M. Perl, L. Berard, T. Kosatsky, J. Hockin, E.C.D. Todo and R. Remis, An outbreak of toxic encephalopathy caused by eating mussels contaminated with domoic acid. *New Eng. J. Med.* **322**, 1775–1780 (1990).
72. J.-C. Poncer, H. Shinozaki and R. Miles, Dual modulation of synaptic inhibition by distinct metabotropic glutamate receptors in the rat hippocampus. *J. Physiol.* **485**, 121–134 (1995).
73. S.L.N. Rao, P.R. Adiga and P.S. Sarma, The isolation and characterization of β-N-oxalyl-L-α,β-diaminopropionic acid: a neurotoxin from the seed of Latyrus sativus. *Biochem.* **3**, 432–436 (1964).
74. M.B. Robinson, J.D. Sinor, L.A. Dowd and J.F. Kerwin, Jr., Subtypes of sodium-dependent high-affinity L-[^{3}H]glutamate transport activity: Pharmacologic specificity and regulation by sodium and potassium. *J. Neurochem.* **60**, 167–179 (1993).
75. D.B. Sattelle, M-I. Sepulveda, H. Shinozaki and M. Ishida, Actions of acromelic acid on nervous system of L-glutamate receptors. *Arch. Insect Biochem. Physiol.* **25**, 87–94 (1994).
76. D.D. Schoepp and P.J. Conn, Metabotropic glutamate receptors in brain function and pathology. *Trends Pharmacol. Sci.* **14**, 13–20 (1993).
77. M.J. Sheardown, E.Ø. Nielsen, A.J. Hansen, P. Jacobsen and T. Honoré, 2,3-Dihydroxy-6-nitro-7-sulfamoylbenzo(F)quinoxaline: a neuroprotectant for cerebral ischemia. *Science* **247**, 571–574 (1990).
78. H. Shinozaki, Discovery of novel actions of kainic acid and related compounds. In *Kainic Aid as a Tool in Neurobiology*, E.G. McGeer, J.W. Olney and P.L. McGeer, eds., Raven Press, New York, pp. 17–35 (1978).
79. H. Shinozaki, The pharmacology of the excitatory neuromuscular junction in the crayfish. *Prog. Neurobiol.* **14**, 121–155 (1980).
80. H. Shinozaki, Pharmacology of the glutamate receptor. *Prog. Neurobiol.* **30**, 399–435 (1988).
81. H. Shinozaki, Kainic acid receptor agonists, In *Excitatory Amino Acid Receptors*, P. Krogsgaard-Larsen and J.J. Hansen, eds., Ellis Horwood, Chichester, pp. 261–291 (1991).
82. H. Shinozaki, Neuron damage induced by some potent kainoids and neuroprotective action of new agonists for metabotropic glutamate receptors. *Eur. Neurol.* **34**(suppl.3), 2–9 (1994).
83. H. Shinozaki and M. Ishida, Theanine as a glutamate antagonist at a crayfish neuromuscular junction. *Brain Res.* **151**, 215–219 (1978).
84. H. Shinozaki and M. Ishida, Inhibitory actions of tuberostemonine on the excitatory transmission at the crayfish neuromuscular junction. *Brain Res.* **334**, 33–40 (1985).
85. H. Shinozaki and M. Ishida, Stizolobic acid, a competitive antagonist of the quisqualate type receptor at the crayfish neuromuscular junction. *Brain Res.* **451**, 353–356 (1988a).
86. H. Shinozaki and M. Ishida, Novel excitatory amino acid related compounds of natural origin. In *Neurotox '88: Molecular Basis of Drugs and Pesticide Action*, G.G. Lunt, ed., Elsevier, Amsterdam, pp. 91–104 (1988b).
87. H. Shinozaki, M. Ishida and T. Okamoto, Acromelic acid, a novel excitatory amino acid from a poisonous mushroom: effects on the crayfish neuromuscular junction. *Brain Res.* **399**, 395–398 (1986).

88. H. Shinozaki, M. Ishida, K. Shimamoto and Y. Ohfune, A conformationally restricted analogue of L-glutamate, the *threo*-folded form of L-α-(carboxycyclopropyl)glycine, activates the NMDA-type receptor more markedly than NMDA in the isolated rat spinal cord. *Brain Res.* **480**, 355–359 (1989a).
89. H. Shinozaki, M. Ishida, K. Shimamoto and Y. Ohfune, Potent NMDA-like actions and potentiation of glutamate responses by conformational variants of a glutamate analogue in the rat spinal cord. *Br. J. Pharmacol.* **98**, 1213–1224 (1989b).
90. H. Shinozaki, M. Ishida, Y. Gotoh and S. Kwak, Specific lesions of rat spinal interneurons induced by systemic administration of acromelic acid, a new potent kainate analogue. *Brain Res.* **503**, 330–333 (1989c).
91. H. Shinozaki, M. Ishida, Y. Gotoh and S. Kwak, Acromelic acid as a tool for the study of specific neurone damages. In *Amino Acids: Chemistry, Biology and Medicine*, G. Lubec and G.A. Rosentahal, eds., ESCOM, Leiden, pp. 281–293 (1990).
92. H. Shinozaki, M. Ishida, S. Kwak and T. Nakajima, Use of acromelic acid for porduction of rat spinal lesions. In *Methods in Neuroscience*, P.M. Conn, ed., Vol. 7, Academic Press, New York, pp. 38–57 (1991).
93. H. Shinozaki and S. Konishi, Actions of several anthelmintics and insecticides on rat cortical neurones. *Brain Res.* **24**, 368–371 (1970).
94. H. Shinozaki and I. Shibuya, A new potent excitant, quisqualic acid: effects on crayfish neuromuscular junction. *Neuropharmacol.* **13**, 665–672 (1974).
95. H. Shinozaki and I. Shibuya, Effects of kainic acid analogues on crayfish opener muscle. *Neuropharmacol.* **15**, 145–147 (1976).
96. A.L. Smith and R.A.J. Mallhinney, Effects of acromelic acid A on the binding of [^{3}H]-kainic acid and [^{3}H]-AMPA to rat brain synaptic plasma membranes. *Br. J. Pharmacol.* **105**, 83–86 (1992).
97. B. Sommer, K. Keinänen, T.A. Verdoorn, W. Wisden, N. Bunashev, A. Herb, M. Kohler, T. Takagi, B. Sakmann and P.H. Seeburg, Flip and flop: a cell-specific functional switch in glutamate-operated channels of the CNS. *Science* **249**, 1580–1585 (1990).
98. P. Spencer, P.B. Nunn, J. Hugon, A.C. Ludolph, S.M. Ross, D.N. Roy and R.C. Robertson, Guam amyotrophic lateral scleosis-Parkinsonism-Dementia linked to a plant excitant neurotoxin. *Science* **237**, 517–522 (1987).
99. G.R. Stewart, C.F. Zorumski, M.T. Price and J.W. Olney, Domoic acid: a dementia-inducing excitotoxic food poison with kainic acid receptor specificity. *Exp. Neurol.* **110**, 127–138 (1990).
100. T.W. Stone and N.R. Burton, NMDA receptors and ligands in the vertebrate CNS. *Prog. Neurobiol.* **30**, 333–368 (1988).
101. T. Takemoto, Isolation and structural identification of naturally occurring excitatory amino acids. In *Kainic Acid as a Tool in Neurobiology*, E.G. McGeer, J.W. Olney and P.L. McGeer, eds., Raven Press, New York, pp. 1–15 (1978).
102. J.S. Teitelbaum, R.J. Zatotte, S. Carpenter, D. Gendron, A.C. Evans, A. Gjedde and N.R. Cashman, Neurologic sequelae of domoic acid intoxication due to the ingestion of contaminated mussels. *New Eng. J. Med.* **322**, 1781–1787 (1990).
103. L. Tryphonas and F. Iverson, Neuropathology of excitatory neurotoxins: the domoic acid model. *Toxi. Pathol.* **18**, 165–169 (1990).
104. J.H. Weiss and D.W. Choi, β-N-methylamino-L-alanine neurotoxicity: Requirement for bicarbonate as a cofactor. *Science* **241**, 973–975 (1988).
105. D.F. Wozniak, G.R. Stewart, J.P. Miller and J.W. Olney, Age-related sensitivity to kainate neurotoxicity. *Exp. Neurol.* **114**, 250–253 (1991).
106. A.B. Young and G.E. Fagg, Excitatory amino acid receptors in the brain: membrane binding and receptor autoradiographic approaches. *Trends in Pharmacol. Sci.* **11**, 126–133 (1990).

9 PATHOPHARMACOLOGICAL EVALUATION OF BAKUMONDO-TO (MAI-MEN-DONG-TANG) AS A CURATIVE FOR CHRONIC INFLAMMATORY AIRWAY DISEASES

TAKESHI MIYATA, YOICHIRO ISOHAMA, SHUSHENG TAI, HIROFUMI KAI and KAZUO TAKAHAMA

Department of Pharmacological Sciences, Faculty of Pharmaceutical Sciences, Kumamoto University, Kumamoto 862-0973, Japan

INTRODUCTION

Although coughing is an important respiratory defense reflex (Banner, 1986), a prolonged, uncontrolled or non-productive cough should be treated. Many antitussives, including narcotic and non-narcotic types, have been widely used for such coughs. However, their insufficient efficacy and/or risk of creating a dependency have led to an increasing usage of traditional Chinese medicines in clinics and hospitals, because the Chinese medicines tend to have minor side effects and sometimes exhibit remarkable efficacy which may not be achieved by synthetic medicinal drugs.

Bakumondo-to (Mai-men-dong-tang), a traditional Chinese medicinal prescription consisting of 6 herbs (*Ophiopogonis tuber, Pinelliae tuber, Zizyphi fructus, Glycyrrhizae radix, Ginseng radix and Oryzae fructus*), has been used for the treatment of bronchitis and pharyngitis accompanying severe dry cough. The antitussive activity of this medicine has not yet been evaluated with modern pharmacological techniques. In the present study, we examined the antitussive and related actions and properties of Bakumondo-to and compared them with those of codeine.

The following factors were considered in this study: 1). Since a cough can be induced by various stimuli, we tried several methods to induce the cough reflex in order to evaluate antitussive activity and to clarify the characteristics of the activity. 2). Since Chinese medicines generally have weak side effects but tend to show relatively potent therapeutic effects especially in the pathogenic state, we used animals with bronchitis produced by long-term exposure to SO_2 (Miyata *et al.*, 1987, 1989) or animals with coughs facilitated by an angiotensin converting enzyme (ACE) inhibitor (Fuchikami *et al.*, 1990; Takahama *et al.*, 1993). 3). Mucociliary clearance as well as cough is an important pulmonary defense mechanism that serves to remove inhaled substances from the lung (Miyata *et al.*, 1988). We studied the effects of Bakumondo-to on airway secretion, because most synthetic antitussives and narcotic antitussives such as codeine disturb the airway mucociliary clearance through an inhibition of airway secretion. 4). Finally, we studied the effects of Bakumondo-to on tracheal mucociliary transport velocity (MCTV) to examine the total effect of the drug on mucociliary clearance.

1. ANTITUSSIVE ACTIVITIES

1. Effects on Cough Responses Induced by Mechanical Stimulation

Mechanical stimulation of the bronchus activates the irritant receptors, i.e., the Aσ fibers (Widdicombe, 1954). The results of a comparative study of the antitussive effects of Bakumondo-to, codeine and eprazinone, a non-narcotic synthetic antitussive, on mechanically induced coughs in normal and bronchitic guinea pigs are shown in Figure 1. Typical cough recordings are presented in Figure 2.

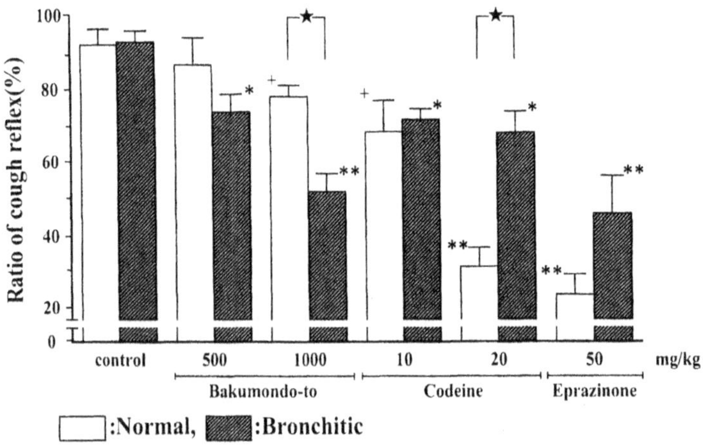

Figure 1 Comparison of antitussive effects of Bakumondo-to, codeine and eprazinone on the cough responses induced by mechanical stimulation to the tracheal mucosa in normal and bronchitic guinea pigs.
$^{+}$: $p < 0.1$, *: $p < 0.05$, **: $p < 0.01$ *vs.* the control groups. ★ : $p < 0.01$ *vs* the normal groups.

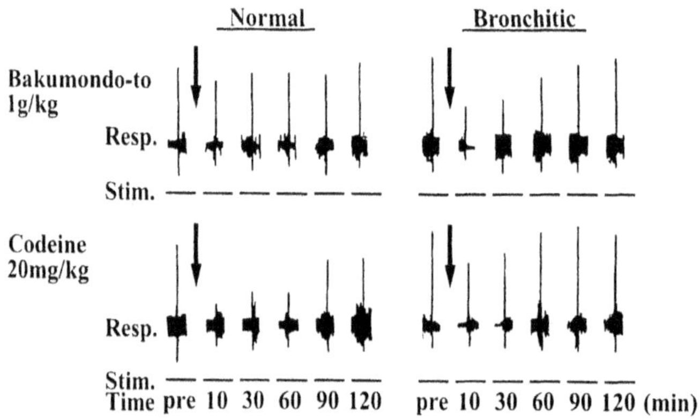

Figure 2 Typical cough recordings of differential effects of Bakumondo-to and codeine on the cough responses induced by mechanical stimulation of tracheal mucosa a normal and bronchitic guinea pigs.
Needle-like deflection in each recording represents a cough response. Resp: respiration.

In normal animals, Bakumondo-to had no effect at the doses of 0.5 and 1 g/kg, while it showed significant and dose-dependent actions causing 26% and 48% reductions in the amplitude of cough responses in bronchitic animals at the same dose, respectively. Codeine reduced the amplitude of cough responses in normal animals by 32% and 70% at doses of 10 mg/kg and 20 mg/kg, respectively. In bronchitic animals, the antitussive effect of codeine was weak and not dose-dependent, and there was a significant difference between the effects of codeine on normal and bronchitic animals only at the dose 20 mg/kg. The effects of eprazinone were similar to those of codeine. Thus, Bakumondo-to showed selectively potent antitussive action in bronchitic animals (Figures 1 and 2).

2 Effects on Cough Responses Induced by Chemical Stimulation

Citric acid-induced coughs

Citric acid, as a chemical stimulant, is also known to act mainly on Aσ-fibers. The effects of 1 g/kg of Bakumondo-to and 20 mg/kg of codeine on citric acid-induced coughs in normal and bronchitic animals are shown in Table 1. Similar to its effect on mechanically induced coughs, the antitussive effect of Bakumondo-to on citric acid-induced coughs was stronger in bronchitic animals than in normal animals. The effects of codeine were the opposite(Table1).

Capsaicin- and substance P-induced cough:

Capsaicin stimulates afferent C fibers (Coleridge *et al.*, 1984) and releases sensory neuropeptides such as substance P and other tachykinins (Serra *et al.*, 1988), and consequently produces coughs in animals (Kohrogi *et al.*, 1988; Takahama *et al.*, 1993) and humans (Fuller *et al.*, 1985). Inhaled substance P has also been reported to cause coughs by stimulating afferent C fibers (Kohrogi *et al.*, 1989). Bakumondo-to dose-dependently decreased the number of capsaicin- and substance P-induced coughs by 45% to 50% at 0.5 g/kg to 1 g/kg. Codeine, at 20 mg/kg, also

Table 1 Effects of Bakumondo-to and codeine on the cough reflex induced in guinea pigs by citric acid inhalation.

Drug		Effect (% change) Dose Normal	Bronchitic
Control	(CMC-Na)	99.3 ± 12.4	93.1 ± 9.8
Bakumondo-to	1.0 g/kg	59.5 ± 16.9	49.1 ± 12.7*
Codeine	20 mg/kg	25.7 ± 8.7*	54.5 ± 25.0

The values are expressed as % change (mean ±S.E.) of pre-administration control in amplitude of cough responses.
*: $p < 0.05$, **: $p < 0.01$ *vs.* the control group

significantly decreased the number of coughs by 65% to 70%. There was no significant difference between the effect of 1 g/kg of Bakumondo-to and that of 20 mg/kg of codeine (Figure 3).

3. Effects on Cough Responses Facilitated by Enalapril

ACE inhibitors are increasingly used to treat both hypertension and congestive cardiac failure, but persistent dry cough has arisen as an adverse side effect of this treatment (Seseko and Kaneko, 1985). During coughing, large fluctuations in intrathoracic pressure are produced, and consequently, hemodynamic and reflex cardiovascular changes occur. Thus, the use of ACE inhibitors sometimes presents a serious problem in patients with hypertension and heart failure. In our experiments, an oral dose of 10 mg/kg of enalapril, a representative ACE inhibitor, did not induce coughs in guinea pigs within an observation period of 120 min after administration. The number of capsaicin-induced coughs, however, significantly increased from 4.5 ± 0.4 to 7.1 ± 0.6 when a prior administration of the same dose of enalapril was given (Figure 4). Bakumondo-to (1 g/kg) completely inhibited the enalapril-induced facilitation of cough responses. Codeine (20 mg/kg) tended to reduce the facilitation in some animals tested, but the effect was uncertain and insignificant (Figure 4).

We did not study the effect of Bakumondo-to on coughs induced by the ACE inhibitor alone, because neither enalapril alone nor other ACE inhibitors tested caused reproducible cough responses in normal guineapigs (Takahama *et al.*, 1993). Nevertheless, Bakumondo-to might be clinically effective on ACE inhibitor-

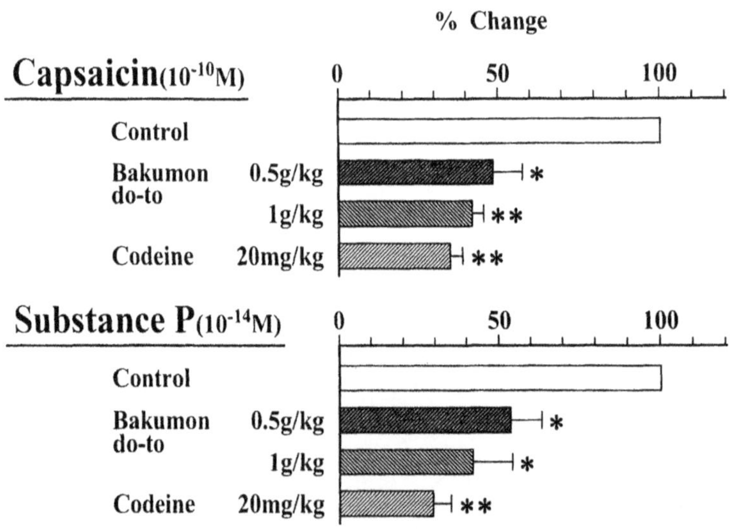

Figure 3 Effects of Bakumondo-to and codeine on the cough reflex of guinea pigs induced by capsaicin and substance P.
Values are expressed as control % change in the number of cough responses during a15 min period. *: $p < 0.05$, **: $p < 0.01$ *vs.* the control groups.

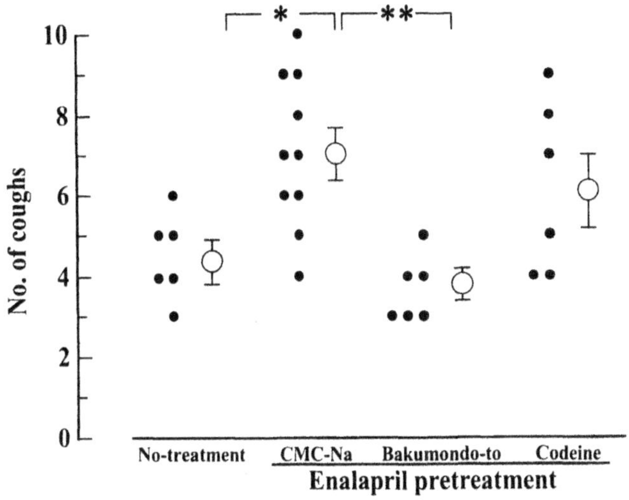

Figure 4 Differences in antitussive effects of Bakumondo-to (1 g/kg) and codeine (20 mg/kg) on the augmentation of capsaicin-induced cough by enalapril in guinea pigs. 10 mg/kg of enalapril, an ACE inhibitor which augments cough responses, was administered p.o. 90 min before the administration of CMC-Na, Bakumondo-to or codeine.
*: $p < 0.05$, **: $p < 0.01$.

induced coughs, since it showed a powerful antitussive action on the augmentation of capsaicin-induced coughs by enalapril.

2. CHARACTERISTICS AND MECHANISMS OF ANTITUSSIVE ACTION OF BAKUMONDO-TO

Cough can be induced by various mechanisms, e.g., by the irritation of rapidly adapting (irritant) receptors with vagal myelinated A-type fibers (Widdicombe, 1954) and/or C fiber receptors with unmyelinated fibers (Coleridge *et al.*, 1976), which are located within and immediately below the epithelial cell lining. Mechanical stimulants and chemical stimulants such as citric acid act mainly on Aσ-fibers, whereas capsaicin and tachykinin mainly act on C fibers.

Bakumondo-to did not depress coughs induced by a mechanical stimulation and/or by citric acid but reduced the coughs induced by capsaicin and substance P, whereas codeine showed potent antitussive actions on mechanically induced as well as chemically induced coughs. There was no significant difference between the antitussive activities of Bakumondo-to and codeine at the doses tested on capsaicin- and substance P-induced coughs.

As shown in Figure 5, which presents the summarized spectra of the effects of Bakumondo-to and codeine on the cough responses induced by various substances in normal guinea pigs, Bakumondo-to but not codeine significantly suppressed coughs induced by neurokinin A (another tachykinin), and bradykinin, which is known to release natural tachykinins from C fiber terminals in the airway.

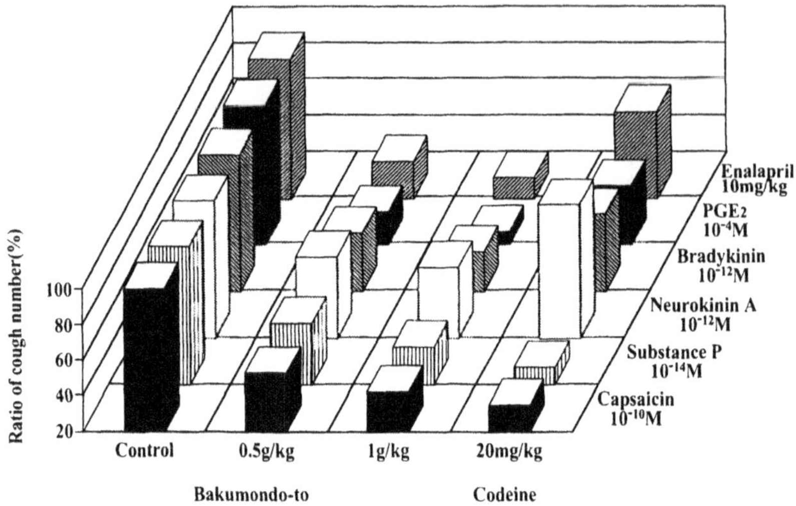

Figure 5 Spectra of antitussive effects of Bakumondo-to and codeine on the cough reflex induced by various substances in guinea pigs.

It is well known that tachykinins are degraded by their cleavage enzyme neutral endopeptidase (NEP). In addition, increased concentrations of bradykinin and/or substance P (Fuller, 1987) in the lung and the subsequent bronchial hyperactivity (Kaufman *et al.*, 1980) have been suggested as the mechanism for ACE inhibitor-induced cough. Interestingly, pretreatment with phosphoramidon, an NEP inhibitor, significantly increased the number of coughs induced by substance P and capsaicin (Takahama *et al.*, 1995). These results are in accord with other experimental findings that tachykinins- and capsaicin-induced coughs were more easily produced in bronchitic guinea pigs than in normal guinea pigs (Figure 6).

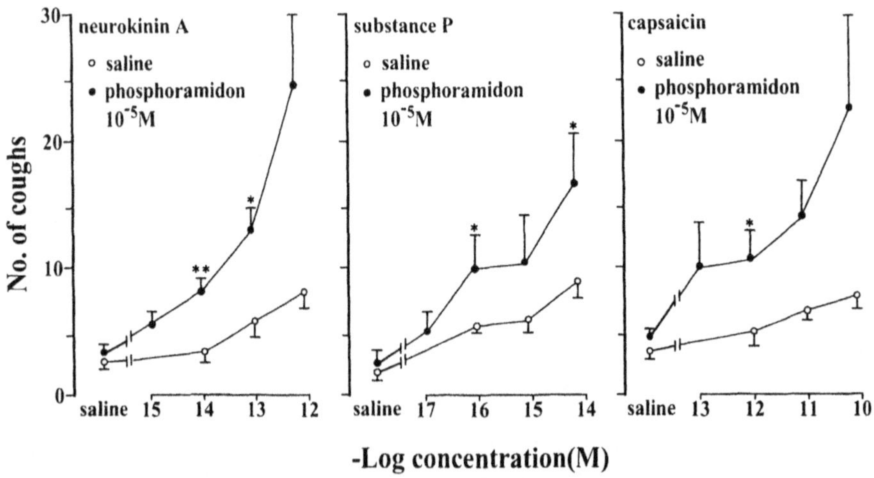

Figure 6 Augmenting effect of phosphoramidon on the coughs induced by neurokinin A, substance P and capsaicin. *: $p < 0.05$, **: $p < 0.01$ *vs.* the control groups.

Bakumondo-to and ophiopogonin, a steroid saponin extracted and purified from Bakumondo (*Ophiopogonis tuber*) (Figure 7) also suppressed tachykinins- and phosphoramidon-induced coughs while codeine was nearly ineffective (Figures 8

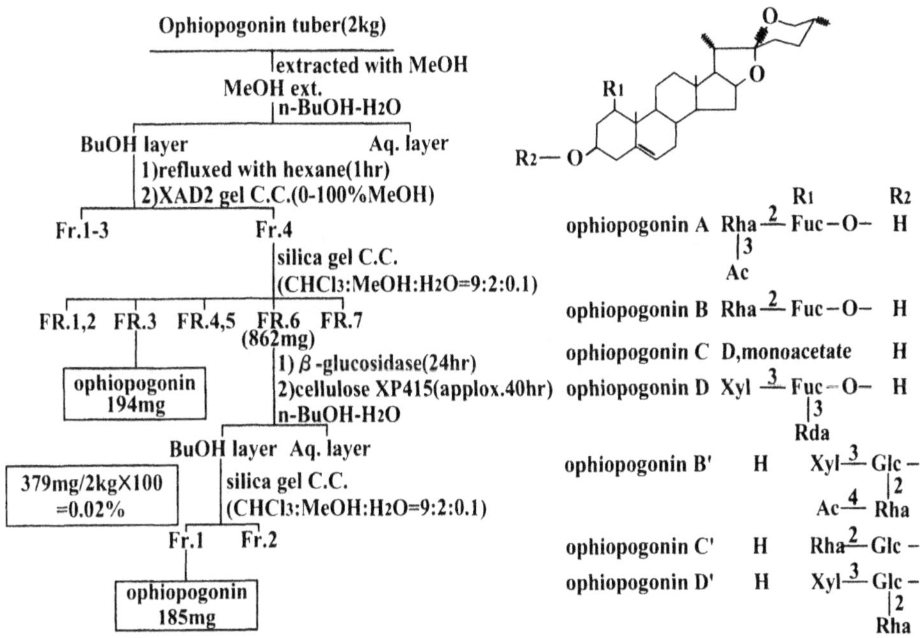

Figure 7 Isolation procedure of ophiopogonin from *Ophiopogonis tuber*.

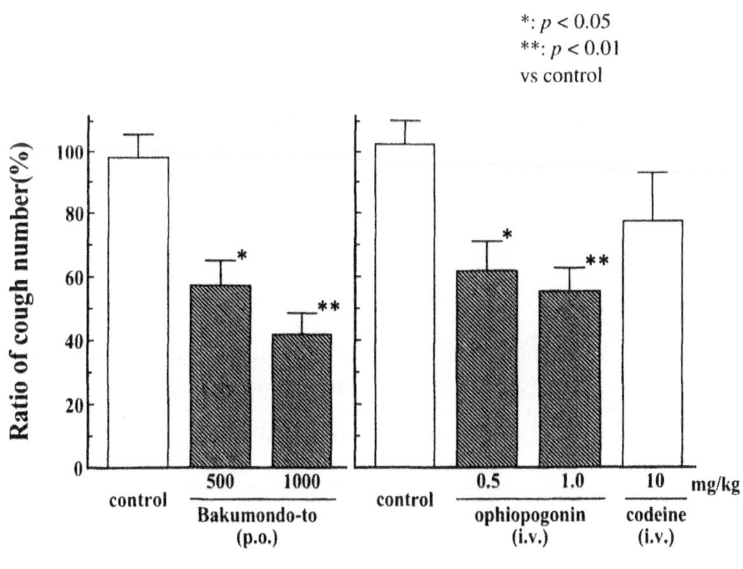

Figure 8 Effects of Bakumondo-to, ophiopogonin and codeine on substance P-induced coughs in bronchitic guinea pigs.

and 9). In this context, we found a difference in NEP-like activities in the airway between normal and bronchitic guinea pigs. The airway NEP-like activities in the bronchitic animals decreased to 1/8–1/9 of those in normal healthy animals, and treatment with Bakumondo-to for 7 days prevented this reduction of NEP-like activities (Table 2).

It is likely, therefore, that Bakumondo-to produces its antitussive action through a peripheral mechanism such as by an inhibition of the excitability of C fibers in the airway, although it is unknown whether or not the Aσ fibers and C fibers ascending from the respiratory tract converge at the site within the cough center consisting of the vagal sensory nucleus (N. Solitarius), autonomic nucleus (N. Dorsalis) and motor nucleus (N. Ambiguus) and reticular parbocellular nucleus in the medulla. In this regard, we have confirmed that Bakumondo-to acts neither on the neuronal activities of the cough center in the medulla nor on the activities of cortices.

It is likely that part of the peripheral mechanism of action of Bakumondo-to is attributable to the depression of the tracheal and bronchial hyperactivity associated with an inflammation, because Bakumondo-to showed potent antitussive action in

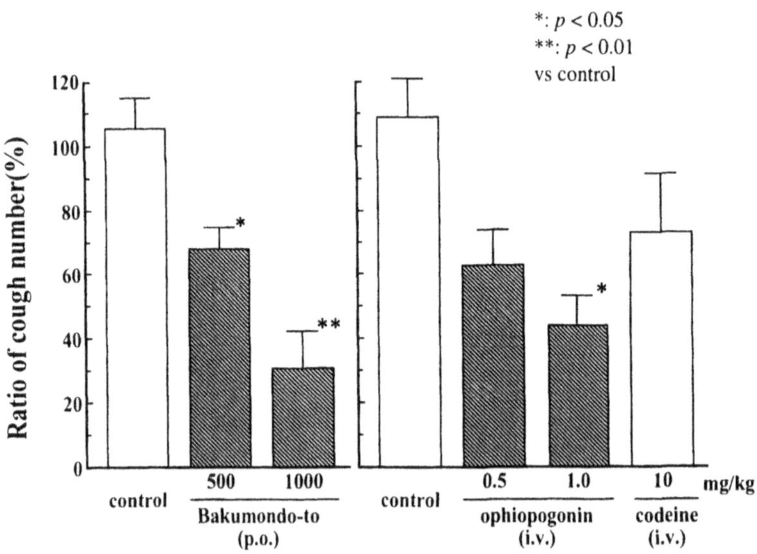

Figure 9 Comparison of antitussive effects of Bakumondo-to, ophiopogonin and codeine on the cough responses induced by phosphoramidon (10^{-5} M) in guinea pigs.

Table 2 Effect of repeated administration of Bakumondo-to (1 g/kg) on SO_2-exposure induced reduction of NEP-like activities in the trachea and bronchi in guinea pigs.

	NEP-like activity (fmol·.min^{-1}·mg protein^{-1}) (n) Trachea	Bronchi
Normal	324.1 ± 80.9 (6)	317.8 ± 123.7 (4)
Bronchitic control	49.1 ± 24.9 (6)*	35.8 ± 35.8 (4)†
Bakumondo-to	247.2 ± 101.4 (4)	128.5 ± 36.2 (4)

†: $p < 0.01$, *: $p < 0.05$ *vs.* the normal groups.

bronchitic animals. Indeed, we found that the bronchitic animals had bronchial hyperactivity with the increased afferent discharges of the superior laryngeal nerve (Miyata *et al.,* 1989). We also found that the repeated administration of Bakumondo-to significantly suppressed the increased afferent discharge, whereas codeine significantly augmented the increased discharge of the nerve (Figure 10). These findings indicate that the inflammation of the trachea and bronchus reduces the threshold for cough responses. In this regard, substance P reportedly causes many pro-inflammatory reactions in the airway, i.e., bronchoconstricton, an increase in vascular permeability and an increase in tracheal and bronchial secretion (Borson, 1991). As presented in Figure 11 and described later, Bakumondo-to significantly inhibited hypersensitive bronchoconstriction and airway hypersecretion by hydrogen peroxide (Miyata *et al.*, 1995). Investigations of the antitussive effects of the crude drug components of Bakumondo-to revealed that Bakumondo (*Ophiopogonis tuber*), Hange (*Pinellia tuber*) and Kanzo (*Glycyrrhiza radix*) were effective, that Ninjin (*Ginseng radix*) was slightly effective, and that Taiso (*Zizyphi fructus*) and Koubei (*Oryzae fructus*) were not effective at all (Figure 12). These results suggested that only the crude components with anti-inflammatory actions have antitussive activities.

In contrast to Bakumondo-to, codeine showed a significantly weaker antitussive action in bronchitic animals than in normal animals. Since inflammation causes the bronchial hyperactivity, codeine may peripherally act to facilitate cough responses by releasing histamine and other mediators. In accord with clinical findings, codeine had no significant effect on the facilitation of cough responses by enalapril. This finding may also be interpreted by its peripheral action.

Unlike codeine, Bakumondo-to and ophiopogonin as a major active component of Bakumondo-to thus have notable antitussive activities on coughs associated with bronchitis and coughs increased by ACE-inhibitor treatment, and these activities

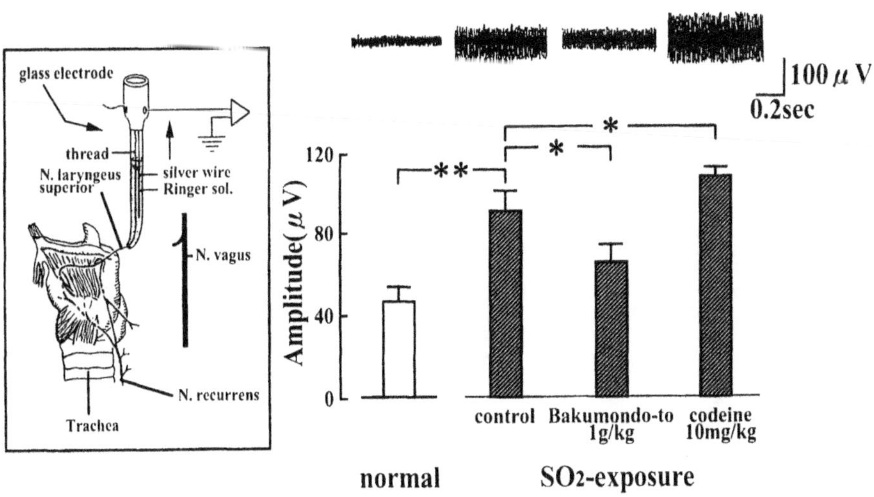

Figure 10 Effects of repeated administrations of Bakumondo-to and codeine on spontaneous afferent discharges of the superior laryngeal nerve in SO_2-exposed guinea pigs. *: $p < 0.05$ *vs.* control bronchitic group, **: $p < 0.01$ *vs.* normal group.

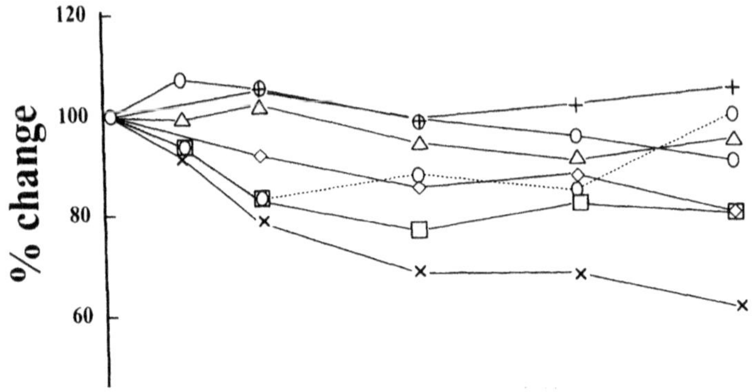
120
100
80
60
% change

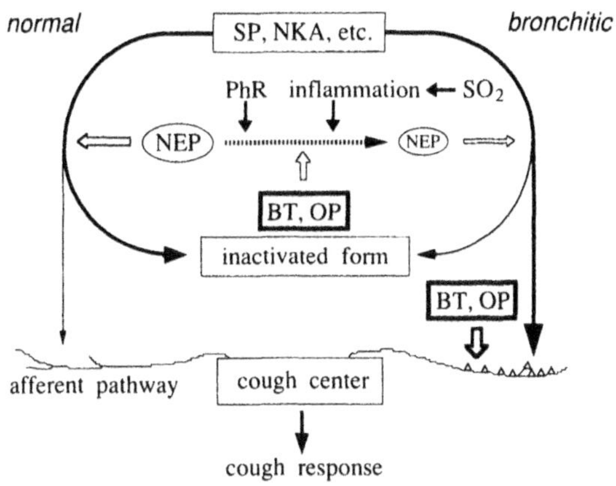

Figure 13 The possible mechanism of the antitussive action of Bakumondo-to and ophiopogonin. NEP: neutral endopeptidase, PhR: Phosphoramidon, BT: Bakumondo-to, OP: ophiopogonin.

3. MUCOREGULATORY AND SECRETOMOTOR ACTIONS

Abnormal and excessive production of airway mucus is a characteristic feature of many chronic inflammatory lung diseases. Although coughs are primarily triggered by the stimulation of airway Aσ and C fibers, an increase in mucous secretion and adhesion of viscous sputa to the airway wall also significantly affect cough responses. Coughs are often alleviated by the expectoration of viscous sputa. In the symptomatic treatment of coughs, even antitussives possessing potent antitussive activity may cause severe adverse effects if they decrease the normal secretion of airway tract fluid and increase sputum viscosity and therefore inhibit the expectoration of sputa. It is thus essential that the productive cough (wet cough) as a protective reflex should not be inhibited and that airway cleaning should be facilitated.

Traditional Chinese medicines may be useful for ameliorating overall defects in airway disorders because the medicines are composed of various herbs with weak but ubiquitous pharmacological activities. Our previous study (Miyamoto *et al.*, 1990) showed that Seihai-to (Qingfei-tang, composed of 16 herbs and described in Wang Bin Hui Chun, a medical text published in 1587 in China), inhibited the release of a slow-reacting substance of anaphylaxis from passively sensitized guinea pig lung after antigen challenge. Seihai-to also suppressed the chemiluminescence of oxygen radicals when healthy human leukocytes were stimulated by opsonized zymosan. In normal rabbits, Seihai-to increased the output volume and fatty acid contents in airway secretions. In bronchitic rabbits, a 6-week administration of Seihai-to restored the decreased amount of saturated fatty acid in the sputa, and histological examinations revealed an amelioration of the inflammation of lung tissues. In pigeons, Seihai-to facilitated tracheal mucociliary transport. Accordingly, Seihai-to seems to exert effectiveness via multiple mechanisms.

Based on these findings and from polypharmacological points of view, we evaluated the effects of Bakumondo-to on airway secretion.

1. Actions on High Molecular Weight Glycoconjugate Secretion from Tracheal Epithelial Cells

The normal human airway is covered with high molecular weight glycoconjugates (HMWG), which consist of O-glycosylated tandem repeats with a high percentage of seerine and threonine. The airway mucous glycoproteins have a variety of protective functions, including entrapment and clearance of exogenous materials by ciliary transport, as well as humidification and lubrication of the airway mucosa (Rose *et al.,* 1992). Excessive mucous production, however, characterizes airway diseases, such as chronic bronchitis, asthma, cystic fibrosis, and bronchiectasis.

In these diseases, the influx of polymorphonuclear leucocytes (PMNs) into the airways and the release of substance P and neurokinin A from the peripheral endings of primary sensory neurons have also been observed. Numerous studies have shown that substance P induces a series of neutrophil responses such as adhesion to endothelial cells, chemotaxis, and the release of inflammatory mediators that triggers and amplifies tissue injury and inflammation (Umeno *et al.,* 1990; Wiedermann *et al.,* 1989; Ohlen *et al.,* 1989; Marasco *et al.,* 1981). Substance P itself can induce mucus secretion in airway submucosal gland mucous cells (Lundgren *et al.,* 1989; Rogers *et al.,* 1989), but not in airway epithelial cells *in vitro.* It is therefore suspected that substance P may affect the mucus secretion of airway epithelial cells through the activation of PMNs.

Thus, we first examined whether PMNs activated by substance P can affect the secretion of HMWG, which is used as a marker of mucus, from hamster tracheal epithelial cells in culture (Yoshitake *et al.,* 1995a, 1995b). Hamster tracheal epithelial cells have been shown to be morphologically and biochemically similar to tracheal goblet cells (Kim *et al.,* 1985; Kim 1991). The secretion of HMWG from cultured cells was stimulated by neutrophil elastase, cathepsin G, ATP, protein kinase C activators, and mechanical stimulation (Kai *et al.,* 1994). Based on these findings, the hamster tracheal epithelial cell is a suitable model to determine mucus secretion by the tracheal surface secretary cells (Miyata *et al.,* 1996).

In our study, the activity of lactose dehydrogenase (LDH) released into the medium as an assessment of cellular integrity did not exceed 1% of the total cell content in all experiments. Figure 14 shows the effect of 10 M of substance P on HMWG secretion from hamster tracheal epithelial cells and PMNs co-culture. According to a previous report (Marasco *et al.,* 1981), we used 10 μm of substance P as a concentration to induce PMNs activation. In the present study, substance P or PMNs alone did not affect HMWG secretion by hamster tracheal cells in culture (Figure 17). As shown in Figure 14, however, substance P-activated PMNs (10^{-6} cells/ml) significantly increased the amount of HMWG released into 30-min challenged.

As expectorant-antitussives, drugs active on airway secretion have been proposed and used to cleanse and clear the respiratory tract for several centuries in many countries. On the basis of the mechanism of this action, the mucoactive drugs were classified into several groups (Braga *et al.,* 1989). Some mucoactive drugs possess direct effects on the production of airway secretions or on changes in their compositions, resulting in increased effectiveness of mucociliary clearance. Other mucoactive drugs do not have a specific action on mucus, but produce a general

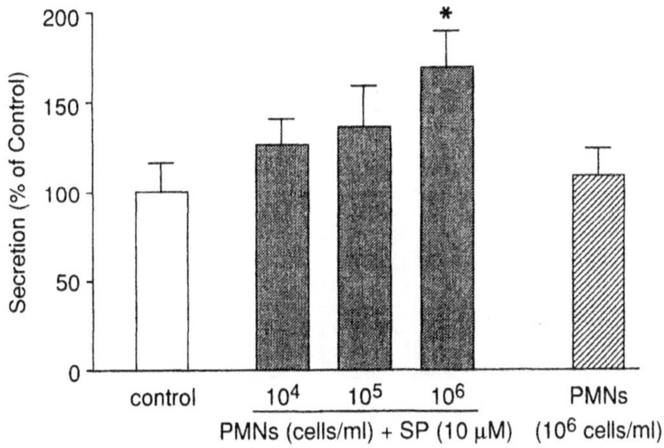

Figure 14 Effect of substance P on high molecular weight glycoconjugate secretion from tracheal epithelial cells and polymorphonuclear leukocyte co-incubation.
*: $p < 0.05$ *vs.* the control group (n = 5).

benefit in airway structure and function which secondarily lead to correction of the pathophysiologic mechanisms that result in abnormal secretions associated with severe cough attacks. From this standpoint we have evaluated the mucoregulatory action of Bakumondo-to using the above-described culture systems of hamster tracheal epithelial cells and co-culture of the cells with PMNs activated by substance P or by other stimulants to simulate the airway inflammatory state.

In the concentrations tested, the Bakumondo-to filtrate, the hydrophobic fraction containing flavonoid and saponins, and the hydrophilic fraction containing sugars and peptides (prepared by the procedure shown in Figure 15) did not produce any effect on HMWG secretion from cultured tracheal epithelial cells (Figure 16).

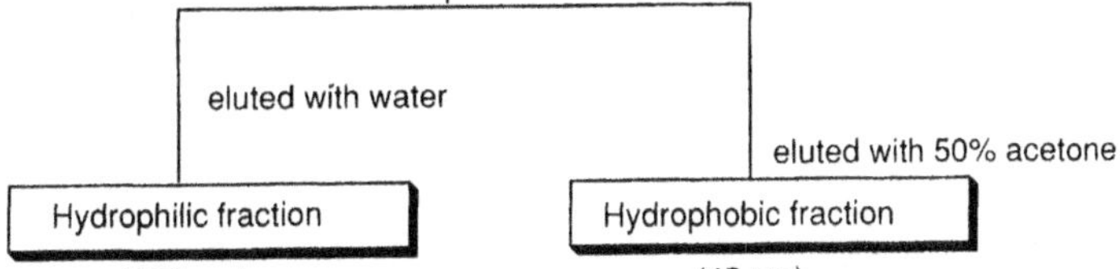

Figure 15 Preparation of Bakumondo-to hydrophobic and hydrophilic fractions. The indicated values represent the yields of the fractions.

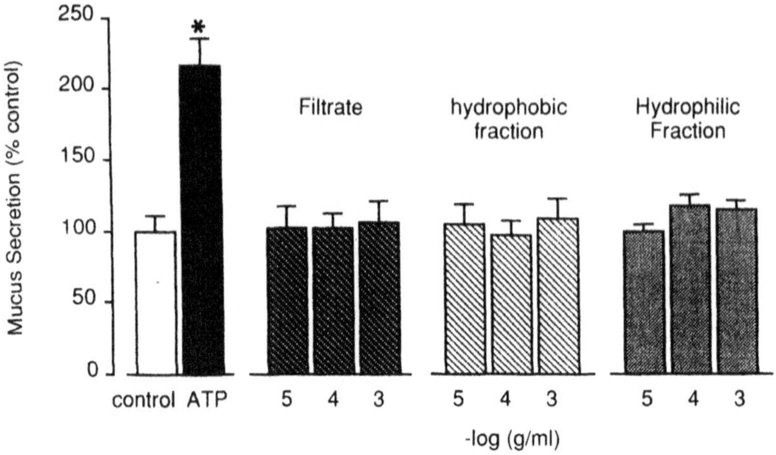

Figure 16 Effect of Bakumando-to filtrate and hydrophobic and hydrophilic fractions on high molecular weight glycoconjugate secretion from tracheal epithelial cell culture. *: $p < 0.05$ *vs.* the control group (n = 4).

However, Bakumondo-to filtrate, in a concentration of 1 mg/ml, significantly suppressed the substance P-induced secretion of HMWG from tracheal epithelial cells and PMNs co-culture (Figure 17).

As a source of the HMWG released by substance P-induced PMNs, apical cell surface membrane-bound mucins may be potent candidates, because it has been shown that some of the HMWG released by human neutrophil elastase in cultured hamster tracheal epithelial cells are apical cell surface membrane-bound mucins (Kim *et al.*, 1987).

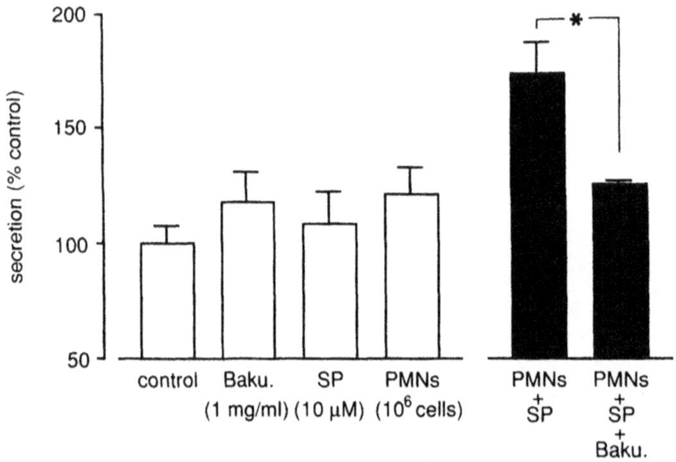

Figure 17 Effects of Bakumondo-to filtrate on substance P-induced high molecular weight glycoconjugate secretion from tracheal epithelial cells and polymorphonuclear leuukocyte co-incubation. SP: substance P, PMNs: polymorphonuclear leukocytes, BT: Bakumondo-to filtrate. *: $p < 0.05$ (n = 5).

The activity of activated PMNs in releasing and degrading mucin may play a significant role in certain pathological conditions. During airway injury, secretory cells are the major airway epithelial cells that proliferate and eventually reconstitute the injury area (Keenan *et al.*, 1982). Thus, it is possible that during injury, reactive oxygen species and elastases generated and released by PMNs at the inflammatory site act by removing the protective layer of mucins from the surface of the secretory cell, which in turn provides a signal for secretory cell proliferation followed by maturation (Kim *et al.*, 1987).

Although the suppressive effect of Bakumondo-to on the increased release of HMWG by substance P-activated PMNs is an *in vitro* finding, we speculate that it is one of the possible mechanisms of effect of the Chinese medicine on chronic airway diseases.

2. Actions on Pulmonary Surfactant Secretion

Pulmonary surfactant secreted from alveolar type II cells lowers the surface tension at the air-liquid interface in the lung and provides alveolar stability. It has been clearly demonstrated that, in addition to this vital role, surfactants are also important in airway mucociliary clearance (De Santis *et al.*, 1994). Several studies have suggested that the presence of phospholipids in the airway may modify the clearance of mucus and make expectoration easy (Allegra *et al.*, 1985; Morgenroth and Bolz, 1985; Gehr *et al.*, 1990; Girod *et al.*, 1991; Rubin *et al.*, 1992a, 1992b). The existence of surfactant films has been demonstrated in the airways of several species by electron microscopy (Gehr *et al.*, 1990; Mercer *et al.*, 1992) and by in situ surface tension measurement (Gehr *et al.*, 1990; Schuch *et al.*, 1990). We found an inhibitory action of inflammatory mediators on mucociliary transport (Tai *et al.*, 1995) and confirmed a protective effect of surface active phospholipids on an acid-induced inhibition of mucociliary transport in pigeons (Kai *et al.*, 1989).

In light of these findings and those of several reports on pulmonary surfactants in mucociliary clearance, it is reasonable to study the effects of mucoactive drugs on pulmonary surfactant secretion in alveolar type II cells. The secretion of phosphatidylcholine (PC), a major surfactant phospholipid, has been shown to be influenced by a variety of physiological and pharmacological agents in alveolar type II cells (Wright and Dobbs, 1991; Kai *et al.*, 1996a, 1996b). We also found that both β_1- and β_2-adrenoceptor subtypes mediated phosphatidylcholine secretion (Kai *et al.*, 1992), and confirmed that both β_1- and β_2-adrenoceptor genes were expressed in rat alveolar type II cells (Isohama *et al.*, 1995).

To simulate airway inflammatory states, we have used co-culture systems of alveolar type II cells with activated PMNs (Oda *et al.*, 1991; Okumura *et al.*, 1996). Activated PMNs in airway epithelium are thought to be involved in the pathogenesis of many airway diseases. PMNs activated by opsonized zymosan caused a significant increase in phosphatidylcholine secretion. Pretreatment of the culture with the combination of superoxide dismutase and catalase reduced the increase in PC secretion (Oda *et al.*, 1991). These results suggested that activated PMNs stimulated the secretion of pulmonary surfactant, and that the stimulation was partly mediated by oxygen radicals. These systems may be useful to assay the anti-oxidant effects of mucoactive drugs.

To examine the effect of Bakumondo-to on pulmonary surfactant secretion, we cultured adult rat alveolar type II cells in the presence or absence of Bakumondo-to filtrate or its fractions for 90 min. To assess cellular integrity, lactate dehydrogenase (LDH) activity in the cells and medium was measured. The LDH activity released into the medium did not exceed 1% of the total cell content in all experiments.

In the cells treated with 0.1 μm of Bakumondo-to filtrate, the basal secretion rate of PC was similar to that in control cells treated with vehicle. In contrast, the secretion was significantly enhanced in the cells treated with 1 or 10 μm of the filtrate, compared to the control cells (Figure 18), which is similar to the effects of β_2 adrenoceptor stimulants (Kai *et al.*, 1992). However, the effect of Bakumondo-to filtrate was not augmented when the concentration was further increased (Figure 18). To examine whether the enhancing effect of Bakumondo-to filtrate on pulmonary surfactant reflects the action of Bakumondo-to in whole animals, we applied the secretion from rats administered 1 g/mg of Bakumondo-to. The serum from rats administered Bakumondo-to strongly enhanced the basal secretion of PC from the cultured alveolar type II cells. The serum from control rats administered vehicle manifested a similar but less significant effect (Figure 19). Our further studies revealed that the Bakumondo-to hydrophobic fraction (containing mainly flavonoids and saponins) increased the PC secretion, and the Bakumondo-to hydrophilic fraction (containing mainly sugars and peptides) suppressed the PC secretion (Figure 18). These contradictory and reciprocal effects of the two fractions might explain the limited highest response induced by whole Bakumondo-to filtrate, namely the ceiling phenomenon. Moreover, the hydrophobic fraction significantly increased the content of intracellular adenosine 3', 5'-cyclic monophosphate (cAMP) levels (Figure 20). Since its enhancing effect on secretion could be inhibited by pretreatment with H-89 (a proteinkinase-A inhibitor shown in Figure 20), but not by propranolol (a β-adrenoceptor antagonist) (Figure 21), the hydrophobic fraction might induce enhanced PC secretion mostly by increasing intracellular cAMP levels followed by the activation

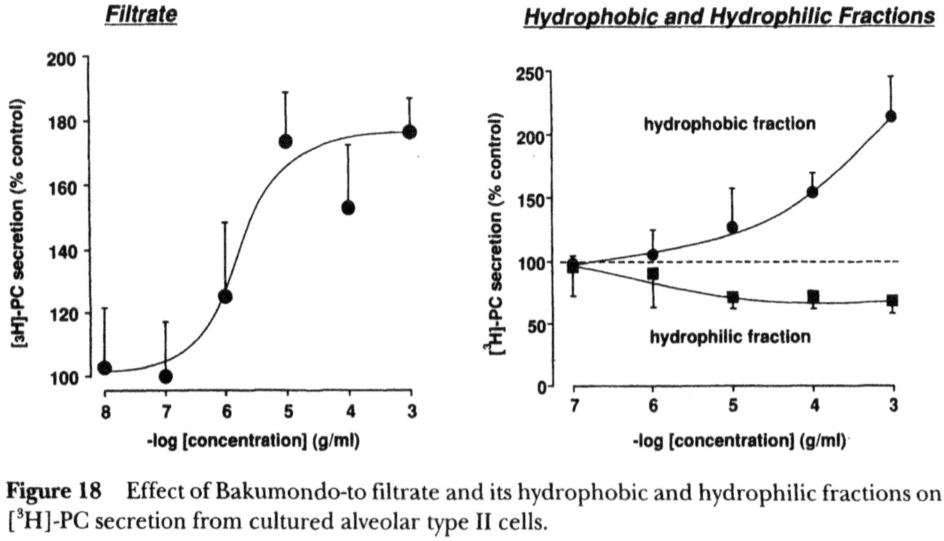

Figure 18 Effect of Bakumondo-to filtrate and its hydrophobic and hydrophilic fractions on [^{3}H]-PC secretion from cultured alveolar type II cells.

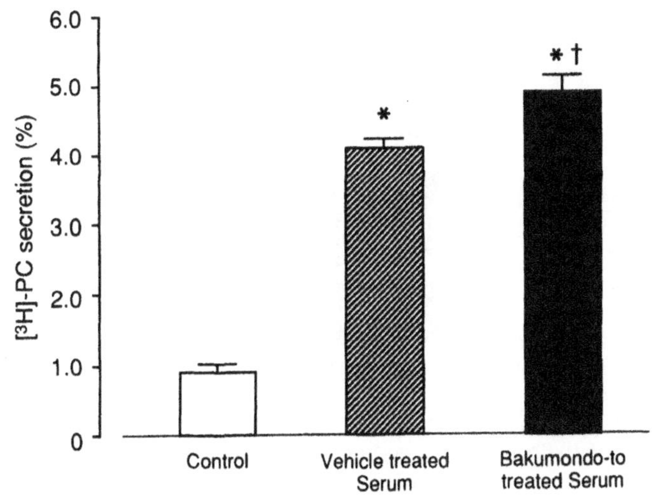

Figure 19 Effect of sera from rats treated with vehicle and/or Bakumondo-to on [^{3}H]-PC secretion from cultured alveolar type II cells. Alveolar type II cells cultured for 18 h were incubated with 50 ml/1ml DMEM/well of sera from rats treated with distilled water or Bakumondo-to. Each value represents the mean ± S.E. of 4 experiments. *: $p < 0.01$ *vs.* the control, †: $p < 0.05$ *vs.* vehicle-treated serum.

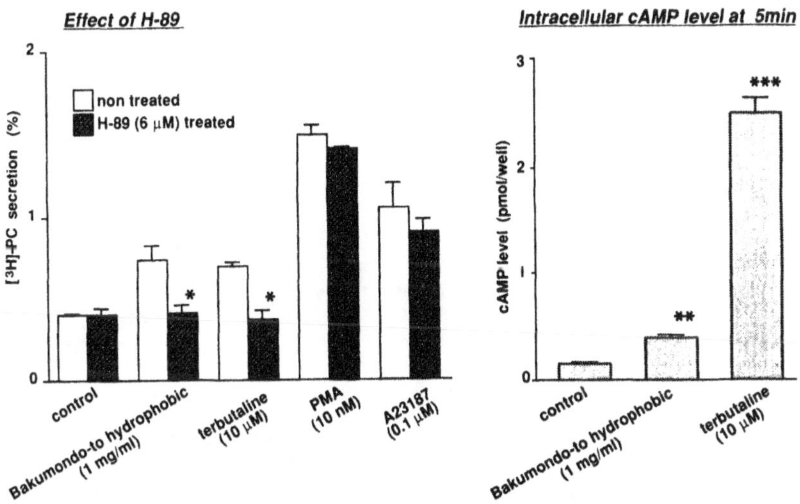

Figure 20 Effect of the hydrophobic fraction from Bakumondo-to on intracellular cAMP level in cultured alveolar type II cells.

of proteinkinase-A. The phosphatidylcholine secretion induced by a β-adrenergic mechanism is controlled by a cascade consisting of receptors, G-proteins, adenylate cyclases and cAMP-dependent protein kinases, one of which is kinase-A. Since dexamethasone treatment is known to sensitize cAMP formation by altering the adenylate cyclase catalyst (Michel *et al.*, 1994), it is likely that the effect of the

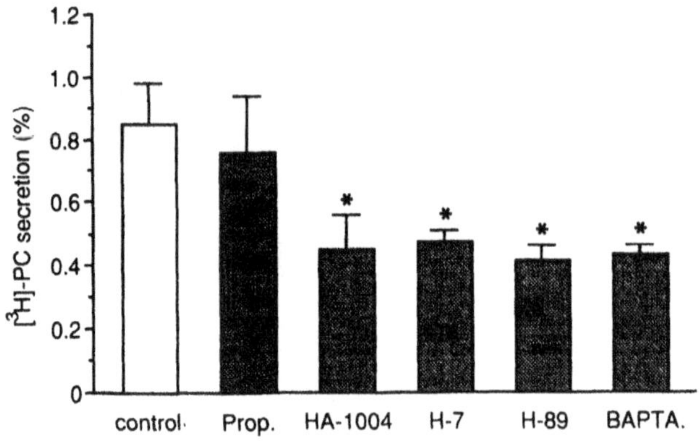

Figure 21 Effects of various inhibitors on [^{3}H]-PC secretion from cultured alveolar type II cells increased by the hydrophobic fraction of Bakumondo-to. Cultured alveolar type II cells were pretreated with Prop (Propranolol, 1 μM), HA-1004 (100 μM), H-7 (100 μM), H-89 (6 μM) and BAPTA (BAPTA-AM, 5 μM) for 30 min, and incubated with the hydrophobic fraction (1 mg/ml) of Bakumondo-to for 90 min. Data are expressed as secretion percentage; each bar represents the mean ± S.E. of 4 experiments. *: $p < 0.05$ *vs.* the control.

hydrophobic fraction of Bakumondo-to was produced through a dexamethasone-like action. Furthermore, we observed that the hydrophobic fraction-induced secretion could be inhibited by pretreatment with BAPT-AM, an intracellular Ca^{2+} chelating agent (Figure 22), and by pretreatment with H-7 (a kinase C inhibitor) or HA-1004 (a nonselective inhibitor of protein kinase C, G and A) (Figure 21). By means of Fara2 fluorescence determination, we also found that the hydrophobic

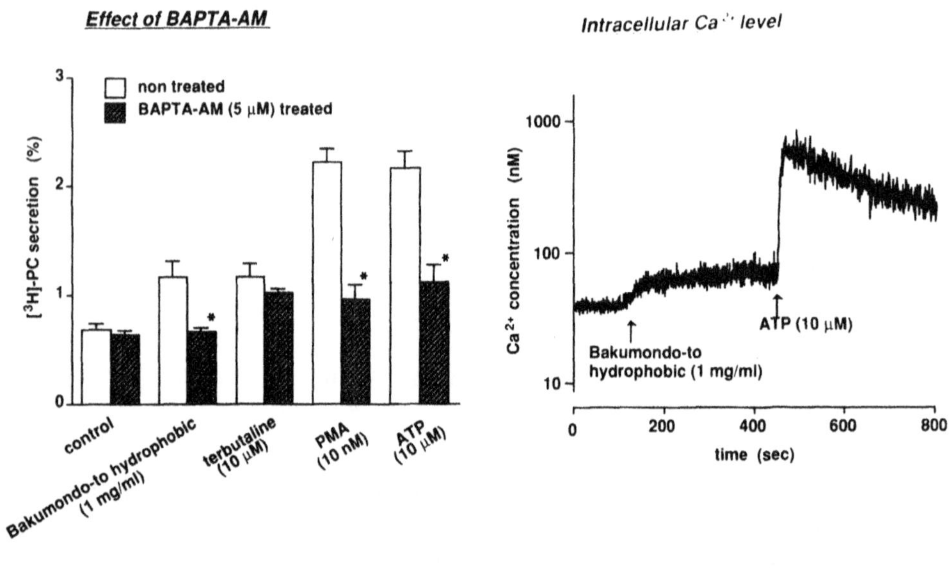

Figure 22 Effect of the hydrophobic fraction of Bakumondo-to on intracellular Ca^{2+} in cultured type II cells.

fraction could significantly and continuously increase intracellular Ca^{2+} levels, although the increase was much less powerful than that produced by adenosine 5'-triphosphate (ATP) (Figure 22). It was suggested therefore that the hydrophobic fraction-induced enhancement of PC secretion is related to the increase of intracellular Ca^{2+} and to the activation of proteinkinase C.

To simulate inflammatory states and to examine the effects of Bakumondo-to and ophiopogonin as an active component on the hypersecretion of pulmonary surfactant in these states, we used co-culture systems of alveolar type II cells with PMNs activated by substance P. The increase in PC secretion from alveolar type II cells depended on the numbers of activated PMNs. Bakumondo-to filtrate (1 mg/ml) and ophiopogonin significantly inhibited the increased secretion (Figures 23 and 24). Elastase and hydrogen peroxide, major mediators derived from activated PMNs, stimulated PC secretion from alveolar type II cells (Figure 25). Although it had no significant effects on the elastase-induced enhancement of PC secretion, Bakumondo-to filtrate significantly inhibited the enhanced secretion by hydrogen peroxide (Figure 25).

These results suggest that Bakumondo-to has a characteristic secretory effect on pulmonary surfactants and normalizes hypersecretion by an inhibition of super oxide (Figure 26).

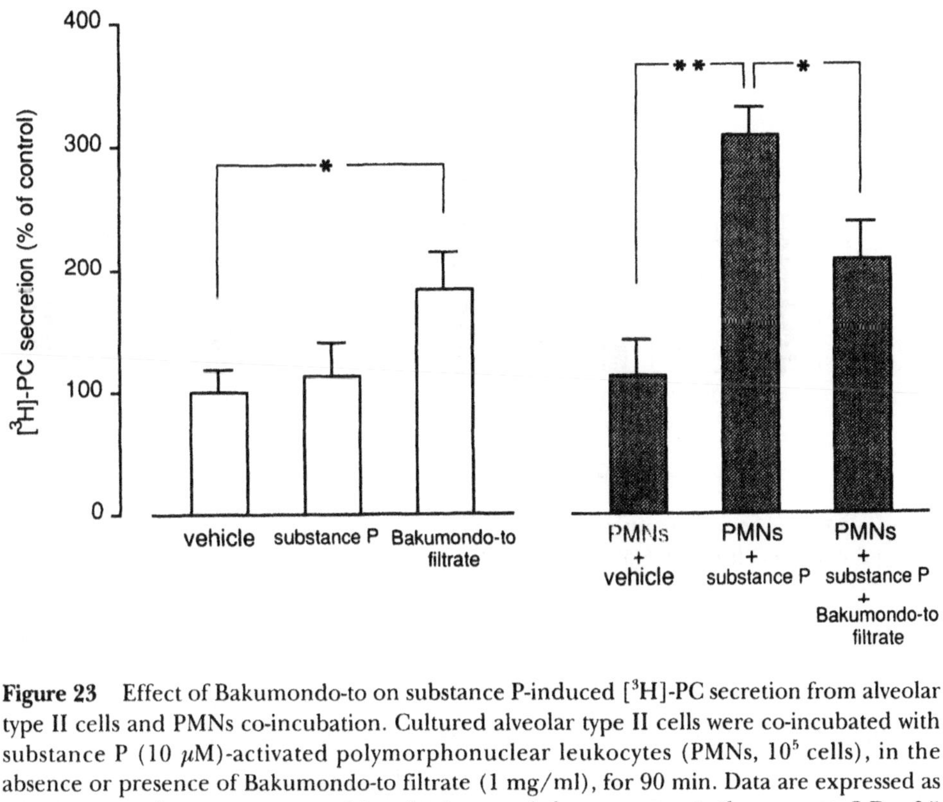

Figure 23 Effect of Bakumondo-to on substance P-induced [^{3}H]-PC secretion from alveolar type II cells and PMNs co-incubation. Cultured alveolar type II cells were co-incubated with substance P (10 μM)-activated polymorphonuclear leukocytes (PMNs, 10^5 cells), in the absence or presence of Bakumondo-to filtrate (1 mg/ml), for 90 min. Data are expressed as percent secretion versus control incubation; each bar represents the mean ± S.E. of 5 experiments. *: $p < 0.05$, **: $p < 0.01$.

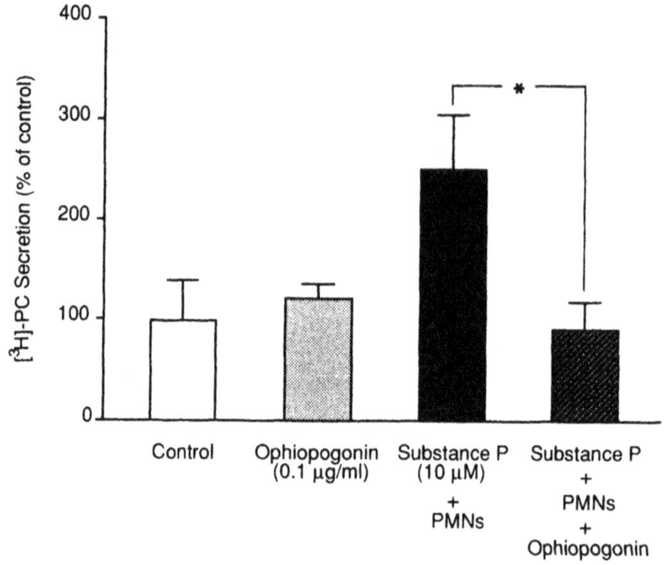

Figure 24 Effect of ophiopogonin on substance P-induced [^{3}H]-PC secretion from alveolar type II cells and PMNs co-incubation.

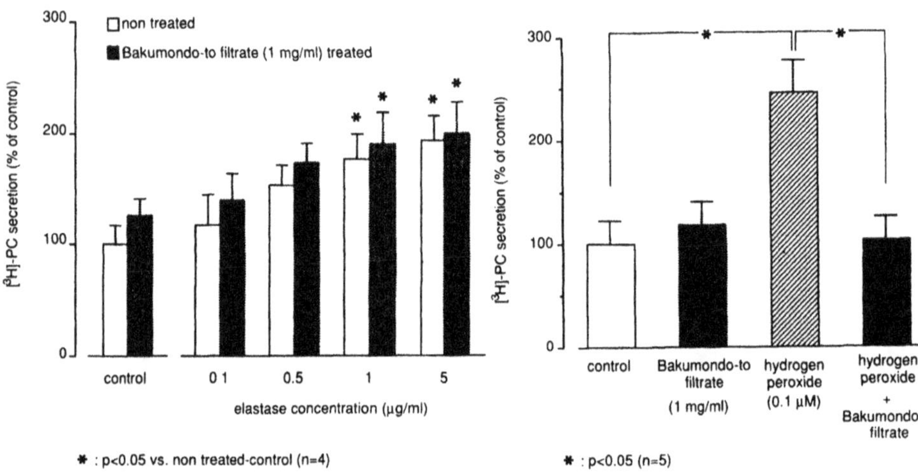

Figure 25 Effects of Bakumondo-to on elastase- and hydrogen peroxide-induced [^{3}H]-PC secretion from cultured alveolar type II cells.

3. Actions of Bakumondo-to on mRNA Expression of β Adrenoceptor in Alveolar Type II Cells

We have been interested in the similarity of the effects of Bakumondo-to and glucocorticoid on airway inflammatory responses. In our previous experiments, dexamethasone was found to increase both the phosphatidylcholine secretion and the intracellular cAMP levels in alveolar type II cells. Furthermore, dexamethosone

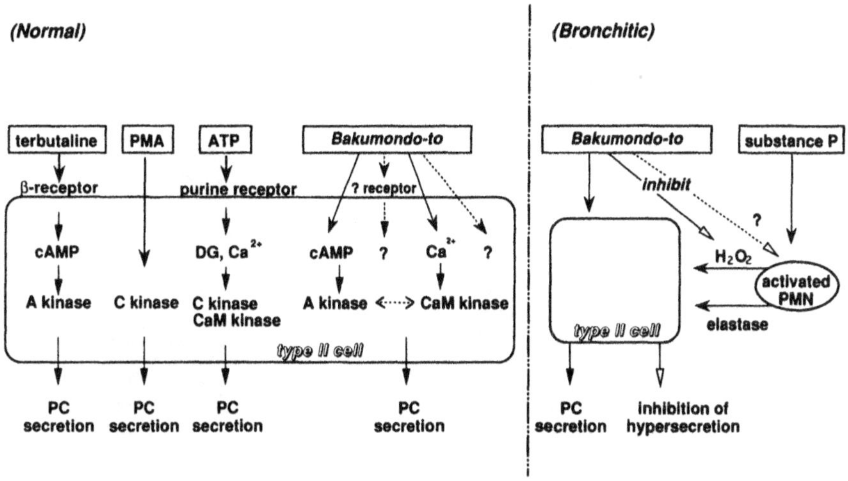

Figure 26 Characteristics and mechanisms of effects manifested by Bakumondo-to.

increased the number of β-adrenoceptors and the mRNA expression of β_2-adrenoceptors in alveolar type II cells. These findings indicated that dexamethasone increased pulmonary surfactant secretion through an enhancement of β_2-adrenoceptor gene expression (Isohama *et al.*, 1996).

Therefore, we also examined the effects of Bakumondo-to on the mRNA expression of β adrenoceptor in alveolar type II cells, using the method described previously (Isohama *et al.*, 1995, 1996). In control cells, the steady-state levels of mRNA for β_1- and β_2-adrenoceptors were 0.66 ± 0.06 and 1.81 ± 0.17 10^{-15} mol/total cellular RNA, respectively. The level of β_2-adrenoceptor mRNA was increased by the treatment with dexamethasone in a concentration-dependent manner. In addition, the increase in the β_2-adrenoceptor mRNA level produced by dexamethasone was significantly suppressed by the simultaneous application of RU-486, a glucocorticoid receptor antagonist. In contrast, the β_1-adrenoceptor mRNA level was not significantly affected by dexamethasone treatment. Unlike dexamethasone, the β_2-adrenoceptor mRNA level was not affected by treatment with Bakumondo-to filtrate. The β_1-adrenoceptor mRNA level, however, was significantly increased by the Bakumondo-to filtrate treatment (Figure 27). Although the mechanism of selective action on β_1-adrenoceptor gene expression in alveolar type II cells remains unclear, the effect may contribute to the effectiveness of Bakumondo-to on chronic airway diseases because both the β_1-and β_2-adrenoceptor subtypes mediate the secretion of phosphatidylcholine (Kai *et al.*, 1992).

4. FACILITATION OF BAKUMONDO-TO ON MUCOCILIARY TRANSPORT FUNCTION

Mucociliary transport of upper and lower airway is an important defense mechanism against inhaled particles, bacteria and viruses (Schuil *et al.*, 1994).

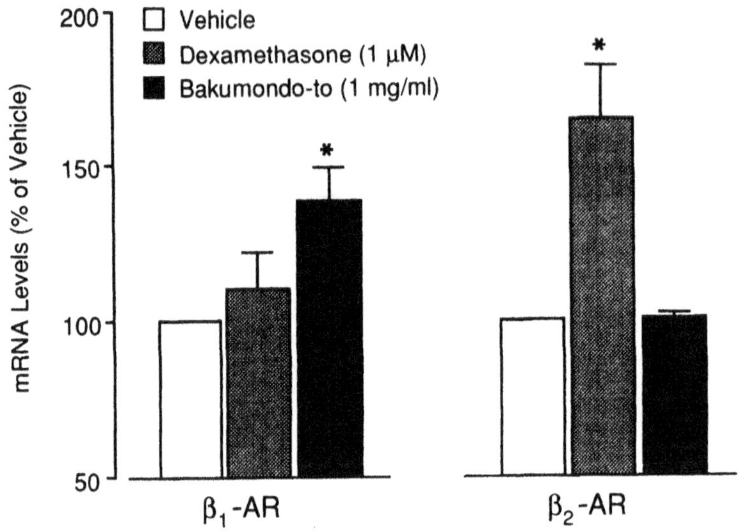

Figure 27 Effect of Bakumondo-to on β-adrenoceptor mRNA expression.

Mucociliary clearance may be impaired in patients with chronic obstructive pulmonary diseases, such as chronic bronchitis, chronic obstructive bronchiolitis and cystic fibrosis (Puchelle *et al.*, 1996). Decreased mucociliary transport function can be mainly attributed to abnormal rheological properties, i.e., an increase in viscosity of airway secretions. Massive accumulated macromolecules, i.e., airway mucus glycoproteins and DNA originated from infecting organisms, inflammatory cells and other exfoliated cells, may be responsible for the viscous purulent secretions (Hodson *et al.*, 1995; Lopez-vidriero and Rud, 1978). Especially, DNA, an extremely viscous polyanion, has been accepted to be present in very large amounts in purulent secretions, but only trace in normal airway secretions (Chernick *et al.*, 1959; Potter *et al.*, 1960).

We have found that DNA concentration-dependently increases the viscosity of airway secretions and significantly decreases MCTV in a dose-dependent manner (Tai *et al.*, submitted to Pharmacology Reviews and Communicasions).

Therefore, we studied the effect of Bakumondo-to on mucociliary transport impaired by DNA. In the study, we used quails as experimental animals. The reason for using quails is based on our histological findings and biochemical study of broncho-alveolar lavages on quails. The histological features of trachea and contents of bronchoalveolar lavage fluid are similar to airway inflammation states in humans (Miyata *et al.*, 1998). So far, we have examined the effects of many mucoactive drugs on mucociliary clearance in the birds (Kase *et al.*, 1977; Miyata *et al.*, 1990, 1978; Inatomi *et al.*, 1980; Miyata *et al.*, 1988, 1989; Kai *et al.*, 1990; Imai *et al.*, 1994; Tai *et al.*, 1996).

Bakumondo-to dose-dependently restored DNA-induced decrease in MCTV (Figure 28). Bakumondo-to may restore the DNA-induced decrease in MCTV through modulation of mucus and PC secretion and subsequent improvement of rheological properties.

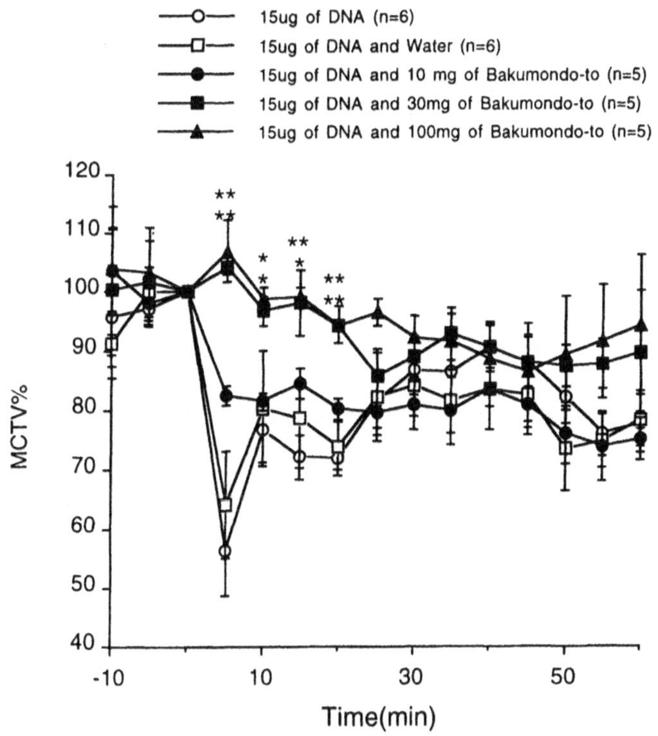

Figure 28 Effect of Bakumondo-to on tracheal mucociliary transport velocity in anaesthetized quails. 15 μg of DNA, which impairs mucociliary transport function, was topically adminstrated 45 min after the administration of Bakumondo-to or water. *: $p < 0.05$, **: $p < 0.01$.

SUMMARY AND CONCLUSION

There is increasing usage of traditional Chinese medicines in clinics and hospitals, because Chinese medicines tend to have moderate side effects and often exhibit remarkable efficacy. Traditional Chinese medicines may be useful drugs for the treatment of overall defects in airway disorders, because the medicines are composed of various herbs with weak but ubiquitous pharmacological activities.

We previously found that codeine was ineffective against both coughs associated with bronchitis and those increased by angiotensin-converting enzyme inhibitors, and that Bakumondo-to, a traditional Chinese medicinal prescription, had marked antitussive effects against codeine-resistant coughs. Since then, we have been investigating the characteristics of the antitussive and airway cleaning effects of this traditional medicine.

First, the effects of Bakumondo-to and the ophiopogonin extracted and purified from Bakumondo (*Ophiopogonis tuber*) on airway-inflammatory mediators, in particular tachykinin and its cleavage enzyme NEP, were investigated. The airway NEP-like activities in the bronchitic animals decreased to 1/8–1/9 of those in normal healthy animals, and treatment with Bakumondo-to for 7 days prevented the reduction of NEP-like activities. Bakumondo-to and ophiopogonin also suppressed tachykinin- and phosphoramidon (an NEP inhibitor)-induced coughing, while codeine was nearly ineffective.

In another experiment, Bakumondo-to stimulated the basic pulmonary surfactant secretion from cultured alveolar type II cells associated with increases in intracellular cyclic AMP and Ca^{2+} levels. It was also found that Bakumondo-to could increase β1 adrenoceptor gene expression in the cultured alveolar type II cells.

Further, a mixed culture of airway epithelial secretory cells and polymorphonuclear leucocytes (PMN) activated by substance P was devised as an *in vitro* model reflecting airway hypersecretion in the case of tachykinin-induced inflammation. The effects of Bakumonto-to in relation to secretion in this model were investigated.

The secretion enhancement of pulmonary surfactants from alveolar type II cells depended on the number of activated PMNs. Bakumondo-to significantly inhibited this secretion and completely inhibited the secretion enhancement of mucus from tracheal epithelial cells stimulated by PMNs. Elastase and hydrogen peroxide, major mediators derived from activated PMNs, stimulated secretion from both kinds of cells. Although it had no significant effects on the elastase-induced hypersecretion, Bakumondo-to significantly inhibited the hypersecretion by hydrogen peroxide.

Finally, Bakumondo-to was found to produce dose-dependent restoration on DNA-induced decrease in MCTV.

Various parameters (airway irritability, mucus production, surfactant phospholipids production, and mucociliary transport) are considered to be important for the dynamics of airway clearance. Pharmacological investigation with appropriate techniques on the ability of an agent to modify these parameters can give us useful information about its mechanism of action. However, since these parameters are closely interconnected, the mechanisms of action of mucoactive antitussive drugs are complicated. This means that the treatment goals, i.e., the facilitation and the control of the airway clearance, cannot always be achieved by the modification of a single parameter, but should more realistically be based on a general renormalization of several parameters. Bakumondo-to may be a good candidate for an ideal mucoactive because it consists of several active components that have various pharmacological effects.

It is therefore suggested that traditional Chinese medicinal prescriptions and their crude drug constituents are useful resources for the development of the next generation of pharmaceuticals contributing to the radical cure of chronic obstructive pulmonary diseases.

REFERENCES

A.S. Banner, Cough: physiology, evaluation, and treatment. *Lung.* **164**, 79–92 (1986).

A. Ohlen, A. Thureson-Klein, L. Lindbom, M.G. Persson and P. Hedqvist, Substance P activates leukocytes and platelets in rabbit microvessels. *Blood Vessels* **26**, 84–94 (1989).

B.K. Rubin, O. Ramirez and M. King, The role of mucus rheology and transport in neonatal respiratory distress syndrome and the effect of surfactant therapy. *Chest* **101**, 1080–1085 (1992a).

B.K. Rubin, A superficial view of mucus and the cystic fibrosis defect. *Pediatr. Pulmonol.* **13**, 4–5 (1992b).

C.J. Wiedermann, F.J. Wiedermann, A. Apperl, G. Kieselbach, G. Konwalinka and H. Braunsteiner, *In vitro* human polymorphonuclear leukocyte chemokinesis and human monocyte chemotaxis are different activities of aminoterminal and carboxyterminal substance P. Naunyn-schmiedeberg's *Arch. Pharmacol.* **340**, 185–190 (1989).

D.B. Borson, Roles of neutral endopeptidase in airways. *Am. J. Physiol.* **260**, 435–440 (1991).

D.F. Rogers, B. Aursudkij and P.J. Barnes Effects of tackykinins on mucus secretion in human bronchi in vitro. *Eur. J. Pharmacol.* **174**, 283–286 (1989).

E. Puchelle, J.M. Zahm, S.D. Bentzmann, C. Grosskopf, S. Shak, D. Mougel and J.M. Polu, Effects of rhDNase on purulent airway secretions in chronic bronchitis. *Eur. Respir. J.* **9**, 765–769 (1985).

G.T. De Sanctis, R.P. Tomkiewicz, B. K. Rubin, S. Schurch and M. King, Exogenous surfactant enhances mucociliary clearance in the anesthetized dog. *Eur. Respir. J.* **7**, 1616–1621 (1994).

H.M. Coleridge, J.C.G. Coleridge, A. Ginzel, D.G. Baker, R.B. Banzett and H.M. Morrison, Stimulation of 'irritant' receptors and afferent C-fiber in the lungs by prostaglandins. *Nature* **264**, 451–453 (1976).

H.M. Widdicombe, Receptors in the trachea and bronchi of the cat. *J. Physiol.* **123**, 415–453 (1954).

H. Kai, M. Saito, K. Furusawa, Y. Oda, Y. Okano, K. Takahama and T. Miyata, Protective effect of surface-active phospholipids against the acid-inducing inhibition of the tracheal mucociliary transport. *Jpn. J. Pharmacol.* **49**, 375–380 (1989).

H. Kai, S. Yamamoto, K. Takahama and T. Miyata, Influence of corticosterone on tracheal mucociliary transport in pigeons. *Jpn. J. Pharmacol.* **52**, 496–499 (1990).

H. Kai, Y. Isohama, K. Takaki, Y. Oda, K. Murahara, K. Takahama and T. Miyata, Both β_1- and β_2-adrenoceptors are involved in mediating phosphatidylcholine secretion in rat type II pneumocyte cultures. *Eur. J. Pharmacol.* **212**, 101–103 (1992).

H. Kai, K. Yoshitake, Y. Isohama, I. Hamamura, K. Takahama and T. Miyata, Involvement of protein kinase C in mucus secretion by hamster tracheal epithelial cells in culture. *Am. J. Physiol.* **267**, L526–L530 (1994).

H. Kai, K. Yoshitake, A. Hisatsune, T. Kido, Y. Isohama, K. Takahama and T. Miyata, Dexamethasone suppresses mucus production and MU-2 and MUC-5AC gene expression by NCI-H292 cells. *Am. J. Physiol.* **271** (*Lung cell. Mol. Physiol.* **15**), L484–488 (1996a).

H. Kai, K. Murahara, Y. Isohama, K. Takahama, Y. Oda, I. Hamamura, K. Yoshitake and T. Miyata, 4-Aminopyridine stimulates phosphatidylcholine secretion in primary cultures of rat type II pneumocytes. *J. Pharm. Pharmacol.* **48**, 53–56 (1996b).

H. Kohrogi, P.D. Graf, K. Sekizawa, D.B. Borson and J.A. Nadel, Neutral endopeptidase inhibitor potentiate substance P- and capsaicin-induced cough in awake guinea-pigs. *J. Clin. Invest.* **82**, 2063–2068 (1988).

H. Kohrogi, J.A. Nadel, B. Malfroy, C. Gorman, R. Bridenbaugh, J.S. Patton and D.B. Borson, Recombinant human enkephalinase (neutral endopeptidase) prevents cough induced by tachykinins in awake guinea-pigs. *J. Clin. Invest.* **84**, 781–786 (1989).

J.C.G. Coleridge and H.M. Coleridge. Afferent vagal C-fibre innervation of the lungs and airways and its functional significance. *Rev. Physiol. Biochem. Pharmacol.* **99**, 1–110 (1984).

J.D. Lundgren, C.J. Wiedermann, C. Logun, J. Plutchok, M. Kaliner and J.H. Shelhamer, Substance P receptor-mediated secretion of respiratory glycoconjugate from feline airway in vitro. *Exp. Lung Res.* **15**, 17–29 (1989).

J. Fuchikami, H. Kai, K. Takahama and T. Miyata, Comparative study of the antitussive activity of Mai-Men-Dong-Tang and codeine in normal and bronchitic guinea-pigs. *Pharmacodynamics & Therapeutics* (Life Sci. Adv.) 2–9, 37–43 (1990).

J. Potter, L.W. Matthews, J. Lemm, S. Spector and W. Reserve, The composition of pulmonary secretions from patients with and without cystic fibrosis. *Am. J. Dis. Child.* **100**, 493–495 (1960).

J.R. Wright and L.G. Dobbs, Regulation of pulmonary surfactant secretion and clearance. *Annu. Rev. Physiol.* **53**, 395–414 (1991).

K.C. Kim, J.I. Rearick, P. Nettesheim and A.M. Jetten, Biochemical characterization of mucous glycoproteins synthesizes and secreted by hamster tracheal cells in primary culture. *J. Biol. Chem.* **260**, 4021–4027 (1985).

K.C. Kim, K. Wasano, R.M. Niles, J.E. Schuster, P.J. Stone and J.S. Brody, Human neutrophil elastase releases cell surface mucins from primary cultures of hamster tracheal epithelial cells. *Proc. Natl. Acad. Sci. USA* **84**, 9304–9308 (1987).

K.C. Kim, Biochemistry and pharmacology of mucin-like glycoproteins produced by cultured airway epithelial cells. *Exp. Lung Res.* **17**, 533–545 (1991).

K.P. Keenan, J.W. Combs and E.M. McDowell, Regeneration of hamster tracheal epithelium after mechanical injury. I. Focal lesions: quantitative morphologic study of cell proliferation. *Vichows Arch. B. Cell. Pathol.* **41**, 193–214 (1982).

K. Miyamoto, K. Fukusawa, A. Kuroiwa, T. Saito, T. Miyata and T. Furukawa, Effects of Qing-Fei-Tang on the Airway Inflammation and Clearance. *Am. J. Chin. Med.* **18**, 5–18 (1990).

K. Morgenroth and J. Bolz, Morphological features of the interaction between mucus and surfactant on the bronchial mucosa. *Respiration* **47**, 225–231 (1985).

K. Takahama, J. Fuchikami, Y. Isohama, H. Kai and T. Miyata, Neurokinin A but not neurokinin B and substance P, induces codeine-resistant coughs in awake guinea-pigs. *Regulatory Peptides* **46**, 236–237 (1993).

K. Takahama, J. Fuchikami, H. Kai, Y. Isohama and T. Miyata, Inhalation of phosphoramidon, a neutral endopeptidase inhibitor, induces coughs in awake guinea-pigs. *Arch. Int. Pharmacodyn. Ther.* **330**, 241–250 (1995).

K. Yoshitake, Y. Isohama, H. Kai, K. Takahama and T. Miyata, Activated polymorphonuclear leucocytes stimulates the loss of cell-associated high molecular weight glycoconjugates from hamster tracheal epithelial cells in culture. *Pharmac. Sci.* **1**, 475–478 (1995a).

K. Yoshitake, Y. Isohama, H. Kai, K. Takahama and T. Miyata, Substance P stimulates the loss of cell-associated high molecular glycoconjugates from cultured hamster tracheal epithelial cells through polymorphonuclear leucocytes activation. *Biochem. Mol. Biol. Int.* **36**, 1009–1016 (1995b).

L. Allegra, R. Bossi and P.C. Braga, Influence of surfactant of mucociliary transport. *Eur. J. Respir. Dis.* **67**, 71–76 (1985).

M.C. Michel, J.K. Knapp and H. Ratjen, Sensitization by dexamethasone of lymphocyte cyclic AMP formation: evidence for increased function of the adenylyl cyclase. *J. Pharmacol.* **113**, 240–246 (1994).

M.C. Rose, Mucins: structure, function, and role in pulmonary diseases. *Am. J. Physiol.* **263** (*Lung Cell. Mol. Physiol.* **7**), L413–429 (1992).

M.C. Serra, F. Bzzoni, B.V. Della, M. Greskowiak and F. Rossi, Activation of human neutrophils by substance P: effects on oxidative metabolism, exocytosis, cytosolic Ca^{2+} concentration and inositol phosphate formation. *J. Immunol.* **141**, 2118–2124 (1988).

M.E. Hodson, Aerosolized dornase alfa (rhDNase) for therapy of cystic fibrosis. *Am J Respir Crit Care Med.* **151**, S70–S74 (1995).

M.P. Kaufman, H.M. Coleridge, J.C.G. Coleridge and D.G. Baker, Bradykinin stimulates afferent vagal C-fibres in intrapulmonary airways of dogs. *J. Appl. Physio.* **48**, 511–517 (1980).

M.T. Lopez-vidriero and L. Reid, Chemical markers of mucous and serum glycoproteins and their relation to viscocity in mucoid and purulent sputum from various hypersecretory diseases. *Am. Rev. Respir. Dis.* **117**, 465–477 (1978).

M. Okumura, M. Tsuruoka, Y. Isohama, H. Kai, K. Takahama and T. Miyata, Activated eosinophils stimulate phosphatidylcholine secretion in primary culture of rat type II pneumocytes. *Biochem. Mol. Biol. Int.* **38**, 571–577 (1996).

N. Inatomi, H. Kuriki, M. Kanno, K. Kawagoe, Y. Nagawa, T. Miyata, K. Takahama and Y. Kase, Pharmacological studies of *trans*-6-hydroxy-5-hydroxymethyl-2-isopropylamino-1,2,3,4-tetrahydro-1-naphthalenol hydrochloride (AA-497), a new ptent bronchodilator. *Arzneim-Forsch/Drug Res.* **30**, 276–285 (1980).

P.C. Braga, I. Ziment and L. Allegra, Classification of agents that act on bronchial mucus. In *Drugs in bronchial mucology*, eds. P.C. Braga and L. Allegra, Raven Press, pp. 59–76 (1989).

P. Gehr, S. Schurch, Y. Berthiaume, Im Hof, V. and Geiser, M. Particle retention in airways by surfactant. *J. Aerosol Med.* **3**, 27–43 (1990).

P.J. Schuil, J.M.E. van Gelder, M. ten Berge, K. Graamans and E.H. Huizing, Histamine and leukotriene C4 effects on *in vitro* ciliary beat frequency of human upper respiratory cilia. *Eur. Arch. Otorhinolaryngol.* **251**, 325–328 (1994).

P.J. Umeno, J.A. Nadel and D.M. McDonald, Neurogenic inflammation of the rat trachea: fate of neutrophils that adhere to venules. *J. Appl. Physiol.* **69**, 2131–2136 (1990).

R.R. Mercer, M.L. Russell and J.D. Crapo, Mucous lining layers in human and rat airways. *Am. Rev. Respir. Dis.* **145**, A355 (1992).

R.W. Fuller, C.S.M. Dixon and P.J. Barnes, Bronchoconstrictior response to inhaled capsaicin in humans. *J. Appl. Physio.* **58**, 1080–1084 (1985).

R.W. Fuller, C.S.M. Dixon, F.M.C. Cuss and P.J. Barnes, Bradykinin-induced bronchoconstriction in humans. *Am. Rev. Respir. Dis.* **135**, 176–180 (1987).

S. Girod, C. Galabert and D. Pierrot, Role of phospholipid lining on respiratory mucus clearance by cough. *J. Appl. Physiol.* **71**, 2262–2266 (1991).

S. Schurch, P. Gehr, V. Im Hof, M. Geiser and F.H.Y. Green, Surfactant displaces particles toward the epithelium in airways and alveoli. *Respir. Physiol.* **80**, 17–32 (1990).

S. Soseko and Y. Kaneko, Cough associated with the use of captopril. *Arch. Intern. Med.* **145**, 1524 (1985).

S. Tai, H. Kai, Y. Isohama, K. Takahama and T. Miyata, Effect of leukotriene D4 on tracheal mucociliary transport velocity in quails. *Jpn. J. Pharmacol.* **70**, 195–197 (1996).

T. Imai, H. Kai, K. Takahama, T. Miyata, J. Hiroi, K. Shimomura and M. Kohsaka, Effects of anovel orrally-active antiallergic drug, quinotolast (FK021), on airway clearance. *Folia Pharmacol. Jpn.* **104**, 347–355 (1994).

T. Miyata, K. Takahama, A. Hasanat, K. Ikegami, K. Iwasaki, Y. Okano and Y. Kase, Action of 4-amono-α-[(*tert*-butylamino) methyl]-3,5-dichlorobenzylalchol hydrochloride (N-AB 365, clenbuterol) on the respiratory system. *Folia Pharmacol. Japon.* **74**, 573–588 (1978).

T. Miyata, H. Kai, K. Furusawa, H. Nakamura, H. Saitou, Y. Okano and K. Takahama, Secretomotor and mucolytic effects of mabutenol, a new bronchodilator. *Arch. Int. Pharmacodyn.* **288**, 147–159 (1987).

T. Miyata, N. Matsumoto, H. Yuki, K. Takahama, Y. Okano and H. Kai, The effect of N-acetylneuramic acid on the mucociliary transport impaired by cigarette smoke. *Arch. Int. Pharmacodyn.* **296**, 202–209 (1988).

T. Miyata, N. Matsumoto, H. Yuki, Y. Oda, K. Takahama and H. Kai, Effects of anticholinergic bronchodilators on mucociliary transport and airway secretion. *Jpn. J. Pharmacol.* **51**, 11–15 (1989a).

T. Miyata, J. Fuchikami, H. Kai and K. Takahama, Comparison of antitussive effects of Bakumondo-to and codeine in bronchitic animals. Japan. *J. Thoracic. Disease.* **27**, 1157–1162 (1989).

T. Miyata, T. Ishii, N. Sugiyama, Y. Okano, N. Nishi, K. Takahama, S. Ogasawara, Y. Oda, K. Yokoyama, Y. Murata and H. Kai, Effect of N—acetylneuraminic acid on respiratory tract secretion and inflammation in the bronchitic rabbit. *Arch. Int. Pharmacodyn. Ther.* **304**, 277–289 (1990).

T. Miyata, Pathopharmacological evaluation of Bakumondo-to (Mai men dong tang) as a curative for chronic inflammatory airway disease – A new concept. *Kampo and Immuno-allergy.* **9**, 122–144 (1995).

T. Miyata, H. Kai, Y. Isohama and K. Takahama, Current status of mucus drug research: Strategies and Problems. *Eur. Respir. Rev.* **11**, 480–491 (1996).

W.A. Marasco, H.J. Showell and E.L. Becker, Substance P binds to the formylpeptide chemotaxis receptor on the rabbit neutrophil. Biochem. *Biophys. Res. Commun.* **99**, 1065–1072 (1981).

W.S. Chernick and Barbero, G.J. Composition of tracheobronchial secretion in cystic fibrosis of the pancreas and bronchietasis. *Pediatrics* **24**, 739–745 (1995).

Y. Isohama, T. Matsuo, H. Kai, K. Takahama and T. Miyata, Changes in β_1- and β_2-adrenoceptor mRNA levels in alveolar type II cells during cultivation. *Biochem. Mol. Biol. Int.* **36**, 561–568 (1995).

Y. Isohama, Y. Kumanda, K. Tanaka, H. Kai, K. Takahama and T. Miyata, Dexamethasone increases β_2-adrenoceptor-regulated phosphatidylcholine secretion in rat alveolar type II cells. *Jpn. J. Pharmacol.* (In press) (1996).

Y. Kase, T. Yakushiji, H. Seo, M. Sakata, G. Kito, K. Takahama and T. Miyata, Pharmacological studies on expectorants. *Folia Pharmacol. Japon.* **73**, 605–624 (1977).

Y. Oda, H. Kai, Y. Isohama, K. Takahama and T. Miyata, Stimulation of pulmonary surfactant secretion by activating neutrophils in rat type II pneumocytes culture. *Life Sci.* **49**, 803–811 (1991).

10 EFFECTS OF TRADITIONAL CHINESE MEDICINES ON BLOOD RHEOLOGY

KAZUO TORIIZUKA[1] and KATSUTOSHI TERASAWA[2]

[1]*Research Division, Oriental Medicine Research Center, The Kitasato Institute 5-9-1 Shirokane, Minato-ku, Tokyo 108-0072, Japan*
[2]*Department of Japanese Oriental Medicine, Faculty of Medicine, Toyama Medical and Pharmaceutical University, 2630 Sugitani, Toyama 930-0194, Japan*

INTRODUCTION

The population of elderly people in Japan has continued to increase over the past several decades. According to the reports of the Japanese Institute of Population Problems, the proportion of the population in Japan aged 65 or older is forecasted to be 25% in the year 2014, and the population aged 75 or older will be radically higher than it is today. Moreover, the structure of diseases has been changing; there is an increase in disorders brought on by stress, cancer, intractable infections and circulatory dysfunction.

This chapter addresses the recent research into the traditional Kampo treatments for peripheral circulation disorders and "oketsu" syndrome. Oketsu syndrome is considered to be blood stasis or a stagnant blood syndrome, which is one of the patho-physiological concepts existing only in Kampo medicine, and is closely related to the microcirculatory diseases. This chapter also addresses the current hemorheological biochemical studies of oketsu syndrome, and the role of Kampo medicine as treatment for microcirculatory disorders.

KAMPO TREATMENT FOR PERIPHERAL CIRCULATION DISORDERS

Peripheral circulation disorders occur in various types of diseases. These pathological conditions can occur in any region of the body, and may manifest as chilliness of the palm and foot (chillphobia), numbness, and Raynaud's syndrome. Of the peripheral circulation diseases, the disorder of artery vessels (arterial thrombosis) is classified as arteriosclerosis, and thromboangiitis obliterans (Buerger's disease) and the disorders of vein vessels are classified as venous thrombosis. The symptoms of sensitivity to cold and chilblains are correlated with peripheral circulation disorders. Kikutani[1] classified the frequently used Kampo formulas for treating the peripheral circulation diseases and/or sensitivity to the cold according to their traditional use, as shown in Table 1.

In general, Kampo medicine is recognized to be effective in the treatment of sensitivity to cold, which can be caused by hypotension or menopause. Peripheral

Table 1 Ethical Kampo formulations, applied to Health Insurance System, for treating the peripheral circulation disorders and 'Hie-sho" (Chillphobia) from Kikutani (1990).[1]

Peripheral Circulation Disorders	'Hie-sho" (Chillphobia)
Sokei-kakketsu-to	Kami-shoyo-san (Jia-wei-xiao-yao-san)
(Shu-jing-huo-xie-tang)	Keishi-bukuryo-gan
thromboangitis obliterans	(Gui-zhi-fu-ling-wan)
	Ryo-kyo-jutsu-kan-to
Toki-shigyaku-ka-goshuyu-shokyo-to	(Ling-jiang-shu-gan-tang)
(Dang-bui-si-ni-jia-wu-zhu-yu-sheng-jiang-tang)	coldness of the lower half of body
frostbite	Shimotsu-to (Si-wu-tang)
Toki-shakuyaku-san (Dang-gui-shao-yao-san)	Shimbu-to (Zhen-wu-tang)
frosbite	Toki-shigyaku-ka-goshuyu-shokyo-to
Unkei-to (Wen-jing-tang)	(Dang-bui-si-ni-jia-wu-zhu-yu-sheng
chilblains	-jiang-tang)
	Toki-shakuyaku-san
	(Dang gui shao-yao-san)

disorders for which Kampo treatment are generally suitable are functional abnormalities where subjective symptoms are predominant. For example, the complaints with menopause are mainly subjective, without objective signs. Kampo medicines are suitable for such symptoms. However, many clinical reports have indicated that Kampo medicines were applicable to the following peripheral microcirculatory disorders.

Thromboangitis Arteriosclerosis

According to *Kampo Shinryo Iten* (ed. by Otsuka *et al.*[2]), Hachimi-jio-gan (Ba-wei-di-huang-wan: *Pilulae octo-medicamentorum rehmanniae*) has therapeutic efficacy for patients with intermittent lameness or claudication. Hoin-to (Bu-yin-tang) is also effective.

Thromboangitis Obliterans (Buerger's Disease)

Kampo Shinryo Iten (Otsuka) describes notable cases of patients with Buerger's disease treated with Teito-to (Di-dang-tang), Daisaiko-to-go-tokakujyoki-to (Da-chai-hu-tang-he-tao-he-cheng-qi-tang: *Formula bupleuri major cum Formula animationis persicae*), and Keishi-bukuryo-gan (Gui-zhi-fu-ling-wan: *Pilulae hoelen et cinnamomi*). Takayama[3] reported that toes with necrosis recovered after the patients were treated with Toki-shigyakuka-goshuyu-shokyo-to (Dang-gui-si-ni-jia-wu-zhu-yu-sheng-jiang-tang: *Formula quadri-frigorium angelicae cum zingibere et evodia*). Osawa[4] suggested that Toki-shigyakuka-goshuyu-shokyo-to, Keishi-kyo-shakuyaku-ka-mao-bushi-saishin-to (Gui-zhi-qu-shao-yao-jia-ma-huang-fu-zi-xi-xin-tang: *Formula cinnamonic exuo paeonia cum asiasari aconiti ephedrae*), and Sokei-kakketsu-to (Shu-jing-huo-xie-tang: *Formula augmentationis orbis et liberationis cruoris*) were useful Kampo formulas for Buerger's disease.

Raynaud's Disease (Raynaud's Syndrome)

Toki-shigyakuka-goshuyu-shokyo-to and Toki-nentsu-to (Dang-gui-zhan-tong-tang) are listed in Kampo Shinryo Iten (Otsuka) as suitable formulations for Raynaud's disease. Iwata[5] reported a comparative study of Western medicines and Kampo treatment for Raynaud's syndrome caused by chainsaw use. Twenty-one pairs of patients matched as to sex, age, history of chainsaw use and treatment period were divided into two groups; one received ordinary Western medicines and the other received Toki-shigyakuka-goshuyu-shokyo-to. After 8 weeks of Kampo medication, significant improvement was observed in subjective symptoms (chilliness of the palm and foot, and palsy of the fingers). Takaguchi[6] used Toki-shigyakuka-goshuyu-shokyo-to, Toki-shakuyaku-san (Dang-gui-shao-yao-san: *Pulvis paeoniae et angelicae*), and Shinbu-to (Zhen-wu-tang: *Formula divinitatis nigrae*) for 17 patients with peripheral microcirculatory diseases, and observed good improvement in 6 cases, improvement in 9 cases, and no change in 2 cases. The conditions of patients with Raynaud's syndrome administered Ninjin-yoei-to (Ren-shen-yang-rong-tang: *Formula ginseng compositae*) for 8 to 12 weeks were eventually improved as shown by the peripheral blood circulation analyzed by sphygmotachograph (digital plethysmograph).[7]

Thrombophlebitis, Phlebitis: Puerperal Lower Extremity Thrombosis

Yakazu[8] and Akiba[9] reported that Keishi-bukuryo-gan had therapeutic efficacy for patients with puerperal lower extremity thrombosis. Kanda[10] observed the clinical efficacy of Keishi-bukuryo-gan for varicose veins of the spermatic cord.

The Autonomic Disorders Associated with Central and Peripheral Neuron Diseases

Patients with spinocerebellar degeneration, amyotropic lateral sclerosis, and sequella of cerebrovascular disease sometimes complain of peripheral coldness and/or coldness of the lower half of the body. A decrease in skin temperature is recognized as a sign of autonomic disorder in nervous diseases. Terasawa *et al.*[11] classified the symptoms of coldness into four types (A–D) by means of the patho-physiological concepts of Kampo medicine. Type A is peripheral coldness associated with lowered body temperature. This type of coldness is suitable for Kampo prescriptions containing Bushi (*Aconite Tuber*) such as Shigyaku-to (Si-ni-tang), Bukuryo-shigyaku-to (Fu-ling-si-ni-tang), Shinbu-to and Bushi-to (Fu-zi-tang). Type A coldness sometimes accompanies hypothyroidism or hypofunction of the adrenal cortex. Type B coldness is localized in the periphery but body temperature is normal. Toki-shakuyaku-san, Ninjin-to (Ren-shen-tang), Toki-shigyakuka-goshuyu-shokyo-to, Kamishoyo-san (Jia-wei-xiao-yao-san: *Pulvis ambulatorius compositae*), Tokakujyoki-to (Tao-he-cheng-qi-tang: *Formula animationis persicae*), and Keishi-bukuryo-gan are generally used for treating type B coldness. Type B is thought to be caused by vasomotor hypofunction. Type C is characterized as coldness of the lower half of the body. The Kampo treatments for type C coldness are Ryokyo-jutsukan-to

(Ling-jiang-shu-gan-tang: *Formula glycyrrhizae atractylodis zingiberis*), Hachimi-gio-gan, Goshaku-san (Wu-ji-san: *Pulvis quinque-hystericorum*), Kamishoyo-san, Tokakujyoki-to, and Keishi-bukuryo-gan. Type D is subjective and objective hemi-coldness of the body, and is thought to be caused by disorders of the central nervous system. Clinical improvements in Type D cases have been observed after the administration of Toki-shigyakuka-goshuyu-shokyo-to, Shigyaku-san and Toki-nentsu-to. To elucidate the effects of Toki-shigyakuka-goshuyu-shokyo-to and Toki (*Angelicae Radix*), which is the main ingredient of its formulation, the acute effects of both herbal medicines were undertaken by measuring deep body temperature (core temperature) by both thermography and plethysmography.[11] The results showed that both Toki-shigyakuka-goshuyu-shokyo-to and Toki (*Angelicae Radix*) improved peripheral circulation in Charcot-marie-Tooth disease patients. The motor nerve conduction velocity of the pretibial nerve gained about 7 m/sec. Clinically, the activity of patients' daily living was improved even during the winter. In one case, the tendency to get chilblains was diminished. The concept of peripheral microcirculation disorders and "hie-sho" (chillphobia) is closely correlated to the oketsu syndrome. (In China, the term "ketsuo" is commonly used instead of the term "oketsu." In Japan the term "ketsuo" is rarely used. Both terms have the same meaning.) In traditional Chinese medicine, "ketsu" (blood) means human red body fluid containing the life energy ("Ki") circulating in the body. A pathological state induced by circulatory failure of this ketsu is called oketsu. Although the definition of oketsu syndrome in terms of Western medicine is difficult, since it is a general name that could be given to a wide range of diseases, it has been recognized clinically as blood stagnation or disorders in peripheral microcirculation. A number of investigators have recently been engaged in studies to elucidate the rheological and hematological aspects of the blood in oketsu syndrome.

RHEOLOGICAL STUDIES OF "OKETSU" SYNDROME

"Oketsu" is one of the patho-physiological concepts. The concept is defined by the five human senses; therefore it has remained difficult to define all aspects of the term by modern scientific methodology. However, concerning blood circulation, the pathological condition of "oketsu" can be regarded as the actual pathological conditions decreasing the cardiac output, changing the function and structure of blood vessels, changing the viscosity of blood, changing the level of coagulation and fibrinolysis, and changing the aggregation ability of platelets. Several investigators are currently studying the factors related to the pathology of "oketsu." Their findings have indicated that the "oketsu" state corresponds to disorders of the following microcirculatory parameters.

Blood Viscosity

Arichi *et al.*[12] observed that the blood viscosity of Tokaku-jyoki-to type patients was high. Terasawa *et al.*[13] presented the diagnostic criteria for "oketsu" syndrome as the results of a multiple regression analyis of over 200 patients' physical data, and proposed an oketsu "score" (Table 2).

According to the criteria, 82 patients were classified into three groups: non-oketsu, mildly affected and severely affected with oketsu syndrome. Blood was withdrawn from the cubital vein into a plastic syringe with EDTA-2K, and viscosities of whole blood and plasma were measured by a cone-plate rotational viscometer. The measurements were carried out at a temperature of 37°C and a cone angle of 1°37′. For the measurement of whole blood viscosity, blood samples were measured at five different points of shear rates ($\dot{\gamma}$) (19.2, 38.4, 76.8, 192.0, 384.0 sec^{-1}). The blood viscosity values obtained were corrected by a regression equation to values at 45% of hematocrit and then compared among the three groups. The results indicated that both whole blood and plasma viscosity were significantly elevated in the patients with oketsu syndrome (Table 3).

As for the rheological aspects of the blood, the blood is considered to be non-Newtonian fluid, i.e., not a simple liquid but a suspension consisting of blood cells and plasma. This view indicates that shear stress changes depend on the shear rate in the microcirculation. In order to elucidate the rheometric characteristics of blood as a non-Newtonian fluid, Casson's equation was employed[14]. Casson's equation is expressed as follows :

$$\sqrt{\tau} = \sqrt{\tau_f} + \sqrt{\eta_c} \cdot \sqrt{\dot{\gamma}}$$

τ : shear stress
η_c : Casson viscosity
$\dot{\gamma}$: shear rate
τ_f : Casson yeild stress

Table 2 Diagnostic criteria for Oketsu syndrome. From Terasawa *et al.* (1986)[13]

Symptoms	Score		Symptoms	Score	
	Male	Female		Male	Female
Dark shade around the eyes	10	10	Tenderness of left navel region	5	5
Pigmentation over the face	2	2	Tenderness of right navel region	10	10
Rough skin	2	5	Tenderness under the navel region	5	5
Purple discoloration of lips	2	2	Tenderness of iliocoecal region	5	2
Purple discoloration of gums	10	5	Tenderness of hypochondrial region	5	5
Purple discoloration of tongue	10	10	Tenderness of sigmoid region	5	5
Telangiectasis, vascular spider	5	5	Hemorrhoids	10	5
Susceptibility to subcutaneous Bleeding	2	10	Dysmenorrhea		10
Redness of palms, palmar Erythema	2	5			

Evaluation 20 points or less: non-oketsu state.
21–39 points: mildly affected oketsu state.
40 points and above: severely affected oketsu state.

Table 3 Comparison of corrected values of blood viscosity and plasma viscosity. From Terasawa *et al.* (1986)[13]

	Non-oketsu group	Mildly affected group	Severely affected group	
Blood viscosity[#1]	N = 26	N = 31	N = 35	
Shear rate				
19.2 sec^{-1}	6.89 ± 0.38	7.11 ± 0.92	7.49 ± 1.17*	(cP)
76.8 sec^{-1}	4.95 ± 0.23	5.30 ± 0.55 **	5.32 ± 0.54 **	(cP)
384.0 sec^{-1}	3.94 ± 0.21	4.15 ± 0.42 *	4.14 ± 0.36*	(cP)
Plasma viscosity[#2]	N = 15	N = 28	N = 29	
	1.35 ± 0.05	1.49 ± 0.16 **	1.49 ± 0.15**	(cP)

The values are expressed as mean and standard deviation.
The asterisks indicate a significant difference from the non-oketsu group;
* $p < 0.05$, ** $p < 0.01$
[#1]; corrected to standard hematocrit of 45%, measured at various shear rates.
[#2]; measured at 384.0 sec^{-1}.

The shear stress of each subject was calculated by using the blood viscosity value at five points of shear rate. The result obtained from each subject showed a good correlation with Casson's equation. Then, by using the approximation method, both the Casson viscosity and Casson yield stress were estimated. This further indicated that the rheometric characteristics of the blood can be explained by two parameters, i.e., Casson viscosity and Casson yield stress. The results of our study revealed that both Casson viscosity and Casson yield stress were significantly elevated in the patients with oketsu syndrome (Table 4).

Casson yield stress is thought to estimate the initiation of blood flow from stasis. Blood is considered an elastic body. An elastic body will start to flow when the shear

Table 4 Comparison of corrected values of Casson viscosity and Casson yield stress. From Terasawa *et al.* (1986).[13]

	Non-oketsu group	Mildly affected group	Severely affected group	
	N = 26	N = 31	N = 35	
Corrected Casson viscosity	3.21 ± 0.19	3.44 ± 0.35***	3.36 ± 0.28**	(cP)
Corrected Casson yield stress	0.141 ± 0.014	0.138 ± 0.023	0.159 ± 0.048*	(dyn/cm^2)

The values are expressed as mean and standard deviation.
The asterisks indicate significant a difference from the non-oketsu group;
* $p < 0.1$, ** $p < 0.05$, *** $p < 0.01$
* also means $p < 0.05$ from the mildly affected group.

high. Terasawa *et al.*[13] presented the diagnostic criteria for "oketsu" syndrome as the results of a multiple regression analyis of over 200 patients' physical data, and proposed an oketsu "score" (Table 2).

According to the criteria, 82 patients were classified into three groups: non-oketsu, mildly affected and severely affected with oketsu syndrome. Blood was withdrawn from the cubital vein into a plastic syringe with EDTA-2K, and viscosities of whole blood and plasma were measured by a cone-plate rotational viscometer. The measurements were carried out at a temperature of 37°C and a cone angle of 1°37′. For the measurement of whole blood viscosity, blood samples were measured at five different points of shear rates ($\dot{\gamma}$) (19.2, 38.4, 76.8, 192.0, 384.0 sec^{-1}). The blood viscosity values obtained were corrected by a regression equation to values at 45% of hematocrit and then compared among the three groups. The results indicated that both whole blood and plasma viscosity were significantly elevated in the patients with oketsu syndrome (Table 3).

As for the rheological aspects of the blood, the blood is considered to be non-Newtonian fluid, i.e., not a simple liquid but a suspension consisting of blood cells and plasma. This view indicates that shear stress changes depend on the shear rate in the microcirculation. In order to elucidate the rheometric characteristics of blood as a non-Newtonian fluid, Casson's equation was employed[14]. Casson's equation is expressed as follows :

$$\sqrt{\tau} = \sqrt{\tau_f} + \sqrt{\eta_c} \cdot \sqrt{\dot{\gamma}}$$

τ : shear stress
η_c : Casson viscosity
$\dot{\gamma}$: shear rate
τ_f : Casson yeild stress

Table 2 Diagnostic criteria for Oketsu syndrome. From Terasawa *et al.* (1986)[13]

Symptoms	Score		Symptoms	Score	
	Male	Female		Male	Female
Dark shade around the eyes	10	10	Tenderness of left navel region	5	5
Pigmentation over the face	2	2	Tenderness of right navel region	10	10
Rough skin	2	5	Tenderness under the navel region	5	5
Purple discoloration of lips	2	2			
Purple discoloration of gums	10	5	Tenderness of iliocoecal region	5	2
Purple discoloration of tongue	10	10	Tenderness of hypochondrial region	5	5
Telangiectasis, vascular spider	5	5			
Susceptibility to subcutaneous Bleeding	2	10	Tenderness of sigmoid region	5	5
			Hemorrhoids	10	5
Redness of palms, palmar Erythema	2	5	Dysmenorrhea		10

Evaluation 20 points or less: non-oketsu state.
21–39 points: mildly affected oketsu state.
40 points and above: severely affected oketsu state.

20 patients and 35 volunteers were evaluated with a video-microscope system.[19] (Figure 1)

The patients and volunteers were classified into three groups according to oketsu score: non-oketsu, mildly affected and severely affected with oketsu syndrome. The results showed that the intravascular erythrocyte aggregation associated with sludging and/or stasis of blood was significantly correlated with the oketsu syndrome. This syndrome was also correlated with such architectural changes as meandering, narrowing, microaneusma and proliferation of microvessels. These abnormal findings in bulber conjunctiva have been reported in arteriosclerotic and diabetic disorders; however, this investigation suggested that these abnormalities can also appear without such diseases.

The pathological physiological condition of oketsu could be paraphrased as "stasis of the blood." The oketsu syndrome corresponds to the concept of microcirculation congestion, but the oketsu syndrome also embraces more complex pathological conditions.[20]

EFFECTS OF "KU-OKETSU-ZAI" (THE KAMPO FORMULATIONS TO EXPEL BLOOD STASIS)

The typical formulations for treating oketsu syndrome are Tokakujyoki-to, Keishi-bukuryo-gan (KBG), and Toki-shakuyaku-san. The following studies were undertaken in an attempt to elucidate the pharmacological effects of these Kampo remedies on blood coagulation, microcirculation, and oketsu syndrome.

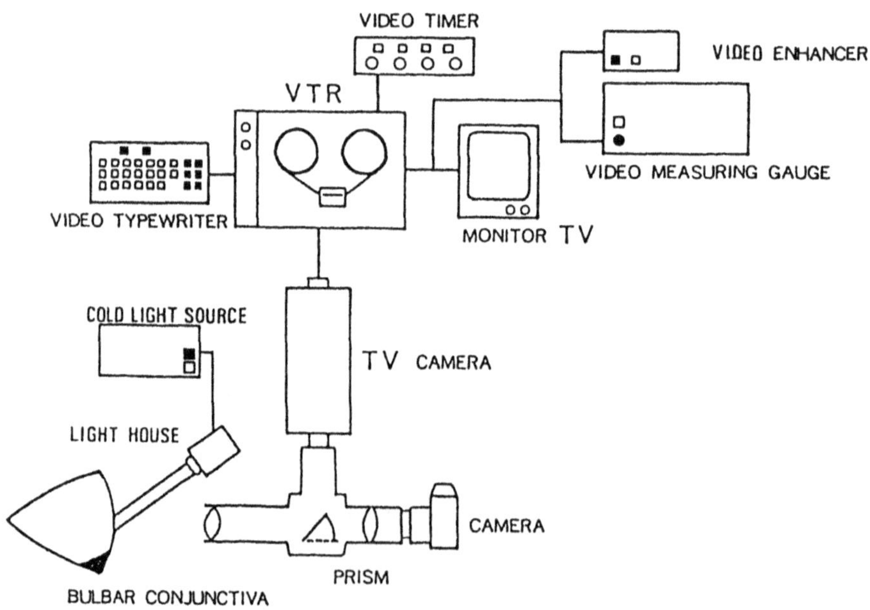

Figure 1 System diagram of the video-microscope system. From Terasawa *et al.* (1988).[19]

Effects of Kampo Medicines on Blood Viscosity

Tani *et al.*[21] evaluated the effects of Kampo medicines on blood viscosity using steroid-treated rats as a model of hyperviscosity. Keishi-bukuryo-gan (KBG) administration was found to suppress hyperviscosity of blood, hyperlipemia, and hypercoagulability. The serum cholesterol level and adrenal weight decreases induced by steroid treatment were also found to be inhibited by KBG administration. The results of a load test with KBG or Toki-shakuyaku-san administered to human volunteers showed that the anti-oketsu remedies had the ability to decrease the blood viscosity and to suppress malondialdehyde (MDA) production, although platelet aggregation showed no change.[22] Tosa *et al.*[23] investigated the effects of two weeks of KBG treatment on blood viscosity, platelet functions and thromboelastogram in ten volunteers, and then followed up for another two weeks. The values of whole blood viscosity decreased significantly and platelet aggregations induced by collagen and ADP were inhibited at two weeks after administration. Thromboxane B_2 synthesis in the platelets was also suppressed after two weeks of KBG treatment (Table 5). There were no significant changes in the thromboelastogram.

In clinical studies of KBG as a treatment for patients with Raynaud's syndrome, the improvement of subjective symptoms was recognized and an increase in peripheral surface temperature was observed by infrared thermogram.[24]

Improvement of Erythrocyte Deformability

Orita *et al.*[16, 17] studied the action of KBG on erythrocyte deformability in young and aged rats. It was observed that KBG caused an increase in erythrocyte deformability in both young and old rats. KBG inhibited the decrease of deformability in aged rats fed a high cholesterol diet. Moreover, the administration of KBG resulted in a decrease in hypotonic hemolysis and a slight decrease of surface negative charge in rat erythrocytes, but did not affect the ATP level or lipid composition of membranes. Kubo *et al.*[25] reported that KBG prevented the formation of hepatic

Table 5 Effect of Keishi-bukuryo-gan extract on thromboxane B_2 formation. From Tosa *et al.* (1987) .[23]

	Threshold dose	
	Collagen	ADP
Before[#1]	56.0 ± 51.9 (ng/ml)	65.1 ± 22.2 (ng/ml)
1 week	57.0 ± 27.5	66.0 ± 18.0
2 week[#2]	27.1 ± 25.4*	33.4 ± 16.0*
3 week	43.0 ± 27.8	52.2 ± 29.6
4 week	40.6 ± 27.0	62.5 ± 19.6

The values are expressed as mean ± standard deviation. * $p < 0.05$,
[#1]; before: at the beginning point of medication,
[#2]; 2 week: at the complete point of medication.

vein thrombosis induced by endotoxin in hyperlipemic rats. In the endotoxin-induced experimental disseminated intravascular coagulation (DIC) model rats, KBG also prevented the decrease of blood platelets and fibrinogen, and prolonged prothrombin time.

Effects on Microcirculation

As described above, to determine the effects of Kampo medicines on the microcirculation of bulbar conjunctiva, a video-microscope system has been used. Hayashi *et al.*[26] reported that the administration of KBG increased the internal diameter of the microvessels and blood flow.

For an investigation of the effect of KBG on the microcirculation of bulbar conjunctiva according to oketsu status, 35 volunteers (6 of whom served as controls) were evaluated.[27] The results indicated that both the blood flow rate and the internal diameter of microvessels increased significantly following the oral administration of KBG, but blood pressure and pulse rate were not changed significantly. In the volunteers with oketsu, there was a significant increase of blood flow volume after the administration of KBG, but it was not significant compared with the changes in the non-oketsu group. The therapeutic effects of KBG and trapidil on microcirculation were investigated in 14 patients with cerebro-spinal vascular diseases with short and long-term administrations of the two medicines.[26] The short-term trial showed that KBG increased the internal diameter and flow rate in the conjuctival venules similar to trapidil. In the long-term trial, the patients were divided into two groups, and KBG and trapidil were administered for four weeks in opposite order in the two groups (cross-over test). KBG increased the flow rate and improved intra-vascular erythrocyte aggregation in the microvessels, and also lowered whole blood viscosity significantly; trapidil did not alter these parameters. (Figure 2)

The results indicated that KBG not only acts to lower blood viscosity and to increase the conjunctival microcirculation, but also has the ability to improve intravascular erythrocyte aggregation; its mode of action on microcirculation was different from that of trapidil, which is generally used to improve microcirculation.

Erythrocyte Aggregation

Erythrocyte aggregation is the reversible adhesion of adjacent erythrocytes induced by the bridging of intercalated macromolecules. Erythrocyte aggregation can also cause a disturbance of cerebral blood flow. Kota *et al.*[28] reported that the severity of the oketsu state is significantly correlated with erythrocyte aggregation, and that the fibrinogen level is one of the important determinants of the aggregation condition in patients with multiple old lacunar infarctions. The effects of KBG on erythrocyte aggregation and on fibrinogen concentration were determined in 23 patients with multiple old lacunar infarction.[29] Erythrocyte aggregation was evaluated by S10, an *in vitro* erythrocyte aggregation index, and the maximum diameter of the column of intravascular erythrocyte aggregation (DEA) was used as

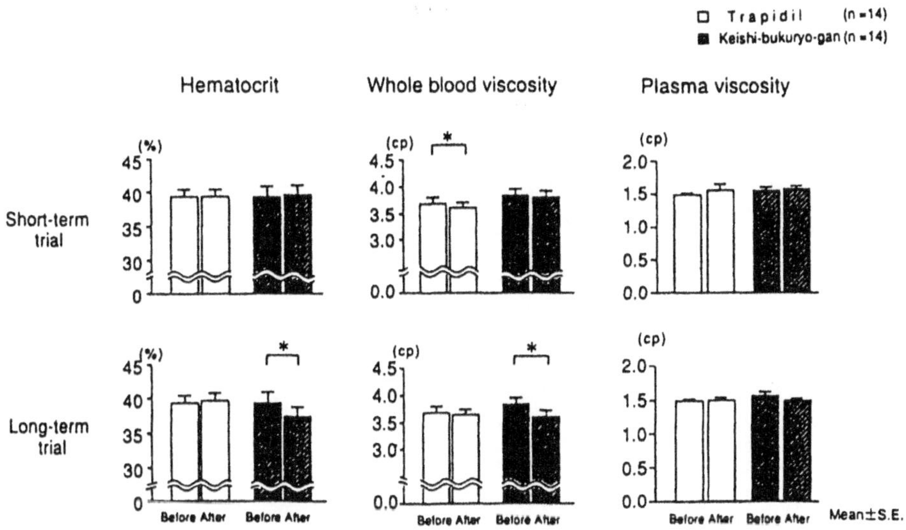

Figure 2 Changes in the rheological parameters after administration of Trapidil and Keishi-bukuryo-gan. From Itoh *et al.* (1992).[27]
Open columns and hatched columns express the mean ± SEM of the patients.
* $p < 0.05$, before *vs.* after each medication.

the *in vivo* erythrocyte aggregation index. The results suggested that KBG had the pharmacological ability to reverse the acceleration of erythrocyte aggregation. Such action was distinct in the more severe oketsu state through the reduction of the fibrinogen level.

According to these studies, the level of oketsu state and the condition of patients were improved depending on the improvement of blood fluidity. These studies provide the empirical concepts of the oketsu syndrome with concrete evidence regarding the abnormalities and disorders of microcirculation. *Moutan Cortex,*[30] *Ginseng Radix,*[31] and some other medicinal plants which are ingredients of oketsu-improving formulations have also been shown to improve the abnormalities of blood flow. These observations provide additional evidence of oketsu-improving formulas ("Ku-oketsu-zai") and may apply to patients with microcirculation disorders.

CONCLUSION

The factors which influence blood viscosity and blood fluidity are shown in Figure 3. Microcirculation disorders and diseases are based on the relationships among these factors.

The pathological condition of oketsu can be paraphrased as "stasis of the blood." An increase in blood viscosity raises peripheral resistance and diminishes cardiac output. Dormandy[32] pointed out that a 10% increase in blood viscosity is associated with a decrease of more than 20% in the total blood flow through a lower limb. In

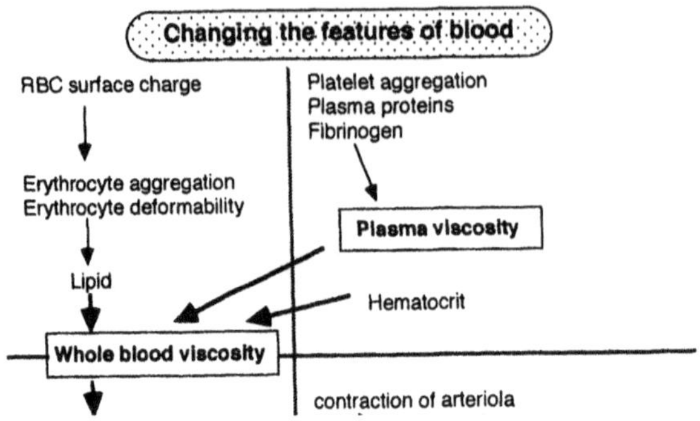
Changing the features of blood
RBC surface charge
Erythrocyte aggregation
Erythrocyte deformability
Lipid
Platelet aggregation
Plasma proteins
Fibrinogen
Plasma viscosity
Hematocrit
Whole blood viscosity
contraction of arteriola

have demonstrated the empirical concept of the "oketsu" syndrome with evidence regarding the abnormalities of microcirculation and/or the peripheral circulatory disorders, including blood viscosity, blood flow in microvessels, erythrocyte aggregation, erythrocyte deformability and platelet function. Pharmacological investigations of "anti-oketsu" Kampo formulations for treating "oketsu" syndrome have demonstrated that these formulations increased the blood flow and improved intra-vascular erythrocyte aggregation in the microvessels, which led to an improved quality of life.

These basic and clinical trials suggest that Kampo medicines have therapeutic efficacy as treatment of peripheral circulatory disorders, and have an important place in the spectrum of medical care for age-related circulation problems and diseases.

REFERENCES

1. T. Kikutani, Treatment of circulatory diseases, (hypertension, hypotension, peripheral circulatory disturbance, hypersensitivity to coldness) by ethical Kampo drugs, applied to health insurance system. *J. Traditional Sino-Japanese Med.* **11**, 445–451 (1990), (Japanese).
2. *Kampo Shinryo Iten,* K. Otsuka, D. Yakazu and T. Shimizu, eds. Nanzan-do Co. Ltd., Tokyo, Japan (1969) (*Natural Healing with Chinese Herbs*, English translation by Hong-Yen-Hsu, Oriental Healing Arts Institute, LA, USA, 1981).
3. H. Takayama, Effective treatment of a case of Buerger's disease with Toki-Shigyaku-ka-Goshuyu-shokyo-to. *J. Traditional Sino-Japanese Med.*, **7** (Supple), 18–20, (1986), (Japanese).
4. T. Osawa, *Thromboangitis obliterans (Buerger's Disease), Practice of Medical Insurance Treatment of Kampo Medicines.* ed. by the Japan Society for Oriental Medicine, pp. 94–96, Tokyo, Japan, (1986) (Japanese)
5. H. Iwata, T. Kasamatsu, K. Miyashita, T. Oshima, T. Yamashita and G. Ichinomiya, Effects of Tsumura Toki-shigyaku-ka-goshuyu-shokyo-to on peripheral circulation. *Medical Consultation and New Remedies* **20**, 2625–2635, (1983) (Japanese).
6. S. Takaguchi, Kampo treatment for peripheral circulatory disturbance. *Kampo Igaku* **6**(6), 18–19 (1982) (Japanese).
7. T. Takemiya, M. Shimizu, H. Yamaguchi and A. Miura, Clinical effects of Ninjin-Yoei-To in patients with peripheral circulatory disturbance. *Pharmacology and Treatment (Yakuri to Chiryo)* **19**, 3801–3808 (1991) (Japanese).
8. D. Yakazu, Kampo therapy on puerperal lower extremity thrombosis. *Jpn. J. of Oriental Med.* **23**, 192–194 (1973) (Japanese).
9. T. Akiba, Effective treatment of a case of thrombophlebitis with Keishi-bukuryo-gan. *J. Traditional Sino-Japanese Med.* **7** (Supple), 24–25 (1986) (Japanese).
10. S. Kanda, A case report. *Kampo Clinic (Kampo Shinryo).* **6**(3), 49–51 (1987) (Japanese).
11. K. Terasawa, A. Imadaya, H. Tosa, T. Ito, I. Horikoshi, K. Toriizuka and S. Honma, A trial of the traditional herbal treatment to the autonomic disorder associated with central and peripheral nervous disease. *Proc. Symp. WAKAN-YAKU,* **15**, 69–76 (1982) (Japanese).
12. S. Arichi, M. Iwanaga, and T. Tani, Blood viscosity of the Tokaku-joki-to-sho, a syndrome used in traditional Chinese medicine. *Med. J. Kinki Univ.* **6**, 403–413 (1981) (Japanese).
13. K. Terasawa, K. Toriizuka, H. Tosa, M. Ueno, T. Hayashi, and M. Shimizu, Rheological studies on "oketsu" syndrome I: The blood viscosity and diagnostic criteria. *J. Med. Pharm. Soc. WAKAN-YAKU* **3**, 98–104 (1986).
14. N. Casson, *Rheology of Disperse Systems* (ed. C.C. Mill), pp. 84, Pergamon Press, London, (1959).
15. K. Terasawa, K. Toriizuka, M. Bandou, A. Imadaya and H. Tosa, Effects of medicinal plants on the metabolism of platelet arachidonic acid – Studies on "oketsu" syndrome, platelet aggregation and changes in malondialdehyde values —. *J. Med. Pharm. Soc. WAKAN-YAKU* **2**, 310–316 (1985).

16. M. Orita (Sakaguchi), H. Abe and S. Arichi, Effects of Keishibukuryo-gan on erythrocyte deformability, *J. Med. Pharm. Soc. for WAKAN-YAKU* **1**, 243–248 (1984) (Japanese).
17. H. Abe, M. Orita, T. Takemura and S. Arichi, Wakan-yaku and hemorheology. *Proc. Symp. WAKAN-YAKU* **16**, 123–125 (1983) (Japanese).
18. T. Tani, M. Iwanaga, T. Ohno and S. Arichi, Effect of crude drugs and their prescriptions on the blood rheology affected by glucocorticoid treatment (I) Effect of betamethasone on rat blood rheology. *Med. J. Kinki Univ.* **9**, 229–238, (1984) (Japanese).
19. K. Terasawa, T. Itoh, Y. Morimoto,Y. Hiyama and H. Tosa, The characteristics of the microcirculation of bulbar conjunctiva in "oketsu" syndrome. *J. Med. Pharm. Soc. WAKAN-YAKU* **5**, 200–205 (1988).
20. Z-Z. Yin, K. Toriizuka and J-C. Cyong, Advances of researches on the blood stasis and antithrombus activity of Chinese drugs. *Kampo and Newest Therapy*, **2**, 155–162 (1993) (Japanese)
21. T. Tani, M. Iwanaga, T. Ohno, M. Higashino, M. Kubo and S. Arichi, Effect of crude drugs and their prescriptions on the blood rheology affected by glucocorticoid treatment (II) Effect of Keishibukuryo-gan on adverse reaction of betamethasone treatment. *Shoyakugaku Zasshi* **38**, 166–174 (1984) (Japanese).
22. K. Toriizuka, Z-T. Zhong, K. Terasawa, M. Okamoto and H. Tosa, Effects of Toki-shakuyaku-san on blood viscosity and platelet functions in normal subjects. *J. Med. Pharm. Soc WAKAN-YAKU* **4**, 20–25 (1987).
23. H. Tosa, K. Toriizuka and K. Terasawa, The effect of Keishi-bukuryo-gan on blood viscosity, platelet functions and blood coagulation in normal subjects, *J. Med. Pharm. Soc. WAKAN-YAKU* **4**, 172–179 (1987).
24. K. Terasawa, H. Matsuda, A. Imadaya, H. Tosa, T. Mitsuma, K. Toriizuka and S. Homma, A study on clinical effects of Keishibukuryo-gan (Kuei-chih-fu-ling-wan) prepared in the hospital pharmacy. *Jpn. J. of Oriental Med.* **35**, 131–136 (1984) (Japanese).
25. M. Kubo, H. Matsuda, T. Nagao, T. Tani, K. Nanba and S. Arichi, Reparatory effects of Keishibukuryo-gan on the full-length figure. *Proc. Symp. WAKAN-YAKU* **16**, 171–182 (1983) (Japanese).
26. T. Hayashi, N. Tsushima, M. Itakura, M. Ho, M. Konishi, T. Tamai, Y. Fujiyoshi and K. Yasunaga, Changes of microcirculation of bulbar conjunctiva and hemorheology on Keishibukuryo-gan Administration. *J. Med. Pharm. Soc. WAKAN-YAKU* **4**, 358–359 (1987) (Japanese).
27. T. Itoh, K. Terasawa, K. Kohta, N. Shibahara, H. Tosa and Y. Hiyama, Effect of Keisi-bukuryo-gan and Trapidil on the microcirculation in patients with cerebro-spinal vascular disease. *J. Med. Pharm. Soc. WAKAN-YAKU* **9**, 40–46 (1992).
28. K. Kohta, Y. Hiyama, K. Terasawa, T. Hamazaki, T. Itoh and H. Tosa, Hemorheological studies of "oketsu" syndrome, Erythrocyte aggregation in "oketsu" syndrome. *J. Med. Pharm. Soc. WAKAN-YAKU* **9**, 221–228 (1992).
29. K. Kohta, H. Hikiami, Y. Shimada, H. Matsuda, T. Hamazaki and K. Terasawa, Effect of Keisi-bukuryo-gan on erythrocyte aggregability in patients with multiple old lacunar infarction. *J. Med. Pharm. Soc. WAKAN-YAKU* **10**, 251–259 (1993).
30. A. Hirai, Study on the mechanisms of anti-platelet and anti-atherosclerotic effects of Sino-Japanese medicines: Focusing arachidonic acid cascade. *J. Traditional Med.* **11**, 16–28 (1994) (Japanese).
31. H. Kaneko, The clinico-pharmacological effects of Korean red ginseng on peripheral circulation disorder. *Taisha* **29** (Supple), 220–227 (1992) (Japanese).
32. J.A. Dormandy, *Br. Med. J.*, **iv**, 716 (1971).
33. L. Dintenfass, Blood viscosity, internal fluidity of the red cell, dynamic coagulation and the critical capillary radius as factors in the physiology and pathology of circulation and microcirculation. *Med. J. Aust.*, **i**, 688 (1968).

11 PHARMACOLOGICAL PROPERTIES OF SOME STRUCTURALLY RELATED INDOLE ALKALOIDS CONTAINED IN THE ASIAN HERBAL MEDICINES, HIRSUTINE AND MITRAGYNINE, WITH SPECIAL REFERENCE TO THEIR CA^{2+} ANTAGONISTIC AND OPIOID-LIKE EFFECTS

KAZUO WATANABE[1], SHINGO YANO[1], SYUNJI HORIE[1], LEONARDO TOMO YAMAMOTO[1], HIROMITSU TAKAYAMA[1], NORIO AIMI[1], SHIN-ICHIRO SAKAI[1], DHAVADEE PONGLUX[2], PAVICH TONGROACH[2], JIE SHAN[3] and PETER K. T. PANG[3]

[1]*Faculty of Pharmaceutical Sciences, Chiba University, 1-33 Yayoi-cho, Inage-ku, Chiba 263-0022, Japan;* [2]*Faculty of Pharmaceutical Sciences, Chulalongkorn University, Bangkok 10330, Thailand;* [3]*Department of Physiology, School of Medicine, University of Alberta, Edmonton, Canada T6G 2H7*

INTRODUCTION

Dried hooks and stems of *Uncaria sinensis* (Oliv.) Havil and *Uncaria rhynchophylla* (Miq.) Jackson are used in Chino-Japanese Kampo medicine for the treatment of patients with hypertension and its associated manifestations such as headache and dizziness. A number of indole alkaloids represented by rhynchophylline and isorhynchophylline have been isolated from these plants. Pharmacological studies have revealed that several alkaloids such as dihydrocorynantheine, hirsutine, hirsuteine, and 3α-dihydrocadambine induce hypotensive and vasodilative effects in anesthetized rats and dogs (Aisaka *et al.*, 1985; Ozaki, 1989, 1990). These findings may indicate the pharmacological basis of the action of *Uncaria* plants in herb prescriptions of the Kampo medicine.

The leaves of the tree called "kratom" in the Thai language (*Mitragyna speciosa*) grown in Thailand are known for their narcotic-like property in traditional use (Suwanlert, 1975; Jansen and Prast, 1988). Mitragynine is a main component of kratom (Jansen and Prast, 1988). Macko *et al.* (1972) reported that mitragynine is comparable to codeine as an analgesic and a cough depressant in rats and mice, and does not induce emesis or dyspnea as does codeine at equivalent doses. Interestingly, the results of that study suggested that mitragynine does not act on opioid receptors, despite the opium-like use of kratom. We have also studied the effects of mitragynine on central nervous systems and observed that mitragynine possesses a marked analgesic effect on phenyl-p-benzoquinone (PBQ) writhing behavior and hot-plate test performances in mice (Watanabe *et al.*, 1992), and the mechanism of this analgesic effect remains to be clarified. The elucidation of the peripheral effect of mitragynine in *in vitro* preparations may help to clarify the analgesic mechanism.

We have studied the pharmacological properties of alkaloids isolated from the Chinese herbal medicine Uncariae Ramulus et Uncus (*Uncaria rhynchophylla*) (Horie *et al.*, 1991; Horiuchi *et al.*, 1988; Ito *et al.*, 1990; Watanabe *et al.*, 1987; Yano *et al.*, 1991). We are interested in the similarity of the chemical structure of mitragynine to those of *Uncaria* alkaloids such as hirsutine and dihydrocorynantheine. In the present study, we reviewed the structure requirements for analgesic action and performed a neuropharmacological analysis of the mode of action of these indole alkaloids with such unique pharmacological properties.

MATERIALS AND METHODS

Hirsutine and dihydrocorynantheine were isolated from the dried hooks and stems of *Uncaria rhynchophylla* (Miq.) in our laboratory (Aimi *et al.*, 1977). Mitragynine was isolated from the extract of leaves of *Mitragyna speciosa* harvested in Bangkok by column chromatography as described previously (Ponglux *et al.*, 1994). The chemical structures of these test compounds are shown in Figure 1.

The pharmacological methods and preparations employed in this investigation were as follows;

1. Whole animal assay for antinociceptive effects: Tail pinch test, hot-plate test, phenyl-*p*-benzoquinone (PBQ) writhing test, formalin-induced pain response.
2. Isolated smooth muscle preparations: Isolated guinea-pig ileum. Isolated vas deferens preparation (epididymal portions). The preparation was transmurally stimulated with a needle-ring platinum electrode. The isometric contraction of the preparation was detected by force transducer and recorded on a pen-writing recorder.
3. Assay of cellular Ca^{2+} movement in isolated organs and cultured cells:
   The intracellular Ca^{2+} level was recorded by using a CAF-100 fluorimeter

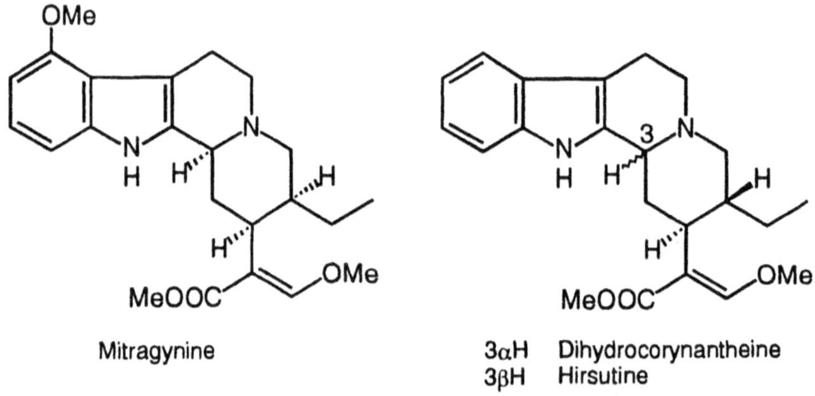

Figure 1 Chemical structures of mitragynine, hirsutine and dihydrocorynantheine.

(Spectroscopic, Tokyo, Japan) with fura-2 fluorescent dye. The activity of Ca^{2+} channels (T-type and L-type) was measured by the whole cell patch-clamp technique with the use of an Axopatch-1B patch-clamp amplifier (Axon Instruments) in neuroblastoma cells.

RESULTS AND DISCUSSION

1. Ca^{2+} Channel Blocking Effects of Hirsutine

1.1. Assay on rat aorta smooth muscle contraction

The effect of hirsutine on norepinephrine (10^{-7} M)-induced contractions was measured by stepwise increasing doses of hirsutine. Hirsutine (10^{-6} M to 3×10^{-5} M) produced a dose-dependent inhibition of the contractions induced by norepinephrine (Figure 2). The dilative effect was not affected by the removal of the endothelium in the aorta strips. The presence of propranolol (10^{-6} M) hardly affected the vasodilatation by hirsutine. The effect of hirsutine on high K^+-induced contractions was measured in a similar way. The sustained contractile response to a high K^+-concentration was dose-dependently inhibited by the treatment with hirsutine, regardless of the absence of the endothelium.

The effect of hirsutine on Ca^{2+}-induced contractions was measured by cumulatively increasing concentrations of Ca^{2+}-depolarized aorta strips without the endothelium. Hirsutine (10^{-6} M to 3×10^{-5} M) dose-dependently shifted the cumulative dose–response curve for Ca^{2+} to the right (Figure 3). The pA_2 value (mean ± S.E.) was calculated from the dose–response curve for Ca^{2+}; it was 6.6 ± 0.1 (n = 4) for hirsutine, and 8.0 ± 0.1 (n = 5) for diltiazem (Yano *et al.*, 1991).

1.2 Effects of hirsutine on cytosolic Ca^{2+} level in vascular smooth muscle

The effects of hirsutine on cytosolic Ca^{2+} ($[Ca^{2+}]i$) in smooth muscle cells of the isolated rat aorta were studied by using fura-2. Hirsutine (3×10^{-6} – 3×10^{-5} M) produced a dose-dependent reduction of $[Ca^{2+}]i$ in the aorta treated with either noradrenaline or high K^+ solution, with similar potency (Figure 4). In addition, the

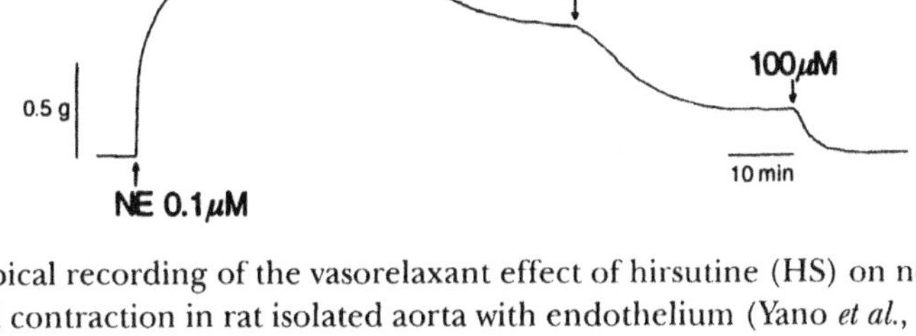

Figure 2 Typical recording of the vasorelaxant effect of hirsutine (HS) on norepinephrine (NE)-induced contraction in rat isolated aorta with endothelium (Yano *et al.*, 1991).

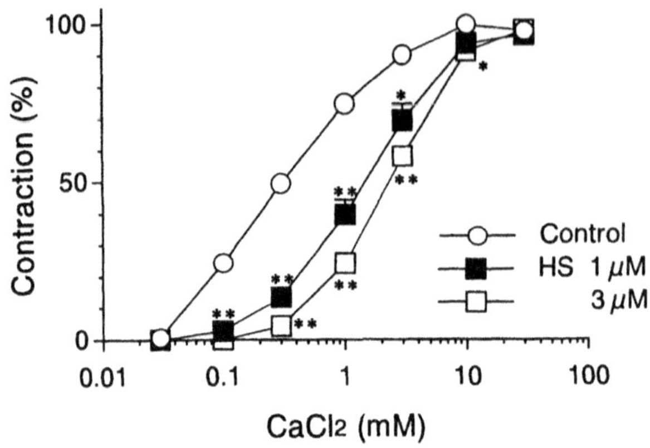

Figure 3 Effect of hirsutine (HS) on Ca^{2+}-induced contraction in rat aorta strip depolarized by KCl (Yano *et al.*, 1991).

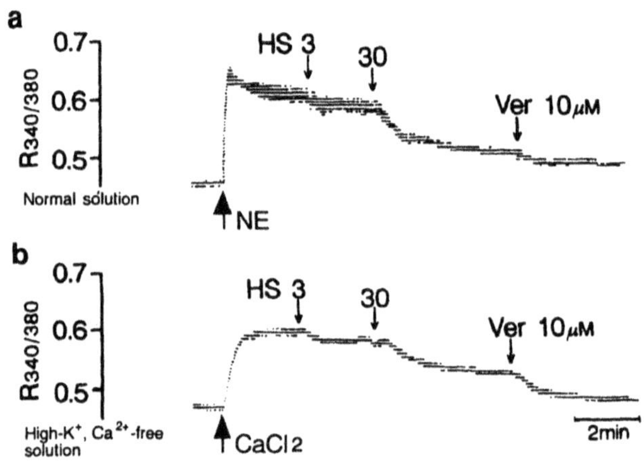

Figure 4 Typical recordings of the effects of hirsutine (HS) and verapamil (Ver) on R340/380 increased by norepinephrine (100 nM) or 60 mM K^+ solution in rat aorta strips without endothelium. The norepinephrine experiment was conducted in solution with a normal amount of Ca^{2+}, and the high K^+ experiment was conducted by adding $CaCl_2$ at 2 mM to 60 mM K^+, Ca^{2+}-free solution. Each trace shown is typical of 6 experiments. (Horie *et al.*, 1992).

effects of hirsutine on intracellular Ca^{2+} store was studied in the experiment with caffeine. Hirsutine was found to inhibit intracellular Ca^{2+} release at the concentration sufficient to block Ca^{2+} influx. These results show that hirsutine predominantly inhibits Ca^{2+} influx through the Ca^{2+} channel, as is also the case with Ca^{2+} release from intracellular Ca^{2+} storage sites (Horie *et al.*, 1992).

Many indole alkaloids have been isolated from the *Uncaria* genus, and some of them are known to possess notable pharmacological properties. Rhynchophylline, isorhynchophylline, and hirsuteine produce a decrease in blood pressure or heart rate (Ozaki, 1989). Dihydrocorynantheine, which is a stereoisomer of hirsutine, selectively blocks α_2-adrenoceptors in vascular responses (Ito *et al.*, 1990). Hirsutine possesses an autonomic ganglionic blocking action, antiarrhythmic action, and hypotensive action (Ozaki, 1989). Some of these actions of hirsutine were presumed to be associated with the Ca^{2+} channel blocking action (Yano *et al.*, 1987). Thus, the action of hirsutine on Ca^{2+} channels in the aorta was elucidated in detail in our present study.

The Ca^{2+} channel blocking activity of hirsutine was found to be of a competitive nature, and more potent than those of other naturally occurring products such as rhynchophylline, isorhynchophylline (Yamahara *et al.*, 1987), and alismol, a sesquiterpenoid from *Alismatis* Rhizome (Matsuda *et al.*, 1987), on the basis of hirusutine's pA_2 value. In addition, a considerable amount of hirsutine (about 11%) is known to be contained in the crude base fraction of the extract of *Uncaria* hooks. Taken together, these findings indicated that hirsutine is the most representative constituent of the vasodilative indole alkaloids contained in the *Uncaria* genus.

2. Effects of Mitragynine

2.1. *Assay of antinociceptive action of the crude alkaloid fraction of Mitragyna leaf and mitragynine*

Macko *et al.* (1972) reported that mitragynine had analgesic and cough depressant effects comparable to those of codeine in dogs, and he also reported the absence of antagonism by nalorphine on its analgesic action. We confirmed the antinociceptive action of the crude alkaloid fraction of *Mitragyna speciosa* extract on some animal responses including PBQ writhing, hot-plate, pressure and formalin test behavior in mice (Figure 5). The crude extract showed a distinct

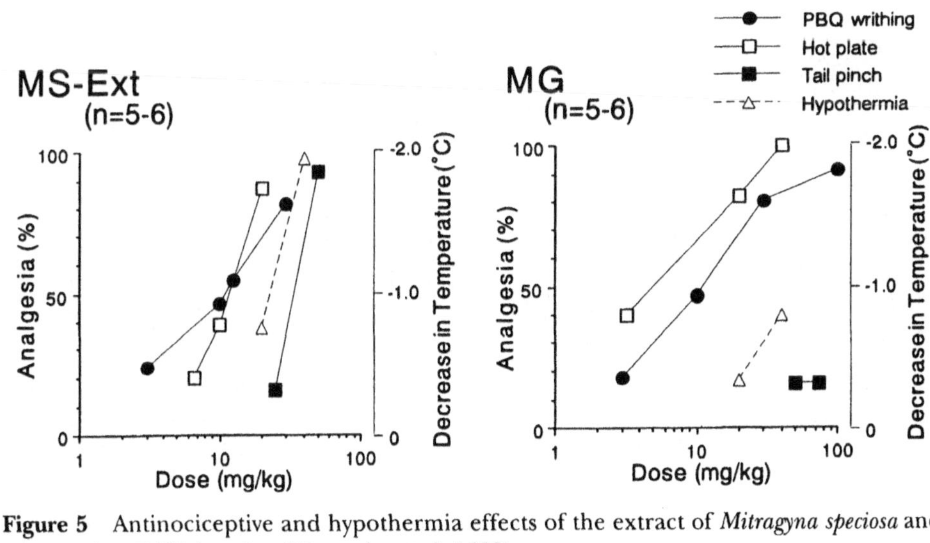

Figure 5 Antinociceptive and hypothermia effects of the extract of *Mitragyna speciosa* and mitragynine (MG) in mice (Watanabe *et al.*, 1992).

antinociceptive activity as assessed by all of these tests in mice, and its potency was about 1/5 that of morphine. Mitragynine also showed a remarkable activity on PBQ writhing in a dose-dependent [3.0–100 mg/kg, per os (p.o.)] manner. In contrast, mitragynine did not show any significant activity on tail pinch behavior (Watanabe *et al.*, 1992).

The pain response induced by a plantar injection of formalin in mice is well known as a convenient test method for the differentiation of two kinds of pain responses, the immediate neurogenic pain response and the delayed inflammatory pain response. As compared with the control responses, mitragynine and the crude extract both showed a marked inhibition of the responses at the dose range of 30–60 mg/kg, p.o. (Figure 6). These results indicate that the type of analgesia provided by the test drugs is different from that provided by anti-inflammatory analgesics, and is associated with the group of opiate-like analgesics.

2.2. *Effect of mitragynine on electrically stimulated contraction of isolated guinea-pig ileum: evidence for the action of opioid receptor*

Although the antinociceptive action of mitragynine seemed to resemble that of morphine, there was no published direct evidence of the opioid effect of mitragynine before our present research. On the contrary, Macko *et al.* (1972) suggested a lack of nalorphine antagonism against mitragynine analgesia in animal experiments. Thus, we studied the opioid action of mitragynine on ileum smooth muscle preparations.

The effects of mitragynine and morphine on contractions of the guinea pig ileum evoked by a single-pulse electrical transmural stimulation were studied. The mean amplitude of ileum contraction evoked by electrical stimulation was about 20% of the maximal response to acetylcholine (3×10^{-6} M). This contraction was abolished by tetrodotoxin (10^{-6} M) and by atropine (10^{-7} M). However,

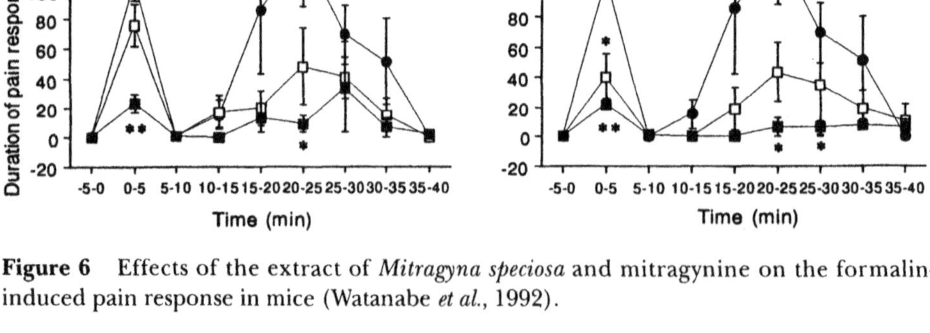

Figure 6 Effects of the extract of *Mitragyna speciosa* and mitragynine on the formalin-induced pain response in mice (Watanabe *et al.*, 1992).

hexamethonium (10^{-4} M) did not affect the contraction (6 ± 3% inhibition) (Watanabe *et al.*, 1997).

As illustrated in Figure 7, mitragynine and morphine inhibited the electrically stimulated contraction in a dose-dependent manner; their pD_2 values were 6.91 ± 0.04 and 7.68 ± 0.11, respectively.

To investigate the involvement of opioid receptors in the inhibitory effect of mitragynine, the effect of naloxone on the decreased contraction was examined (Figure 7). Naloxone (10^{-8} M to 3×10^{-7} M) reversed the inhibitory effect of mitragynine on the electrically stimulated twitch contraction in a dose-dependent manner (Yamamoto *et al.*, 1996). The same effect of naloxone was obtained in the experiment with morphine.

In contrast, naloxone did not show any effect on the inhibition induced by clonidine, an α_2-adrenoceptor agonist, or nicardipine, an L-type Ca^{2+} channel antagonist. Yohimbine, an α_2-antagonist, inhibited the clonidine effect.

As shown in Figure 8, naloxone (3×10^{-9} M to 3×10^{-8} M) shifted the concentration–response curves for mitragynine and morphine to the right in the electrically stimulated ileum preparations. The pA_2 and 95% confidence limits for the effect of naloxone were 8.77 ± 0.04 (n = 3 – 6) for mitragynine and 9.00 ± 0.04 (n = 4) for morphine; these pA_2 values was not significantly different. The slope factors were not significantly different from the unity (0.93 ± 0.29 for mitragynine and 1.72 ± 0.28 for morphine).

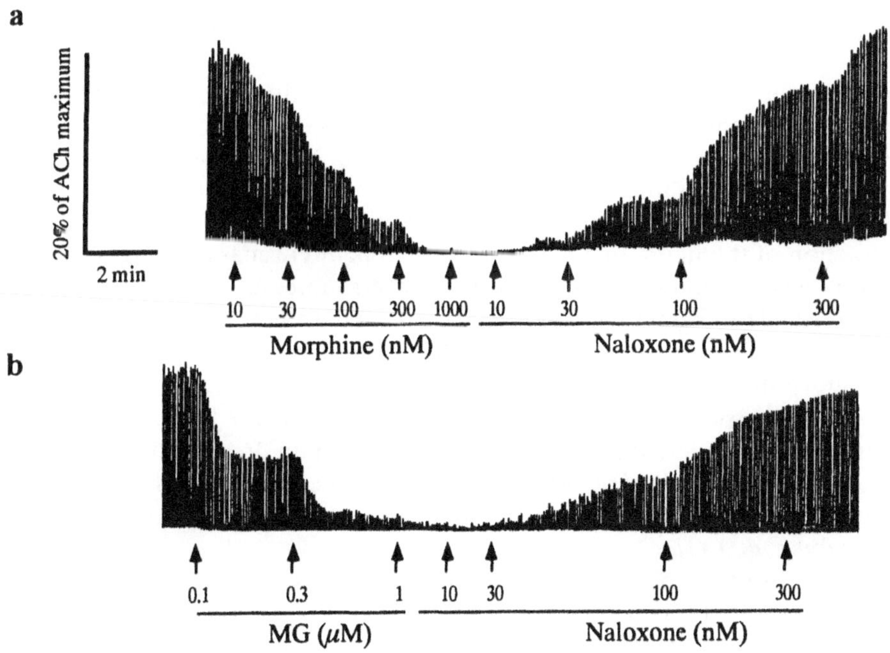

Figure 7 Typical tracings of the effects of naloxone on relaxant responses to (a) morphine or (b) mitragynime (MG) in the guinea pig ileum contracted by electrical transmural stimulation (Watanabe *et al.*, 1997). ACh maximum: the maximum contraction induced by acetylcholine (ACh, 3 μM).

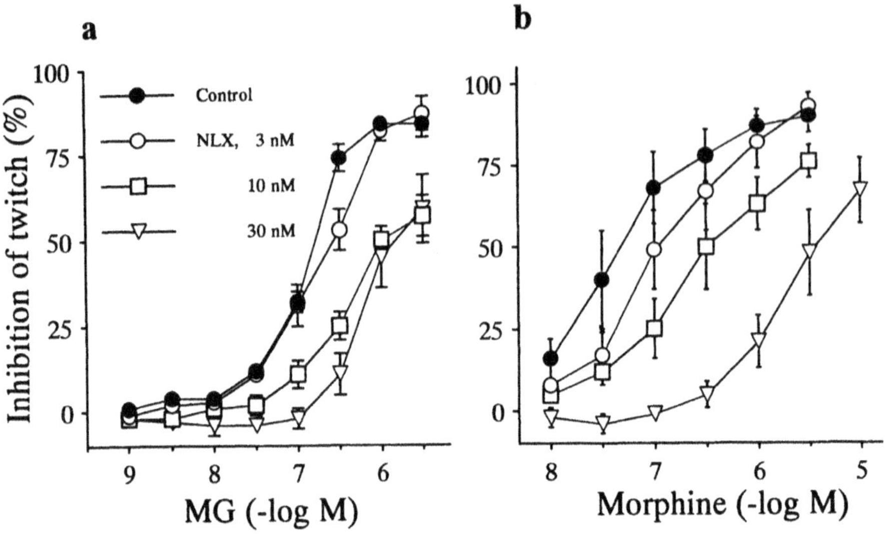

Figure 8 Concentration–response curves for (a) MG and (b) morphine in the electrically stimulated guinea-pig ileum in the absence or the presence of naloxone (3 nM, 10 nM and 30 nM). Responses are expressed as inhibition percentage of the twitch contractions before the agonist addition. Values are means ± S.E.M. (n = 4) (Watanabe *et al.*, 1997).

2.3 Effect of mitragynine on withdrawal contraction after a brief exposure to morphine

Opiate dependency can be accomplished in the guinea pig ileum by a brief incubation with morphine (Lujan and Rodrigues, 1981; Collier *et al.*, 1981; Chahl, 1986; Valeri *et al.*, 1990) In this tissue, the contraction is observed as a washout- or naloxone-induced withdrawal response.

Mitragynine (3×10^{-6} M to 10^{-5} M) inhibited the naloxone-induced withdrawal contraction of the ileum in a concentration-dependent manner (Figure 9). The withdrawal contraction was also abolished by tetrodotoxin (10^{-6} M) and atropine (10^{-6} M), while hexamethonium (10^{-4} M) showed a slight inhibition. Naloxone alone did not show any effect in the guinea pig ileum and it did not show any effect when morphine was replaced by mitragynine (10^{-5} M). The same results were obtained with tissue exposed to mitragynine or morphine for 1 hr before naloxone administration, as seen with the tissue exposed for 5 min. In addition, the addition of morphine or DAMGO (10^{-6} M) in the tissue exposed to morphine (10^{-6} M) for 5 min or 1 hr did not elicit any change in the naloxone-induced withdrawal response (Watanabe *et al.*, 1997).

The results of the present study showed that mitragynine inhibited the nerve-stimulation-elicited twitch contraction through the opioid receptor in the isolated guinea-pig ileum. Mitragynine also inhibited the naloxone-precipitated withdrawal response. Mitragynine thus seems to have two distinct properties: it has an opioid-like effect and an inhibitory effect on the withdrawal response. Matsumoto *et al.* (1996) also showed the naloxone antagonism of the mitragynine-induced antinociceptive effect.

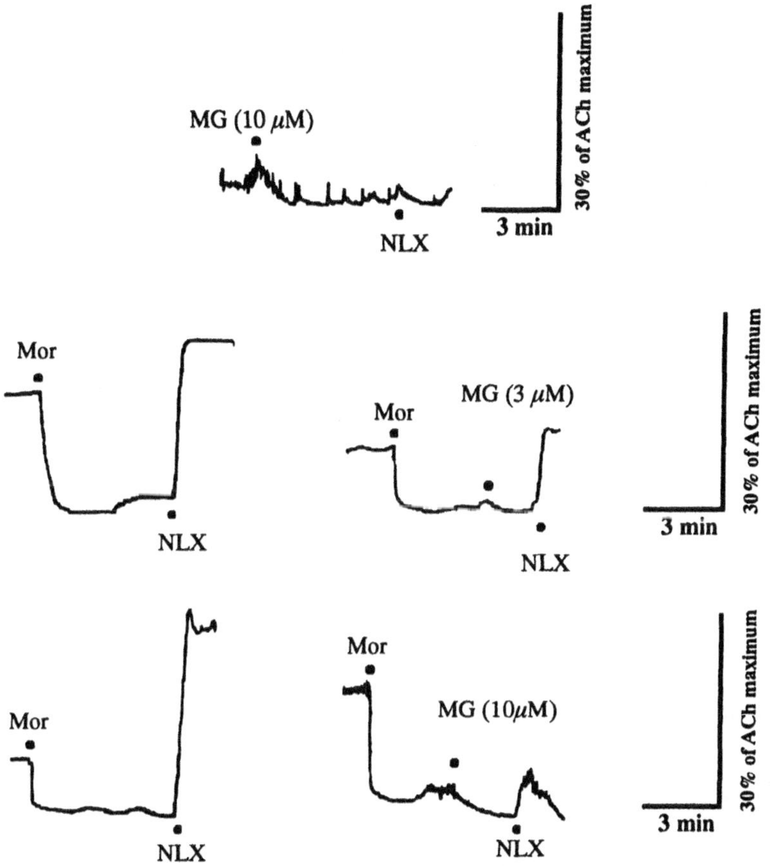

Figure 9 Typical tracings of the effects of mitragynine (MG) and naloxone (NLX, 1 μM)-precipitated withdrawal response after a 5-min exposure of guinea pig ileum to morphine (Mor, 500 nM) (Watanabe *et al.*, 1997)

3. Effect of Mitragynine on Ca^{2+} Movement in Smooth Muscle Cells and Neuroblastoma Cells

Further experiments were conducted to clarify the pharmacological profiles and mechanisms of the inhibitory effects of mitragynine on contractions of the guinea pig vas deferens and ileum with the special reference to the involvement of opioid receptors. We studied the effects of mitragynine on neuronal Ca^{2+} channels with the patch-clamp technique and with the Ca^{2+}-indicative fluorescent dye method, and compared the effect of mitragynine on the neurogenic contraction of smooth muscle with that of morphine.

3.1. Effect of mitragynine on the contraction of rat aorta smooth muscle preparation

With the analogy of Ca^{2+} antagonistic activity of hirsutine and dihydrocorynantheine, the effect of mitragynine on the rat aorta preparation was examined. As shown in Figure 10, a clear shift of the dose–response curve of norepinephrine to the right was observed in this preparation by the doses of 10^{-6} M and 3×10^{-6} M of mitragynine. The same results were obtained in the Ca^{2+}-induced contractions under the K^+-free condition of the nutrient medium. This indicates the Ca^{2+} antagonistic effect of mitragynine on the vascular smooth muscle cells.

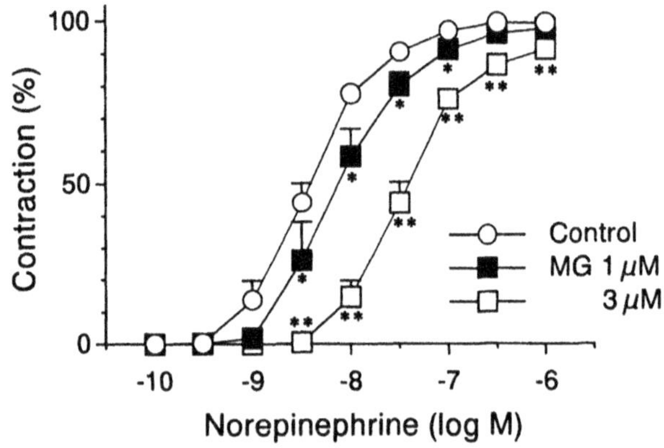

Figure 10 Effect of mitragynine (MG) on norepinephrine-induced contractions in rat aorta strip without endothelium (Watanabe *et al.*, 1992).

3.2. Effects of mitragynine on Ca^{2+} channels in neuroblastoma cells

Figure 11 shows records of the KCl-induced $[Ca^{2+}]i$ increase before and after exposure to mitragynine and after washout of the drug. Mitragynine inhibited the

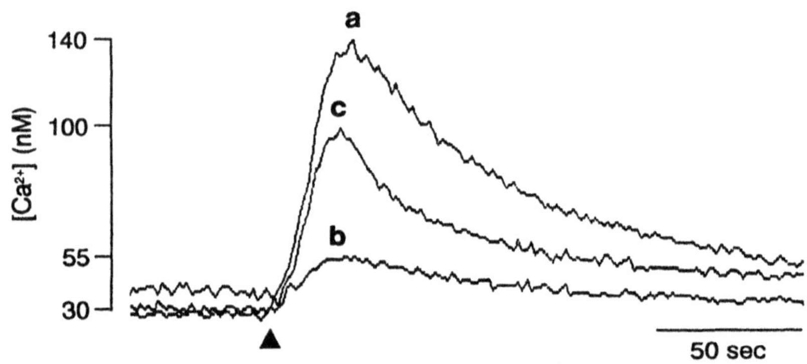

Figure 11 Effect of mitragynine on KCl-induced $[Ca^{2+}]i$ increase in neuroblastoma cells. Three representative records before and after the administration of mitragynine (1 μM) and washout. Original recordings were obtained from 1 typical experiment showing the effect of (a) 15 mM KCl alone, (b) 15 mM KCl after incubation with mitragynine, and (c) 15 mM KCl after washout. These 3 tests were performed sequentially on the same group of cells. ▲: Time when KCl was added. (Horie *et al.*, 1997).

KCl-induced [Ca^{2+}]i increase. This increase was abolished by the washout of mitragynine. In addition, mitragynine inhibited the T-type and L-type Ca^{2+} channel currents in the analysis using the patch-clamp technique.

4. Mode of Action of Mitragynine on the Ca^{2+}-Channel on the Nerve Terminal and Structure Requirement for Analgesic Activity in the Indoloquinolizidine Type Indole Alkaloids

Ca^{2+} channel blockers such as nifedipine have been demonstrated to exhibit analgesic properties in some pain tests (Miranda *et al.*, 1992). The administration of a neuronal Ca^{2+} channel blocker decreases the pain (Chaplan *et al.*, 1994). The blockade of presynaptic Ca^{2+} channels is known to inhibit neurotransmitter release, and this mechanism might produce the inhibition of neuropathic pain. It has been suggested that neuronal Ca^{2+} channels are involved in the genesis of the spontaneous discharge from injured afferents (Xiao and Bennett, 1995). Thus, several studies examined the relationships among analgesics, the blockade of the Ca^{2+} channel and the inhibition of neurotransmitter release. Mitragynine was also found to possess an analgesic activity (Macko *et al.*, 1972; Watanabe *et al.*, 1992). It would be of interest to determine whether the neuronal Ca^{2+} channel-blocking activity of mitragynine is involved in its analgesic effect.

The electrically induced contraction of the guinea pig vas deferens is due to the excitation of postganglionic sympathetic neurons, leading to the co-release of norepinephrine and 3′,5′-adenosine triphosphate (ATP) from the nerve terminal. In the present study, mitragynine almost abolished the electrically induced contraction of the vas deferens at 10 μM, but failed to affect the responses to norepinephrine and ATP at a higher concentration (30 μM). Nicotine and tyramine act on postsynaptic nerve endings, resulting in the release of neurotransmitters; they each elicit the co-release of norepinephrine and ATP from the nerve terminal, resulting in the contraction of the rat vas deferens (Kügelgen and Starke, 1991). As mitragynine inhibited the nicotine- or tyramine-induced contractions, mitragynine may act on the nerve ending to inhibit the release of norepinephrine and ATP. The inhibitory effect of mitragynine on the neuronal Ca^{2+} channels described above may be involved in the inhibition of transmitter release.

Electrical stimulation elicits an ileum contraction by the release of neurotransmitters, such as acetylcholine and substance P, from the nerve endings (Brookes *et al.*, 1991). In the present study, the electrical stimulation-induced contraction was abolished by tetrodotoxin or atropine, but was not affected by hexamethonium. These results show that the electrical stimulation-induced contraction is elicited by acetylcholine released from post-ganglionic cholinergic neurons. Like morphine, mitragynine inhibited the electrically stimulated ileum contraction. However, mitragynine hardly affected the contraction induced by direct stimulation of the receptors on the smooth muscle by acetylcholine or histamine. Taken together, these results indicate that mitragynine acts on the postsynaptic nerves to inhibit the release of neurotransmitters. The involvement of opioid receptors in the effect of mitragynine is suggested by the reversal action of naloxone. The inhibitory effect of mitragynine on ileum contraction is one-sixth as potent as that of morphine. Clear evidence of the involvement of opioid receptors in the inhibitory action of mitragynine was obtained in the present study, which showed that mitragynine inhibits the nerve-stimulation-elicited contraction through the opioid receptor in guinea pig ileum.

We observed that mitragynine stimulated opioid receptors at concentrations less than 1 μM. In addition, mitragynine blocked the Ca^{2+} influx through the neuronal Ca^{2+} channels at concentrations higher than 10 μM. These effects of mitragynine elicited the inhibition of neurotransmitter release from nerve endings, leading to the inhibition of the neurogenic contractions in the vas deferens and ileum (Figure 12). These neuronal Ca^{2+} channel blocking and opioid receptor- stimulating effects may be involved in the analgesic effect of mitragynine.

Another important finding of our study regards the structure requirement for analgesic activity in indoloquinolizidine type indole alkaloids.

We have been studying the structure activity relationship for Ca^{2+} antagonism and analgesic effects in indoloquinolizidine alkaloids derived from natural medicine. As for the Ca^{2+} antagonistic effect of indoloquinolizidine, the stereochemical orientation of the lone pair electrons on the nitrogen atom in the C-ring was found to be very important (Horie *et al.*, unpublished data). Some other substitutions near the nitrogen atom had a significant influence on the Ca^{2+} antagonistic effects among the hirsutine and dihydrocorynantheine analogs.

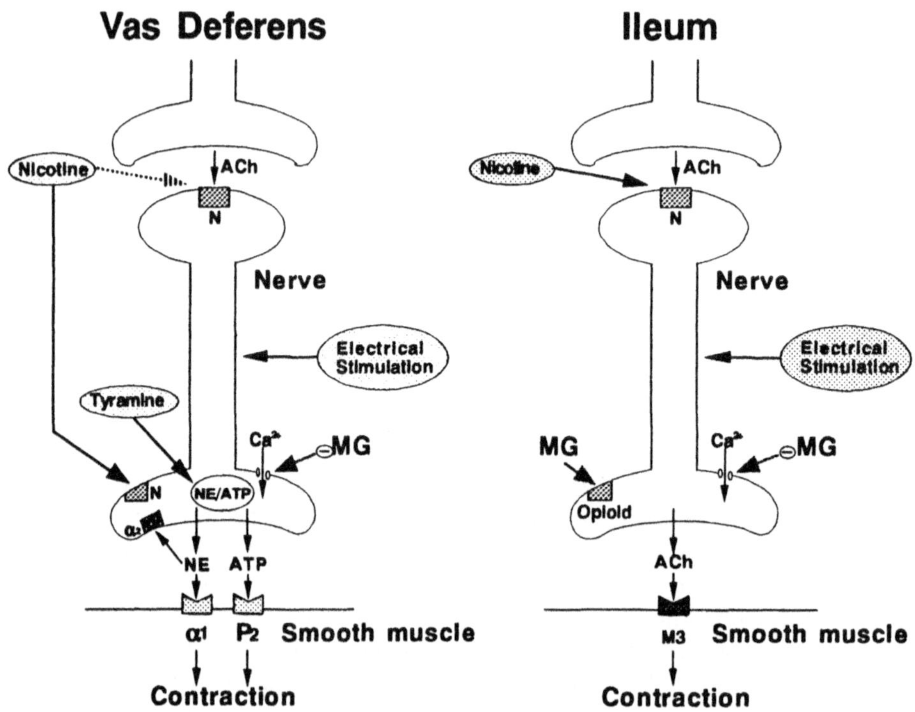

Figure 12 Schematic illustration of the action sites of the inhibitory effect of mitragynine (MG) on neurogenic contractions in the guinea pig vas deferens and ileum (Horie *et al.*, 1997).

Those indoloquinolizidines seemed to have an important role in the therapeutic efficacy of the *Uncaria* hook traditionally used for hypertension or related cardiovascular diseases.

Concerning the opioid-like action of mitragynine, the methoxyl moiety of the A ring of indoloquinolizidines is critically important. None of the indole alkaloids found in the *Uncaria* hook has the indoloquinolizidine constituents which have the methoxyl group in the A-ring. This can be seen in the chemical structures shown in Figure 1. It is important from the therapeutic point of view that this commonly used anti-hypertensive crude drug (*Uncaria* hook) does not contain opiate-like constituents.

SUMMARY AND CONCLUSIONS

Three indole alkaloids (hirsutine, dihydrocorynantheine and mitragynine) are closely related in chemical structure, with an indoloquinolizidine skeleton. Hirsutine and dihydrocorynantheine are components of a Chinese traditional medicine known as *Uncaria* hook, used for hypertension and related symptoms. Mitragynine was isolated from a Thai herbal medicine, *Mitragyna speciosa*, which has been used as an opium substitute in Thailand. Hirsutine was found to have strong hypotensive effects, exhibiting Ca^{2+} channel-blocking activity mainly through the inhibition of the voltage-dependent Ca^{2+} influx.

Mitragynine showed distinct antinociceptive effects in several analgesic assay experiments. In the pharmacological analysis of its actions, mitragynine was found to have a stimulating effect on opioid receptors in the central nervous system, as well as in peripheral nerve terminals in the intestine. The importance of the methoxyl group in the indoloquinolizidine structure was strongly suggested by the opioid effect of the indole alkaloids tested.

In conclusion, we have partly elucidated the mode of action of two popular herbal medicines used in Asia, on the basis of Ca^{2+} antagonistic and opiate receptor actions from the molecular point of view.

ACKNOWLEDGMENTS

This study was supported in part by a grant from the Mombusho International Scientific Research Program: Joint Research (No. 06044035) and by Grants-in-Aid for Scientific Research (No. 07457535, No. 08557128) and for the Encouragement of Young Scientists (No. 09771965) from the Ministry of Education, Sciences, Sports and Culture, Japan. S. Horie is the recipient of a Yamamura Yuichi Memorial WAKAN-YAKU Research Grant, Japan.

REFERENCES

E. Macko, J.A. Weisbach and B. Douglas,. Some observations on the pharmacology of mitragynine. *Arch. Int. Pharmacodyn.* **198**, 145–161 (1972).

D. Ponglux, S. Wongseripipatana, H. Takayama, M. Kikuchi, M. Kurihara, M. Kitajima, N. Aimi and S. Sakai, A new indole alkaloid, 7 α-hydroxy-7H-mitragynine, from *Mitragyna speciosa* in Thailand. *Planta Med.* **60**, 580–581 (1994).

H.F. Miranda, D. Bustamante, V. Kramer, T. Pelissier, H. Saavedra, C. Paeile, E. Fernandez and G. Pinardi, Antinociceptive effects of Ca^{2+} channel blockers. *Eur. J. Pharmacol.* **217**, 137–141 (1992).

H.O.J. Collier, N.J. Cuthbert and D.L. Francis, Model of opiate dependence in the guinea-pig isolated ileum. *Br. J. Pharmacol.* **73**, 921–932 (1981).

H. Horiuchi, S. Yano, K. Watanabe, E. Yamanaka, N. Aimi and S. Sakai, Structure specificities of yohimbine and indoloquinolizidine derivatives in blocking alpha-adrenoceptor and calcium channel. *Res. Comm. Chem. Pathol. Pharmacol.* **59**, 407–410 (1988).

H. Matsuda, G. Kobayashi, J. Yamahara, H. Fujimura, K. Kurahashi and M. Fujiwara, Effects of alismol isolated from *Alismatis* Rhizome on Ca-induced contraction in the rabbit thoracic aorta. *Life Sci.* **41**, 1845–1852 (1987).

I. Von Kügelgen and K. Starke, Noradrenaline-ATP co-transmission in the sympathetic neurons system. *TiPS* **12**, 319–324 (1991).

J. Yamahara, S. Miki, H. Matsuda, G. Kobayashi and H. Fujimura, Screening test for calcium antagonist in natural products. The active principles of Uncaria Ramulus et Uncus. *Folia Pharmacol. Japon.* **90**, 133–139 (1987).

K.L.R. Jansen and C.J. Prast, Ethnopharmacology of kratom and the *Mitragyna* alkaloids. *J. Ethnopharmacol.* **23**, 115–119 (1988).

K. Aisawa, Y. Hattori, T. Kihara, T. Ishihara, K. Endo and H. Hikino, Hypotensive action of 3α-dihydrocadambine, an indole alkaloid glycoside of *Uncaria* hooks. *Planta Med.* **51**, 424–427 (1985).

K. Matsumoto, M. Mizowaki, T. Suchitra, H. Takayama, S. Sakai, N. Aimi and H. Watanabe, Antinociceptive action of mitragynine in mice: evidence for the involvement of supraspinal opioid receptors. *Life Sci.* **59**, 1149–1155 (1996).

K. Watanabe, S. Yano, S. Horie and L.T. Yamamoto, Inhibitory effect of mitragynine, an alkaloid with analgesic effect from Thai medicinal plant, on electrically stimulated contraction of isolated guinea-pig ileum through the opioid receptor. *Life Sci.* **60**, 933–942 (1997).

K. Watanabe, S. Yano, S. Horie, S. Sakai, H. Takayama, D. Ponglux and S. Wongseripipatana, Pharmacological profiles of "Kratom" (*Mitragyna speciosa*), a Thai medical plant with special reference to its analgesic activity. In P. Tongroach, H. Watanabe, D. Ponglux, U. Suvanakoot and N. Ruangrungsi (Eds.), *Advance in Research on Pharmacologically Active Substances from Natural Products.* Chiang Mai University, Chiang Mai, Thailand, pp. 125–132 (1992).

K. Watanabe, S. Yano, H. Horiuchi, E. Yamanaka, N. Aimi and S. Sakai, Ca^{2+} channel-blocking effect of the yohimbine derivatives, 14β-benzoyloxyyohimbine and 14β-nitrobenzoyloxyyohimbine. *J. Pharm. Pharmacol.* **39**, 439–443 (1987).

L.A. Chahl, Withdrawal responses of guinea-pig isolated ileum following brief exposure to opiates and opioid peptides. *Naunyn-Schmiedeberg's Arch. Pharmacol.* **333**, 387–392 (1986).

L.T. Yamamoto, S. Horie, S. Yano, H. Takayama, S. Sakai, N. Aimi, D. Ponglux and K. Watanabe, Opioid-like effect of mitragynine pseudoindoxyl, an analgesic compound related to a Thai medicinal plant, in the isolated guinea-pig ileum. *Jpn. J. Pharmacol.* **71** (Supplement I), 608 (1996).

M. Lujan and R. Rodrigues, Pharmacological characterization of opiate physical dependence in the isolated ileum of the guinea-pig. *Br. J. Pharmacol.* **73**, 859–866 (1981).

N. Aimi, E. Yamanaka, N. Shinma, M. Fujiu, J. Kurita, S. Sakai and J. Haginiwa, Studies on plants containing indole alkaloids, VI. Minor bases of *Ankara rhynchophylla* Miq. *Chem. Pharm. Bull.* **25**, 2067–2077 (1977).

P. Valeri, B. Martinelli, L.A. Morrone and C. Severini, Reproducible withdrawal contractions of isolated guinea-pig ileum after brief morphine exposure: Effects of clonidine and nifedipine. *J. Pharm. Pharmacol.* **42**, 115–120 (1990).

S.J.H. Brookes, P.A. Steele and M. Costa, Identification and immunohistochemistry of cholinergic and non-cholinergic circular muscle motor neurons in the guinea-pig small intestine. *Neuroscience* **42**, 863–879 (1991).

S.R. Chaplan, J.W. Pogrel and T.L. Yaksh, Role of voltage dependent calcium channel subtypes in experimental tactile allodynia. *J. Pharmacol. Exp. Ther.* **269**, 1117–1123 (1994).

S. Horie, S. Yano, N. Aimi, S. Sakai and K. Watanabe, Effects of hirsutine, an antihypertensive indole alkaloid from *Uncaria rhynchophylla*, on intracellular calcium in rat thoracic aorta. *Life Sci.* **50**, 491–498 (1992).

S. Horie, L.T. Yamamoto, Y. Futagami, S. Yano, H. Takayama, S. Sakai, N. Aimi, D. Ponglux, J. Shan, P.K.T. Pang and K. Watanabe, Analgesic, neuronal Ca^{2+} channel-blocking and smooth muscle relaxant activities of mitragynine, an indole-alkaloid, from the Thai folk medicine 'kratom'. *J. Trad. Med.* **12**, 366–367 (1995).

S. Yano, H. Horiuchi, S. Horie, N. Aimi, S. Sakai and K. Watanabe, Ca^{2+} channel blocking effects of hirsutine, an indole alkaloid from *Uncaria* Genus, in the isolated rat aorta. *Planta Med.* **57**, 399–510 (1991).

W.-H. Xiao and G.J. Bennett, Synthetic ω-conopeptides applied to the site of nerve injury suppress neuropathic pains in rats. *J. Pharmacol. Exp. Ther.* **274**, 666–672 (1995).

Y. Ito, S. Yano, K. Watanabe, E. Yamanaka, N. Aimi and S. Sakai, Structure-activity relationship of yohimbine and its related analogs in blocking alpha-1 and alpha-2 adrenoceptors: A comparative study of cardiovascular activities. *Chem. Pharm. Bull.* **38**, 1702–1706 (1990).

Y. Ozaki, Pharmacological studies of indole alkaloids obtained from domestic plants, *Uncaria rhynchophylla* MIQ. and *Amsonia elliptica* ROEM. *Folia Pharmacol. Japon.* **94**, 17–26 (1989).

Y. Ozaki, Vasodilative effects of indole alkaloids obtained from domestic plants, *Uncaria rhynchophylla* MIQ. and *Amsonia elliptica* ROEM. *Folia Pharmacol. Japon.* **95**, 47–54 (1990).

12 THE ROLE OF BIOACTIVE POLYSACCHARIDES IN KAMPO MEDICINES

HARUKI YAMADA

Oriental Medicine Research Center, The Kitasato Institute, 5-9-1, Shirokane, Minato-ku, Tokyo 108-0072, Japan

The use of kampo medicines began more than 2000 years ago, and thus much cumulative experience and knowledge regarding their use have been acquired. Kampo medicines are often formulations consisting of component herbs, and each formulation is known by a characteristic name. The component herbs of kampo medicines contain substances with both low and high molecular weight. Typical of the former are alkaloids, terpenoids, saponins and flavonoids, whereas proteins, tannins and polysaccharides are representative of the latter group. Although biologically active low molecular weight substances in kampo medicines have been studied, they cannot account for all of the clinical effects achieved. Pharmacological activities have often been observed in fractions with high molecular weight from boiled water extracts of the herbs prescribed in kampo medicines. Of the fractions with a high molecular weight, polysaccharides have been shown to possess immunomodulating, anticoagulative, antiemetic, anti-ulcer and hypoglycemic activities.[1,2] Therefore, the elucidation of the role of polysaccharides in kampo medicines may provide an important basis to evaluate the efficacy of a formulated kampo preparation.

The present chapter deals with non-clinical studies of immunostimulating polysaccharides using a particular kampo medicine, Juzen-Taiho-To (Chinese name = Shi-Quan-Da-Bu-Tang), as an example, and recent advances in the study of bioactive polysaccharides in component Chinese and Japanese herbs, using an anti-polysaccharide antibody.

PHARMACOLOGICAL STUDY OF JUZEN-TAIHO-TO

Juzen-Taiho-To (JTT) has been administered to patients with anemia, anorexia or a debilitated general condition caused by surgery, chronic disease or childbirth. The clinical effects observed suggest that JTT enhances immune responses and improves the functioning of hemopoietic systems. JTT consists of ten component herbs: Astragali radix (roots of *Astragalus membranaceus* Bunge), Cinnamomi cortex (bark of *Cinnamomum cassia* Blume), Rehmanniae radix (roots of *Rehmmania glutinasa* Libosch var. *purpurea* Makino), Paeoniae radix (rhizome of *Paeonia lactiflora* Pall), Cnidii rhizoma (rhizome of *Cnidium officinale* Makino), Atractylodis lanceae rhizoma (rhizome of *Atractylodes lancea* DC.), Angelicae radix (root of *Angelica actutiloba* Kitagawa), Ginseng radix (root of *Panax ginseng* C.A. Meyer),

Hoelen (fungus of *Poria cocos* Wolf) and Glycyrrhizae radix (roots of *Glycyrrhiza uralensis* Fisch *et* DC.).

Towards a determination of the clinical efficacy of JTT, the pharmacological effects of JTT on immune and hemopoietic systems have been studied. JTT influences the specifically immune system; it significantly enhanced the sheep red blood cell (SRBC) response,[3] phagocytosis[4] and mitogenic activities[5] against spleen B cells in mice. The peripheral blood cell count in patients treated with JTT (orally) and the anti-cancer drug mitomycin C concomitantly, was higher than that of patients administered the anti-cancer drug alone.[6] Concomitant administrations of JTT with mitomycin C produced significantly extended survival periods in P-388 tumor-bearing mice compared with those treated with mitomycin C alone.[7] JTT also facilitated hematopoietic recovery from mitomycin C-induced bone marrow or radiation injuries.[8, 9] These immuno- and hematopoietic-stimulating activities may be closely related to the clinical effects of JTT. Therefore, the efficacy of JTT can be further evaluated if immuno-modulating substances are isolated and characterized from the preparation. TJ-48, a spray-dried powder of JTT extract preparation, enhances antibody and interleukin-2 (IL-2) production as well as promoting granulocyte-macrophage colony-forming cells (GM-CFC), anti-complementary and mitogenic activites. To isolate these immunopharmacologically active principles, TJ-48 was fractionated into low and high molecular weight fractions. Although the methanol-soluble (F-1) and dialyzable (F3) fractions showed GM-CFC enhancing activities, and the methanol/water-insoluble (F-2) fraction showed anti-complementary activity, the crude polysaccharide fraction (F-5) enhanced antibody and IL-2 production beside showing anti-complementary and mitogenic activities (Figure 1).[10–14] Of the TJ-48-derived fractions, only F-5 (300 mg/kg) displayed an anti-SRBC antibody response *in vivo* when the polysaccharide fraction (F-5) was administered intraperitoneally or orally to normal Balb/c mice (7 weeks old).[14] TJ-48 has been used clinically for the treatment of immuno-compromised patients. It is known that the functions of T cells are reduced with aging. Although aged (6-month-old) Balb/c mice produced a lower level of anti-SRBC-IgG in comparison with young (8-week-old) Balb/c mice, the anti-SRBC-IgG response of the aged mice was stimulated significantly when F-5 or TJ-48 was orally administered to the aged mice from 6 days before immunization (Figure 2).[14] These findings suggest that polysaccharides in F-5 might enhance the function of helper T cells, corresponding to class switching of the antibody in the aged mice. Reticuloendothelial cells such as macrophages are known to contribute to the antibody response, and the response was accelerated by the administration of carrageenan, which inhibits the function of macrophages. When Balb/c mice (8 weeks old) were intraperitoneally injected with *i*-carrageenan before immunization, the anti-SRBC-IgM and IgG responses were significantly increased in comparison with those of normal mice.[14] Oral administrations of TJ-48 (1 g/kg) or F-5 (500 mg/kg) to the *i*-carrageenan-treated mice reduced the level of the anti-SRBC antibody response to near that of normal mice.[14] *cis*-Diaminedichloroplatinum (CDDP) is a widely used anti-tumor drug; however, it has severe adverse effects such as immuno-toxicity, renal toxicity, and an emetic effect. TJ-48 increased the survival rate and anti-SRBC response of CDDP-administered mice.[15] Although the production of anti-SRBC-IgM in the CDDP-treated mice was decreased in comparison with that of normal mice, the anti-SRBC-

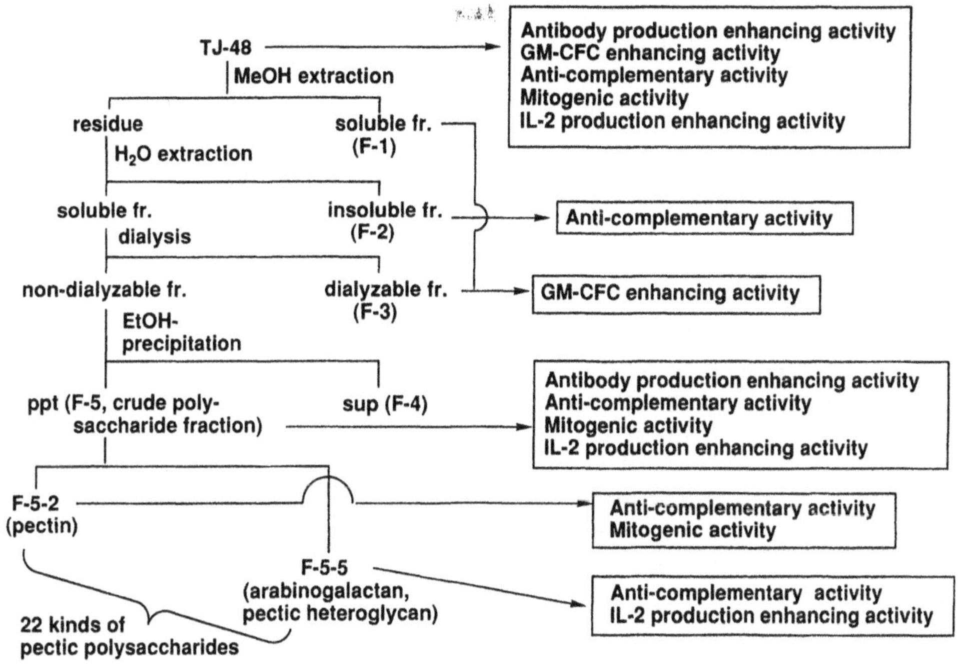

Figure 1 Fractionation and *in vitro* immuno-modulating activities of Juzen-Taiho-To extract preparation (TJ-48).

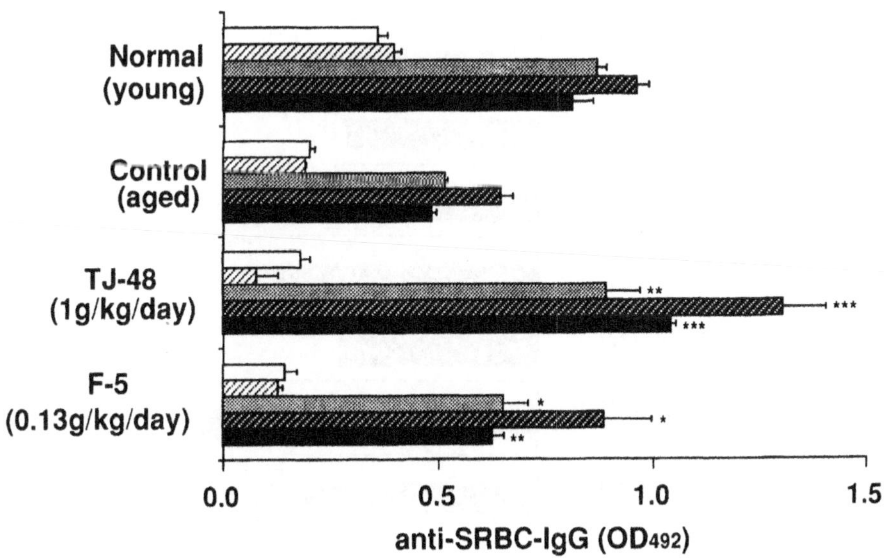

Figure 2 Effect of oral administration of polysaccharide fraction (F-5) from TJ-48 on anti-SRBC antibody response of aged Balb/c mice. Titers of anti-SRBC-IgG in plasma were measured at 7(□), 8(▨), 11 (■), 13 (▨) and 15 days (■) after stimulation. Each value represents the mean ± S.E. of 5 mice, and was compared with that in the control mice. * $p < 0.1$, ** $p < 0.05$, *** $p < 0.001$.

IgM response was increased to the level of normal mice when F-5 was orally administered to the CDDP-treated mice before the first injection of CDDP (Figure 3).[14] The anti-SRBC-response was also stimulated by oral administrations of the other fractions: the methanol-soluble fraction (F-1) and a mixture of water-soluble dialyzable (F-3) and ethanol-soluble non-dialyzable (F-4) fractions from TJ-48, indicating that the response was due to the combination effect of several components containing polysaccharides.[14] The survival rate of CDDP-treated ICR mice was increased by oral administration of F-5, whereas the mortality of the CDDP-treated mice was not changed by oral administration of the methanol-soluble fraction (F-1) (Figure 4).[16] When F-1 is deleted from TJ-48, the remaining fraction containing F-2, F-3, F-4 and F-5 shows a more potent protective effect against the lethal toxicity of CDDP. The oral administration of F-5 (130 mg/kg/day) as well as that of TJ-48 (1 g/kg/day) decreased the level of blood urea nitrogen (BUN) in the plasma of CDDP-treated Balb/c mice (Figure 5).[16] Other fractions, i.e., F-1 and a mixture of F-3 and F-4, seem to be more effective than F-5. These results suggest that the protective effect of TJ-48 against the renal toxicity of CDDP also involves the combination effect of low-molecular- weight substances and pectic polysaccharides. Sugiyama *et al.* recently reported that Angelicae Radix, Ginseng and Hoelen significantly reduced the nephrotoxicity of CDDP in the component herbs of JTT.[17] They also isolated sodium malate from Angelicae Radix as an active gradient reducing the bone marrow toxicity of CDDP. When sodium malate (p.o.) and CDDP (i.p.) with a molar ratio over 0.5:1 was used, the renal, hepatic and gastrointestinal toxicities of CDDP were suppressed almost

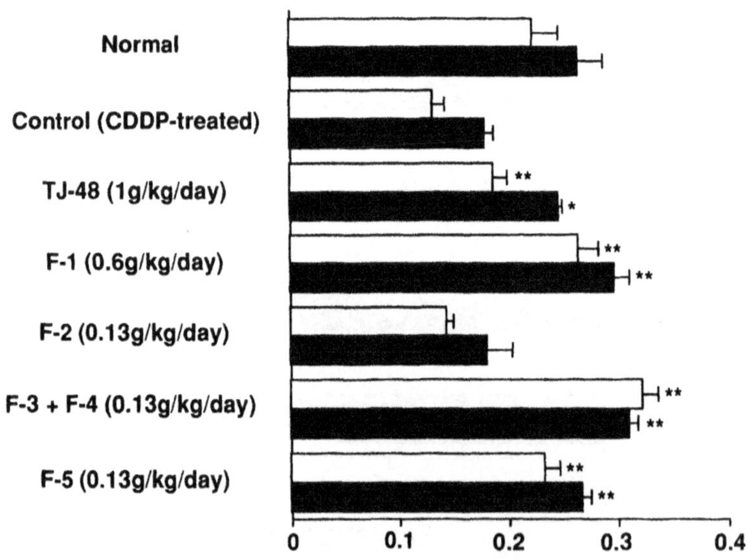

Figure 3 Effect of oral administration of the subfractions from TJ-48 on anti-SRBC-antibody response of *cis*-CDDP-treated Balb/c mice. The titer of anti-SRBC-IgM was measured at 6 (□) and 7 days (■) after immunization. Each value represents the mean ± S.E. from 5 mice, and was compared with the *cis*-CDDP-treated control mice. * $p < 0.05$, ** $p < 0.01$.

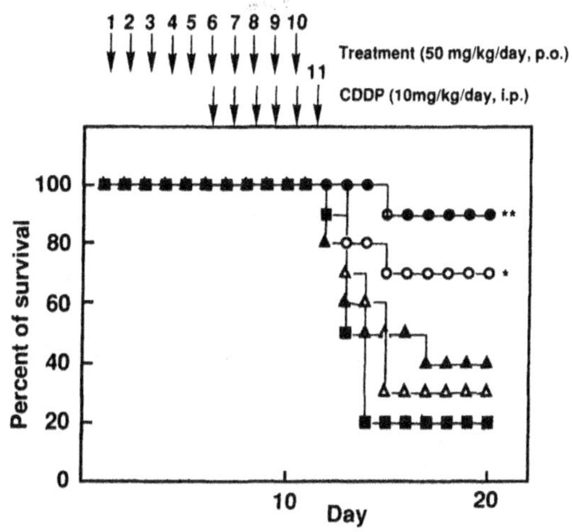

Figure 4 Effect of oral administration of franctions from TJ-48 on the mortality of CDDP-treated ICR mice. ICR mice, which were treated with CDDP (10 mg/kg/day, i.p.,) from 6–11 days after the first oral administration of samples, were administered orally with fraction (50 mg/kg/day) every day at days 1–15. ICR mice were administered with (■), water (control); (▲), TJ-48; (●), F-2~F-5 from which were removed the MeOH-solube substances from TJ-48; (△), MeOH-solube fraction (F-1); and (○), polysaccharide fraction (F-5). Each group ($n = 10$) was compared with the control group by Wilcoxon test. * $p < 0.05$, ** $p < 0.01$. No significance was observed between the survival periods of mice administered F-5 and F-2~F-5.

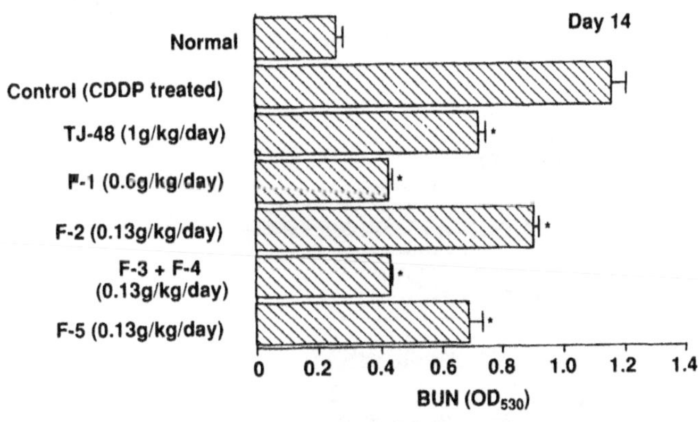

Figure 5 Effect of oral administration of fractions (F-1~F-5) from TJ-48 on the level of blood urea nitrogen (BUN) in CDDP-treated mice. BUN levels were measured at day 14 after the first oral administration. Each value represents the mean ± S.E. of 5 mice, and was compared with that in the control mice. *$p < 0.001$.

completely without affecting the anti-tumor effect of CDDP. These findings suggest that the reducing effect of BUN-level in CDDP-treated mice by the methanol-soluble fraction (F-1) and dialyzable fraction (F-3) from TJ-48 may be caused by sodium malate. However, those investigators[17] used the roots of *Angelica sinensis* (a Chinese herb) as Angelicae Radix, which is different from that of *Angelica acutiloba* (a Japanese herb), in TJ-48 for the preparation of JTT. The possibility exists that the difference in active ingredients arises from the type of Angelicae Radix used as a component herb. The mechanism of action of this toxicity-attenuating effect remains unknown. We recently isolated an anti-ulcer pectic polysaccharide (bupleuran 2IIb) with free radical scavenging activities from Bupleuri Radix (the roots of *Bupleurum falcatum*); this activity may be related to the anti-ulcer effects of the polysaccharide.[18] This finding suggests that the free radical scavenging activity may be involved in the attenuation effects of the polysaccharide fraction on CDDP-induced toxicities.

When F-5 is further fractionated by cetyltrimethylammonium bromide precipitation, four polysaccharide fractions are obtained. The acidic polysaccharide fraction F-5-2 shows anti-complementary and mitogenic activities, while both weakly acidic polysaccharides F-5-3 and F-5-4 have merely anti-complementary effects. The neutral polysaccharide fraction, F-5-5, enhances not only anti-complementary activity but also IL-2 production (Figure 1). F-5-2 at the dose of 100 mg/kg/day also decreased the level of BUN in CDDP-treated mice, but F-5-5 did not show this effect (Figure 6).[16] On further fractionating of the individual acidic and neutral polysaccharide fractions (F-5-2 and F-5-5) *per se* by ion exchange chromatography on DEAE-Sephadex, twenty-two different polysaccharides with varied immunostimulating activities were obtained.[11] The fractionated pectic polysaccharides were classified into three polysaccharide groups: (i) pectins with anti-complementary and mitogenic activities, (ii) pectic arabinogalactans with mere anti-complementary activities, and (iii) pectic heteroglycans with IL-2 enhancement effects.[11, 12] These pectic polysaccharides are composed of two common polysaccharide moieties, the galacturonan [$\alpha(1\to4)$-linked galacturonic acid polymer] and ramified (rhamnogalacturonan core with arabinogalactan and/or

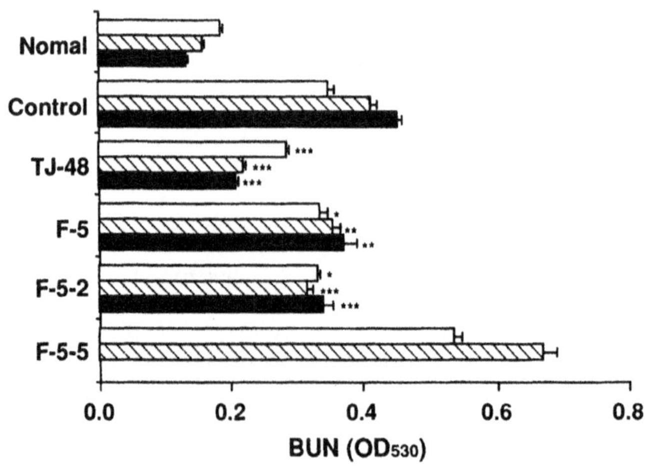

Figure 6 Effect of oral administration of acidic and neutral polysaccharides (F-5-2 and F-5-5) from TJ-48 on the level of BUN in CDDP-treated mice. Levels of BUN were measured at 10 (□), 11 (▨) and 12 days (■) after the first oral administration. Each value represents the mean ± S.E. of 5 mice, and was compared with that in control mice. * $p < 0.1$, ** $p < 0.01$, *** $p < 0.001$.

arabino and galactooligosaccharides side-chains) regions (Figure 7). Pectin and pectic arabinogalactan consist mainly of galacturonan and arabinogalactans, respectively, whereas pectic heteroglycan is composed of ramified and galacturonan regions. However, the neutral sugar side-chain in the heteroglycan consists of glucose and mannose in addition to the arabinose and galactose found in the former two polysaccharides.[11, 12] The ramified region has been suggested to be involved in mitogenic activity.[11]

The acidic pectic polysaccharide fraction (F-5-2) induced the proliferation of B cells as well as TJ-48 (Figure 8), particulary those holding surface IgM- and IgD-positive cells in a manner similar to that of TJ-48 (Table 1).[18] However, another B cell mitogen, lipopolysaccharide (LPS), did not influence the IgD-positive cell count. This result indicates that B cell mitogen in the pectin of TJ-48 is different from LPS. The proliferation induced by F-5-2 was T cell-independent but macrophage-dependent.[19] The macrophages can be substituted for a soluble factor(s) secreted from the macrophages but not for IL-1 (Figure 9).[19] B cell mitogens are generally known to induce the proliferation of B cells and subsequently, their differentiation into plasma cells. However, although F-5-2 induced B cell differentiation, it arrested their development in the intermediate stage of B cell differentiation.[19] The B cells induced by F-5-2 produced IgM antibody in response to IL-6 and an antigen (SRBC) but not in response to IgG antibody (Figure 10).[19] The other cytokines IL-1, 2, 3 and 4 did not produce the antibody. F-5-2 induced the expression of IL-6 receptor as a marker of lymphocyte

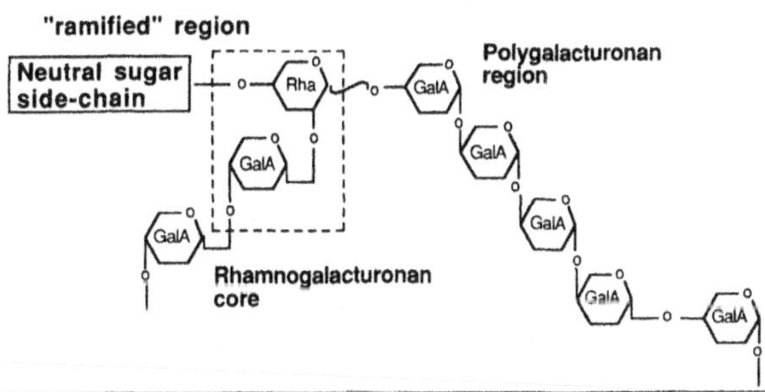

Group	Immuno-modulating activity	"Ramified" region	Polygalacturonan region	Neutral sugar side chain
Pectins (F-5-2)	Anti-complementary activity Mitogenic activity	+	++++	Arabino-oligosaccharides Galacto-oligosaccharides
Pectic arabino-galactans (F-5-5)	Anti-complementary activity	+++	-~+	Arabinogalactans
Pectic heteroglycans (F-5-5)	IL-2 production enhancing activity	++~+	+~++	Neutral side chains consisting of Arabinose, Galactose, Glucose and Mannose

Figure 7 Characterization of immuno-stimulating pectic polysaccharides from Juzen-Taiho-To extract preparation (TJ-48).

Table 1 Fluorocytographic analysis of TJ-48, F-5-2 and LPS-proliferated spleen cells *in vitro.*

Sample	Positive cells (%)								
	Blast cells	Total lg^+	lgM^+	LgG^+	lgD^+	Thy 1.2^+	$L3T4^+$	$Lyt2^+$	L3T4/Lyt2
Control	09.0 ± 1.06	53.8 ± 0.06	51.7 ± 0.55	46.2 ± 4.26	51.5 ± 1.01	41.4 ± 0.40	29.3 ± 0.83	13.6 ± 0.54	2.2 ± 0.15
TJ-48	21.2 ± 2.34	64.5 ± 1.28	62.9 ± 0.73	59.3 ± 1.14	64.1 ± 0.84	29.3 ± 1.14	21.9 ± 0.71	10.4 ± 0.24	2.1 ± 0.12
F-5-2	27.8 ± 2.21	68.6 ± 0.21	66.2 ± 0.80	69.3 ± 0.27	65.9 ± 2.50	29.2 ± 0.21	22.1 ± 0.02	9.8 ± 0.42	2.3 ± 0.10
LPS	36.2 ± 0.54	74.2 ± 0.29	76.6 ± 0.16	76.5 ± 0.49	17.0 ± 0.43	24.4 ± 0.70	18.3 ± 0.58	8.2 ± 2.2	2.2 ± 0.13

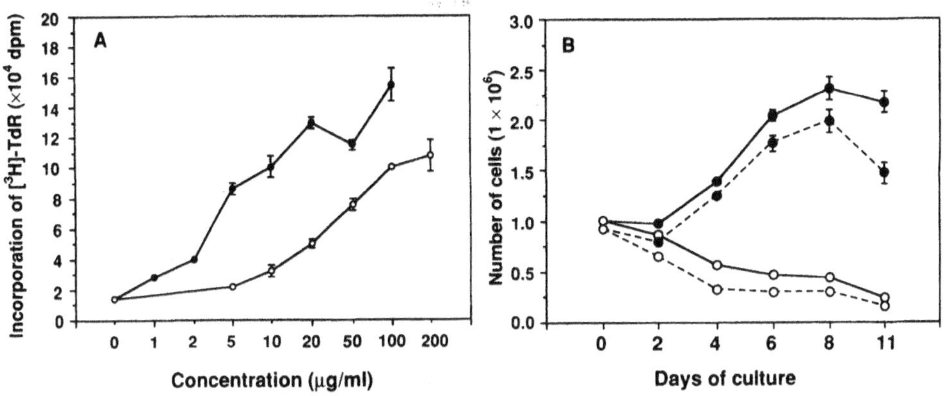

Figure 8 Effect of TJ-48 and F-5-2 on the proliferation of spleen cells.
A: BALB/c spleen cells were cultured with various concentrations of TJ-48 (○) or F-5-2 (●) for 4 days. The proliferation was measured by the incorporation of [^{3}H]-TdR, and the results are expressed as the mean dpm ± S.E. of triplicated cultures.
B: Spleen cells were cultured with F-5-2 (20 mg/ml, ●) or medium alone (○) for the indicated number of days, and the culture medium was changed every 3 days. The cell number was counted using the trypan blue dye-exclusion technique. A solid line indicates the total cell number and a dotted line indicates the number of viable cells.

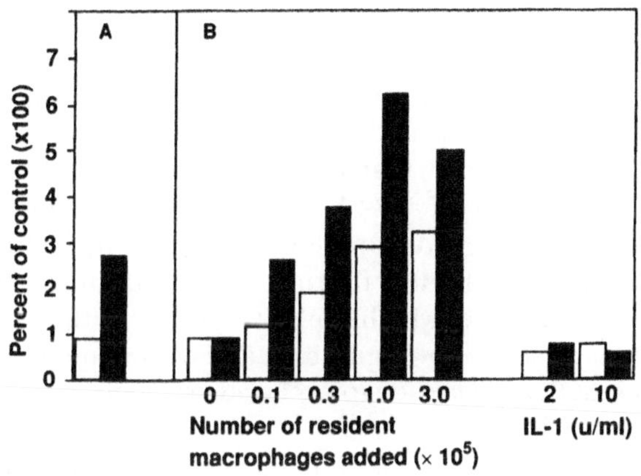

Figure 9 Role of macrophages in F-5-2-induced B-cell proliferation. Cultures were performed using a Millicell system, and the proliferation was assayed using the MTT method.
A: Whole spleen cells (5 × 10^5) from BALB/c *nu/nu* mice were cultured in the inner chamber with (■) or without (□) 20 μg/ml F-5-2 for 3 days.
B: B-cells purified from spleen cells (inner chamber) were cultured with (■) or without (□) 20 μg/ml F-5-2 in the presence or absence of macrophages or IL-1 instead of macrophages (outer chamber). Values are expressed as the means of duplicate cultures.

differentiation not only on IgM$^+$ and IgG$^+$ B cells but also on IgD$^+$ B cells.[19] F-5-2 therefore seems to be a new type of B cell mitogen.

Of the ten component herbs in JTT, Glycyrrhizae Radix contains the most potent mitogenic polysaccharide,[20] and the purified polysaccharide (GR-2IIc) has

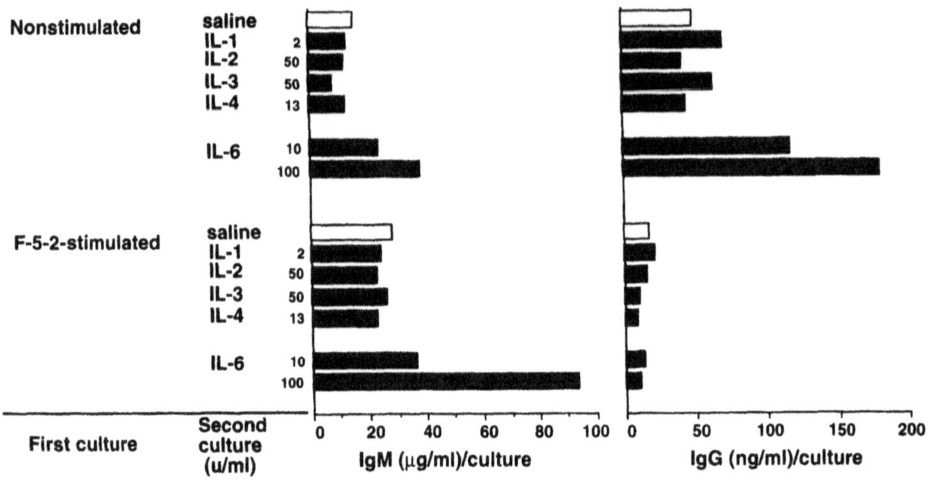

Figure 10 Induction of F-5-2-stimulated cells to antibody-producing cells by cytokines. Spleen cells from BALB/c *nu/nu* mice were pre-cultured with or without 20 μg/ml F-5-2 for 3 days. The cells were harvested and washed twice with culture medium. Viable cells (1×10^6) were cultured with the indicated cytokine for 3 days. The culture supernatants were collected and IgM and IgG were measured by enzyme-linked immunosorbent assay.

chemical properties similar to those of the active polysaccharides in F-5-2,[21, 22] although some other herbs also contain the active polysaccharide fractions.[20] GR-2IIc also plays an important role in the expression of anti-complementary activity in TJ-48.[22] We recently observed that the neutral oligosaccharide chains in the "ramified" region (PG-1c) of GR-2IIc are responsible for the expression of mitogenic activity as well as anti-complementary activity.[22, 23] This observation is consistent with the previous finding that oligosaccharide-alditol fractions consisting mainly of 6-linked galactose, which are obtained from the anti-complementary pectin of *Angelica acutiloba*, show significant anti-complementary activity.[2, 24] Thus, these results suggest that certain oligosaccharide structures may be involved in the expression of the activities.

Since roots have been used in many medicinal herbs, contamination with LPS is possible. LPS, which is a structural component of the outer membranes of Gram-negative bacteria, also has pharmacogical activities such as immunostimulating activity. Therefore, it is important to estimate whether the immunostimulating activity of bioactive polysaccharides from kampo medicines is due to LPS-contamination or the drug itself. The properties of F-5-2 from TJ-48 are distinct from those of LPS: F-5-2 showed similar mitogenic activity on spleen cells from C3H/HeJ mice, which are known to have no sensitity to LPS; F-5-2 required the adherent cells, possibly a factor from those cells, but LPS did not; and the effect of F-5-2 on antibody production was different from that of LPS.[19] Although LPS can be absorbed effectively by polymyxin B-affinity chromatography, acidic polysaccharides are also absorbed to polymyxin B. This method is therefore not available for the removal of LPS from acidic polysaccharide solution.[25] Aida and

Pabst reported the removal of LPS from protein solutions by phase separation using Triton X-114.[26] Bioactive pectic polysaccharides, bupleuran 2IIc, from *Bupleurum falcatum* are positive in the chromogenic Limulus amebocyte lysate (LAL) assay which is sensitive to LPS.[27] However, when the LAL-positive substance was removed from bupleuran 2IIc and its fraction PG-1 by phase separation using Triton X-114, the mitogenic and anti-complementary activities did not change.[27] Therefore, this method is available to estimate whether the activity is due to the polysaccharide itself or to LPS.

TJ-48 and the polysaccharide fraction (F-5) also stimulated the production of growth factors in Peyer's patch cells *in vitro* and *in vivo*, and enhanced the proliferation of bone marrow cells (Figure 11).[28] A part of the growth factors was suggested to be IL-6 and GM-CSF, from the neutralization results using each specific antibody (unpublished). These results suggest that the immunomodulating activities of TJ-48 are expressed mainly as the consequence of a combination of the effects of several different pectic polysaccharides through mucous immunity.

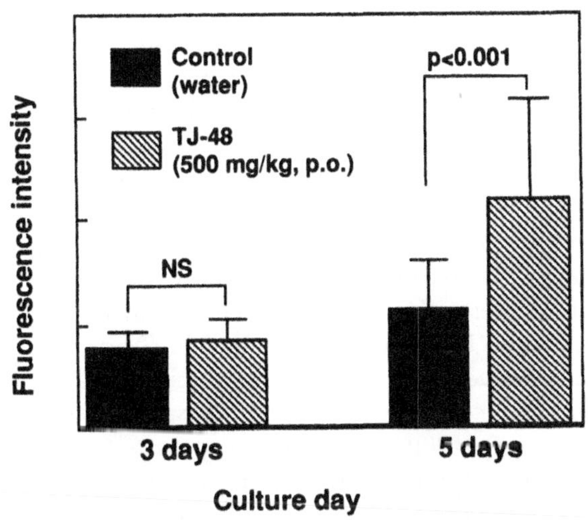

Figure 11 Effect of orally administered Juzen-taiho-to extract preparation on production of hematopoietic factor from Peyer's patch cells. Juzen-taiho-to extract preparation was administered to mice orally at a dose of 500 mg/kg; 18 hrs after administration, the Peyer's patch cells were removed from mice, and then the cells were cultured for 3 or 5 days. The resulting culture supernatant was used for stimulation of bone marrow cells. Proliferative responses of bone marrow cells were measured by Alamar blue reduction assay, and the results are expressed as the mean ± S.D. of bone marrow cell cultures.

PREPARATION OF ANTI-POLYSACCHARIDE ANTIBODY AND ITS APPLICATION

Several pharmacological activities have been observed in pectic polysaccharides such as pectins, pectic arabinogalactans and pectic heteroglycans in addition to low-molecular-weight substances from some kampo formulas and their component herbs. We have isolated several immunostimulating polysaccharides and anti-ulcer polysaccharides from Chinese and Japanese herbs and studied their structures and structure–activity relationship.[1,2] Most of the active polysaccharides obtained in our laboratory have been considered to be pectic polysaccharides consisting of a galacturonan region and a rhamnogalacturonan core with neutral carbohydrate side chains ("ramified" region).[2] Bioactive pectins from medicinal herbs also consist of a unique rhamnogalacturonan-II (RGII)-like region containing 2-keto-3-deoxyocturosonic acid (KDO), which is known as a component of LPS.[29] Several LPS receptor molecules have been identified on the cell surface of lymphocytes, monocytes and macrophages, and one of the LPS receptors has been characterized as having a potential specificity for KDO determinant on LPS.[30] The KDO residues in LPS play a major role in the signal leading to IL-1 production by human monocytes, possibly through LPS receptor.[31] Thus there is a possibility that the RGII-like region containing KDO residues in the pectins are recognized by KDO receptors on the cell surface to act as a trigger for certain biological activities. Because pectic polysaccharides have been considered to be essential for the expression of the clinical effects of some kampo medicines such as Juzen-Taiho-To, their applications to the quality control of kampo medicines and their pharmacokinetics are very important. However, no specific reagent for the detection and quantification of pharmacologically active pectic polysaccharides is available. As described in the previous section, bioactive pectic polysaccharides contain a specific carbohydrate sequence for the expression of the activity. It has been considered that the polysaccharides isolated from natural plants generally have a microheterogeneity due to different biosynthetic processes. The fine structures of these polysaccharides also may be variable due to different harvesting times and cultivation sites. Therefore, the quality control of medicinal herbs and pharmacokinetics using bioactive polysaccharide should be performed for the active parts in the polysaccharide. Because the ramified region is essential for most of the pharmacological activities of the active polysaccharides, the antibody against the ramified region is very useful for several purposes.

Anti-sera against a complement-activating pectin (AR-2IIb), which is purified from the roots of *Angelica acutiloba* Kitagawa,[32,33] were prepared by the immunization of rabbits, and a polyclonal anti-AR-2IIb antibody of the IgG class was purified by affinity chromatography on AR-2IIb-immobilized Sepharose and Protein G-Sepharose.[34] The periodate oxidation of AR-2IIb significantly reduced its inhibitory activity on the reactivity of AR-2IIb to anti-AR-2IIb-IgG, but pronase digestion of AR-2IIb did not affect its inhibitory activity, indicating that the antibody recognizes the carbohydrate moiety as the epitope. Other pharmacologically active pectins from *A. acutiloba, Bupleurum falcatum* and *Glycyrrhiza uralensis* and the complement-activating pectic arabinogalactan from *A. acutiloba* also show significant inhibitory activities on the reactivity of AR-2IIb to anti-AR-2IIb-IgG, but these inhibitory activities are lower than that of AR-2IIb.[34] Other pectins, polygalacturonic acid, arabinogalactan, galactan and araban, tested have negligible inhibitory activity. The endo-$\alpha(1\rightarrow4)$-polygalacturonase digestion of AR-2IIb gives the "ramified" region, KDO-containing region and oligogalacturonides. When these carbohydrate fragments were tested the inhibitory activity by an enzyme-linked immunosorbent assay (ELISA) method, only the ramified region showed potent activity, indicating that it contains an antigenic epitope (Figure 12).[34] Although carboxyl-reduction of galacturonic acid in the "ramified" region decreases the inhibitory activity of the "ramified" region on its reactivity to anti-AR-2IIb, an acidic tetrasaccharide unit in the rhamnogalacturonan core has negligible inhibitory activity (Figure 12).[34] Because α(1→4)-linked polygalacturonic acid does not inhibit the reaction between anti-AR-2IIb and the ramified region of AR-2IIb, anti-AR-2IIb may recognize a neutral carbohydrate moiety containing uronic acid in the region which is the active site in the complement-activating pectin AR-2IIb. Anti-AR-2IIb-IgG may therefore be a useful tool for the quality control and study of the pharmacokinetics of the polysaccharide.

The roots of *Bupleurum falcatum* L. (Japanese name = Saiko) have been used in Chinese and Japanese herbal medicines for the treatment of chronic hepatitis, nephrosis syndrome, and autoimmune diseases. Pharmacologically active polysaccharides having anti-complementary activity,[35] macrophage Fc receptor up-regulating[36] and anti-ulcer activities[37, 38] have been isolated from the hot water extract of the roots of *B. falcatum*, and characterized as pectic polysaccharides.[35, 39–41] Anti-sera against the ramified region (PG-1) of an anti-ulcer polysaccharide (bupleuran 2IIc) which was purified from *B. falcatum* L. were obtained by the immunization of rabbits, and a polyclonal anti-bupleuran 2IIc/PG-1 antibody of the IgG class was purified by affinity chromatography.[42] The antigenic specificity of anti-bupleuran 2IIc/PG-1-IgG was examined by a two-site sandwich ELISA which was developed as an improved method for the microanalysis of bupleuran 2IIc, using a biotinylated antibody (Figure 13).[42] Another pectin with Fc receptor up-regulating activity, bupleuran 2IIb from *B. falcatum*, and anti-complementary pectins from *A. acutiloba* and *G. uralensis* also showed significant reactivity to anti-bupleuran 2IIc/PG-1-IgG, although these reactivities were lower than that of bupleuran 2IIc.[42] Other polysaccharides tested, such as apple pectin, araban, yeast mannan, pulllulan,

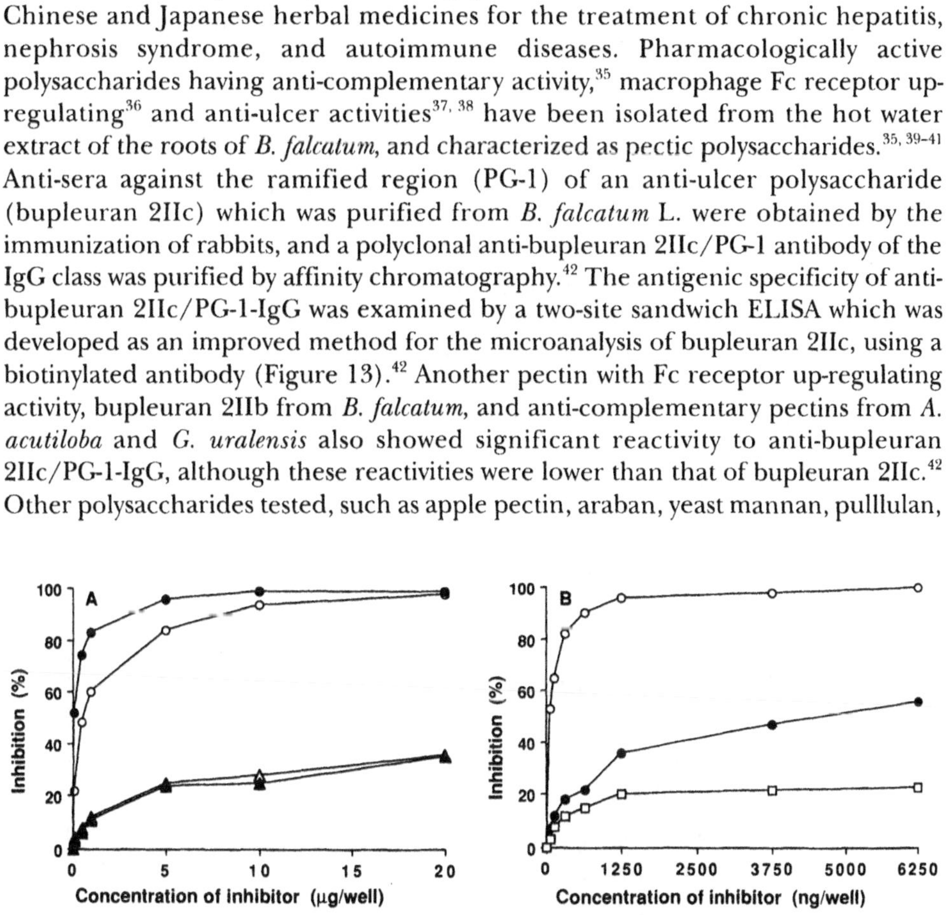

Figure 12 (A) Inhibitory activity of endo-polygalacturonase-digested products from AR-2llb on the reactivity of AR-2llb to anti-AR-2llb-lgG. AR-2llb (1000 ng/well) was coated in the plate. ○, AR-2llb; ●, "ramified" region of AR-2llb; △, KDO-containing region of AR-2llb; ▲, oligogalacturonides from AR-2llb.

(B) Inhibitory activity of carboxyl-reduced "ramified" region and acidic tetrasaccharide on the reactivity of the ramified region to anti-AR-2llb-lgG. Ramified region (125 ng/well) was coated in the plate. ○, ramified region; ●, carboxyl-reduced ramified region; □, methylglycoside of α-L-Rha-(1→4)-α-D-GalA-(1→2)-α-L-Rha-(1→4)-α-D-GalA.

etc. had negligible reactivity (Figure 13).[42] The KDO-containing region and oligogalacturonides which are obtained by the endo-$\alpha(1\rightarrow4)$-polygalacturonase digestion of bupleuran 2IIc were also not significantly recognized by anti-bupleuran 2IIc/PG-1-IgG. These results suggest that the ramified region of bupleuran 2IIc can be detected by sandwich ELISA using anti-bupleuran 2IIc/PG-1-IgG. When bupleuran 2IIc was administered to mice intravenously, the polysaccharide level in the circulation decreased within 24 h and was detected by the two-site sandwich ELISA mainly in the liver.[42] However, the appearance of bupleuran 2IIc from the circulation was delayed by treatment with *i*-carrageenan, which is a representative blocking agent for the phagocytic system (Figure 14).[42] Since kampo medicines are generally given orally to patients, it is necessary to analyze the pharmacokinetics after oral administration of polysaccharides. When the crude polysaccharide fraction (BR-2), containing mainly bupleuran 2IIb and 2IIc from *B. falcatum*, was orally administered to mice, the polysaccharides were detected in the liver and Peyer's patch one week after administration by immunohistochemical staining using biotinylated anti-bupleuran 2IIc/PG-1-IgG.[42] Bupleuran 2IIc/PG-1 was also detected in the liver homogenate one week after oral administration by a sandwich ELISA method (Figure 15).[42] However, when the citrus pectin was orally administered to the mice, the pectin could not be detected in the serum or liver by the same antibody. These results suggest that at least a part of orally administered bupleuran 2IIc is absorbed in the body.

For the study of the pharmacokinetics of pharmacologically active polysaccharides, fluorescent or radio-labeled polysaccharide has been used in order to trace the polysaccharide in the body. However, there are problems in that the uniform labeling of the probe for the polysaccharide is very difficult, and the fluorescence labeling may modify the polysaccharide structure. Two-site sandwich

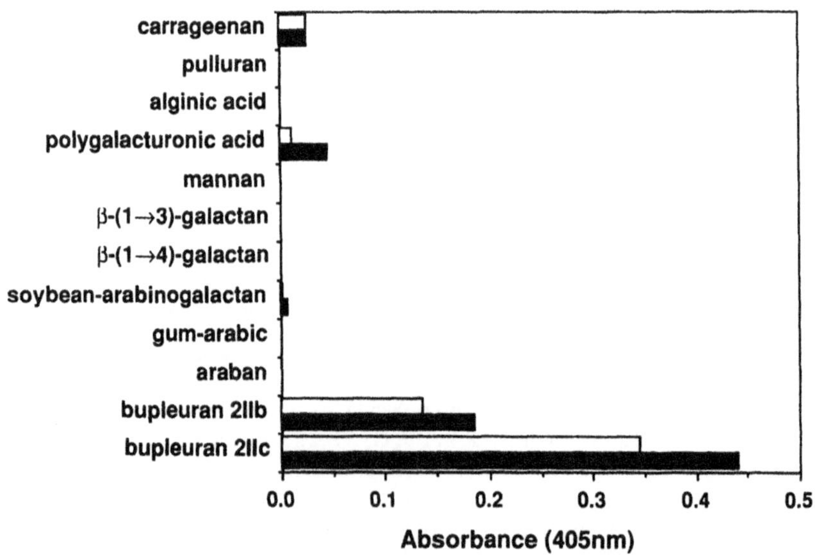

Figure 13 Reactivity of various polysaccharides to anti-bupleuran 2IIc/PG-1-IgG on two-site sandwich ELISA. □, 500 ng/well; ■, 2000 ng/well.

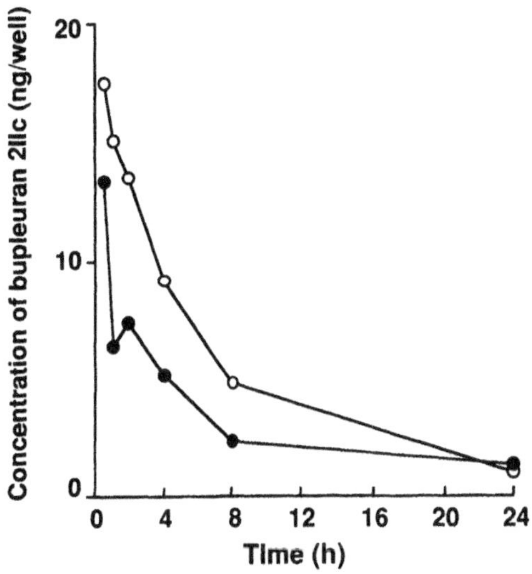

Figure 14 The effect of carrageenan on the clearance of intravenously administered bupleuran 2llc from circulation. ○, carrageenan-treated mice; ●, saline-treated controls. Data are means ± S.D. n = 5.

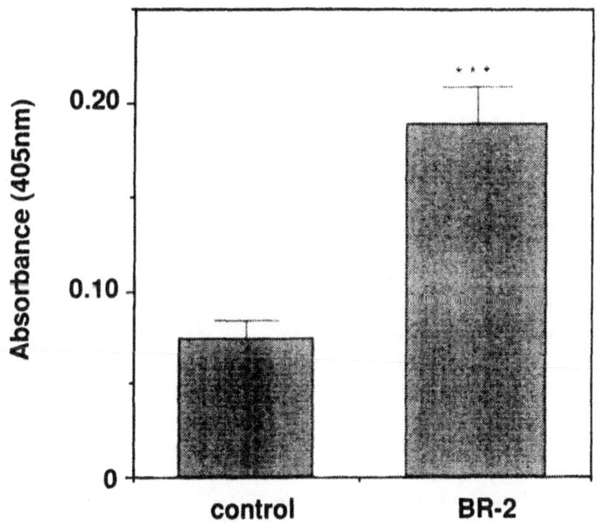

Figure 15 Distribution of orally administered bupleuran 2llc in liver. *** $p < 0.01$. Data are means ± S.D. n = 6.

ELISA using a specific anti-polysaccharide antibody was established as more direct, simple and specific method to detect a certain polysaccharide. Therefore, an antibody against pharmacologically active pectic polysaccharides will be a useful tool for investigating the tissue distribution of the polysaccharide administered and thus the quality control of kampo medicines.

REFERENCES

1. H. Yamada and H. Kiyohara, Bioactive polysaccharides from Chinese herbal medicine. *Abstracts of Chinese Medicine* **3**, 104–124 (1989).
2. H. Yamada, Pectic polysaccharides from Chinese herbs: Structure and biological activity. *Carbohydrate Polymers* **25**, 269–276 (1994).
3. Y. Komatsu, N. Takemoto, H. Maruyama, H. Tsuchiya, M. Aburada, E. Hosoya, S. Shinohara and H. Hamada, Effect of Juzentaihoto on the anti-SRBC response in mice. *Japanese Journal of Inflammation* **6**, 405–408 (1986).
4. H. Maruyama, H. Kawamura, N. Takemoto, Y. Komatsu, M. Aburada and E. Hosoya, Effect of Juzentaihoto on phagocytes. *Japanese Journal of Inflammation* **8**, 461–465 (1988).
5. N. Takemoto, H. Maruyama, H. Kawamura, Y. Komatsu, M. Aburada and E. Hosoya, Mitogenic activity of Juzentaihoto (TJ-48) on murine lymphoid cells. *Japanese Journal of Inflammation* **9**, 137–140 (1989).
6. K. Nabeya and S. Ri, Effect of oriental medical herbs on the restration of the human body before and after operation. *Proceeding of Symposium on WAKAN-YAKU* **16**, 201–206 (1983).
7. M. Aburada, S. Takeda, E. Ito, M. Nakamura and E. Hosoya, Protective effects of Juzentaihoto, dried decoctum of 10 Chinese herbs mixture, upon the adverse effects of mitomycin C in mice. *Journal of Pharmacobio-Dynamics* **6**, 1000–1004 (1983).
8. Y. Ohnishi, R. Yasumizu, H.X. Fan, J. Liu, F. Takao-Liu, Y. Komatsu, E. Hosoya, R. Good and S. Ikehara, Effect of Juzentaihoto (TJ-48), a traditional oriental medicine, on hematopoietic recovery from radiation injury in mice. *Experimental Hematology* **18**, 18–22 (1990).
9. H. Kawamura, H. Maruyama, N. Takemoto, Y. Komatsu, M. Aburada, S. Ikehara and E. Hosoya, Accelerating effect of Japanese Kampo medicine on recovery of murine hematopoietic stem cells after administration of mitomycin C. *International Journal of Immunotherapy* **V**, 35–42 (1989).
10. H. Yamada, H. Kiyohara, J.C. Cyong, N. Takemoto, Y. Komatsu, H. Kawamura, M. Aburada and E. Hosoya, Fractionation and characterization of mitogenic and anti-complementary active fractions from Kampo (Japanese herbal) medicine "Juzen-Taiho-To". *Planta Medica* **56**, 386–391 (1990).
11. H. Kiyohara, N. Takemoto, Y. Komatsu, H. Kawamura, E. Hosoya and H. Yamada, Characterization of mitogenic pectic polysaccharides from Kampo (Japanese herbal) medicine "Junzen-Taiho-To". *Planta Medica* **57**, 254–259 (1991).
12. H. Kiyohara, H. Yamada, N. Takemoto, H. Kawamura, Y. Komatsu and T. Oyama, Characterization of *in vitro* IL-2 production enhancing and anti-complementary pectic polysaccharides from Kampo (Japanese Herbal) medicine "Juzen-Taiho-To". *Phytotherapy Research* **7**, 367–375 (1993) .
13. S. Ikehara, H. Kawamura, Y. Komatsu, H. Yamada, H. Hisha, R. Yasumizu, R. Ohnishi-Inoue, H. Kiyohara, M. Hirano, M. Aburada and R.A. Good Effects of medicinal plants on hematopoietic cells. In *Microbial Infections. Role of biological response modifiers*, eds. H. Friedman, T.M. Klein and H. Yamaguchi, Plenum Press, New York, pp. 319–330 (1992).
14. H. Kiyohara, T. Matsumoto, N. Takemoto, H. Kawamura, Y. Komatsu and H. Yamada, Effect of oral administration of a pectic polysaccharide fraction from a Kampo (Japanese herbal) medicine "Juzen-taiho-to" on antibody response of mice. *Planta Medica* **61**, 429–434 (1995).
15. S. Ebisuno, S. Hirano, T. Ohkawa, H. Maruyama, H. Kawamura and E. Hosoya, Immunomodulating effects of Juzentaihoto in immunosuppressive mice. *Biotheraphy* **4**, 112–116 (1990).
16. H. Kiyohara, T. Matsumoto, Y. Komatsu and H. Yamada, Protective effect of oral administration of a pectic polysaccharide fraction from a Kampo (Japanese herbal) medicine "Juzen-Taiho-To" on adverse effects of *cis*-diaminedichloroplatinum. *Planta Medica* **61**, 531–534 (1995).
17. K. Sugiyama, Effects of Juzen-taiho-to in reducing the side effects of cisplatin. *Journal of Traditional Medicines* **13**, 27–41 (1996).
18. T. Matsumoto, R. Moriguchi and H. Yamada, Role of polymorphonuclear leucocytes and oxygen-derived free radicals in the formation of gastric lesions induced by HCl/ethenol, and a possible mechanism of protection by anti-ulcer polysaccharide. *J. Pharmacy Pharmacology* **45**, 535–539 (1993).
19. N. Takemoto, H. Kiyohara, H. Maruyama, Y. Komatsu, H. Yamada and H. Kawamura, A novel type of B cell mitogen isolated from Juzen-Taiho-To (TJ-48), a Japanese traditional medicine. *International Journal of Immunopharmacology* **16**, 919–929 (1994).
20. H. Yamada, H. Kiyohara, N. Takemoto, J.F. Zhao, H. Kawamura, Y. Komatsu, J.C., Cyong, M. Aburada and E. Hosoya, Mitogenic and complement activating activities of the herbal components of Juzen-Taiho-To. *Planta Medica* **58**, 166–170 (1992).

21. J.F. Zhao, H. Kiyohara, X.B. Sun, T. Matsumoto, J.C. Cyong, H. Yamada, N. Takemoto and H. Kawamura, *In vitro* immunostimulating polysaccharide fractions from the roots of *Glycyrrhiza uralensis Fisch. et* DC. *Phytotherapy Research* **5**, 206–210 (1991).
22. J.F. Zhao, H. Kiyohara, H. Yamada, N. Takemoto and H. Kawamura, Heterogeneity and characterization of mitogenic and anti-complementary pectic polysaccharides from the roots of *Glycyrrhiza uralensis* Fisch. et DC. *Carbohydrate Research* **219**, 149–172 (1991).
23. H. Kiyohara, N. Takemoto, J.F. Zhao, H. Kawamura and H. Yamada, Pectic polysaccharides from roots of *Glycyrrhiza uralensis:* Possible contribution of neutral oligosaccharides in the galacturonase-resistant region to anti-complementary and mitogenic activities. *Planta Medica* **62**, 14–19 (1996).
24. H. Kiyohara, J.C. Cyong and H. Yamada, Relationship between structure and activity of the "ramified" region in anti-complementary pectic polysaccharides from *Angelica acutiloba* Kitagawa. *Carbohydrate Research* **193**, 201–214 (1989).
25. T.E. Karplus, R.J. Ulevitch and C.B. Wilson, A new nethod for reduction of endotoxin contamination from protein solutions. *Journal of Immunological Methods* **105**, 211–220 (1987).
26. Y. Aida and M.J. Pabst, Removal of endotoxin from protein solutions by phase separation using Triton X-114. *Journal of Immunological Methods* **132**, 191–195 (1990).
27. M. Hirano, T. Matsumoto, H. Kiyohara and H. Yamada, Lipopolysacchride-independent limulus amebocyte lysate activating, mitogenic and anti-complementary activities of pectic polysaccharides from Chinese herbs. Planta Medica **60**, 248-252 (1994).
28. T. Hong, T. Matsumoto, H. Kiyohara and H. Yamada, Peyer's patch cells mediated proliferative response of bone marrow cells of Juzen-taiho-to. *Journal of Traditional Medicine* **12**, 412–413 (1995).
29. M. Hirano, H. Kiyohara and H. Yamada, Existence of rhamnogalacturonan II-like region in bioactive pectins from medicinal herbs. *Planta Mededica* **60**, 450–454 (1994).
30. D.C. Morrison, M.G. Lei, T.Y. Chen, M. Flebbe, J. Halling and S. Field, Identification and characterization of mammalian cell membrance receptors for LPS-endotoxin. In *Microbial Infection, Role of biological response modifiers,* eds. H. Friedman, T.M. Klein and H. Yamaguchi, Plenum Press, New York, pp. 23–29 (1992).
31. N. Haeffner-Cavaillon, M. Caroff and J.M. Cavaillon, Interleukin-1 induction by lipopolysaccharide: Structural requirements of the 3-deoxy-D-manno-2-octulosonic acid (KDO). *Molecular Immunology* **26**, 485–494 (1989).
32. H. Kiyohara, H. Yamada, J.C. Cyong and Y. Otsuka, Molecular aggregation and anti-complementary activity of arabinogalactan from *Angelica acutiloba. Journal of Phramacobio-Dynamics* **9**, 339–346 (1986).
33. H. Yamada, H. Kiyohara, J.C. Cyong and Y. Otsuka, Structural characterization of an anti-complementary arabinogalactan form the roots of *Angelica acutiloba Kitagawa. Carbohydrate Research* **159**, 275–291 (1987).
34. N.L. Wang, H. Kiyohara, T. Matsumoto, H. Otsuka, M. Hirano and H. Yamada, Polyclonal antibody against a complement activating pectin from roots of *Angelica acutiloba* Kitagawa. *Planta Medica* **60**, 425–429 (1994).
35. H. Yamada, K.S. Ra, H. Kiyohara, J.C. Cyong and Y. Otsuka, Structural characterization of an anti-complementary pectic polysaccharide from the roots of *Bupleurum falcatum* L. *Carbohydrate Research* **189**, 209–226 (1989).
36. T. Matsumoto, J.C. Cyong, H. Kiyohara, H. Matsui, A. Abe, M. Hirano, H. Danbara and H. Yamada, The pectic polysaccharides from *Bupleurum falcatum* L. enhances immune-complexes binding to peritoneal macrophages through Fc receptor expression. *International Journal of Immunopharmacology* **15**, 683–693 (1993).
37. H. Yamada, X.B. Sun, T. Matsumoto, K.S. Ra, M. Hirano and H. Kiyohara, Purification of anti-ulcer polysaccahrides from the roots of *Bupleurum falcatum* L. *Planta Medica* **57**, 555–559 (1991).
38. X.B. Sun, T. Matsumoto and H. Yamada, Effects of a polysaccharide fraction from the roots of *Bupleurum falcatum* L. on experimental gastric ulcer models in rats and mice. *Journal of Pharmacy & Pharmacology* **43**, 699–704 (1991).
39. H. Yamada, M. Hirano and H. Kiyohara, Partial structure of an anti-ulcer pectic polysaccharide from the roots of *Bupleurum falcatum* L. *Carbohydrate Research* **219**, 173–192 (1991).
40. M. Hirano, H. Kiyohara, T. Matsumotoband H. Yamada, Structural studies of endo-polygalacturonase-resistant fragments in anti-ulcer pectin from the roots of *Bupleurum falcatum* L. *Carbohydrate Research* **251**, 145–162 (1994).

41. T. Matsumoto, M. Hirano, H. Kiyohara and H. Yamada, Characterization of the endo-polygalacturonase-resistant region of the pectin from *Bupleurum falcatum* L. – a polysaccharide with an active function in clearance of immune complexes. *Carbohydrate Research* **270**, 221–229 (1995).
42. M. Sakurai, T. Matsumoto, H. Kiyohara and H. Yamada, Detection and tissue distribution of anti-ulcer polysaccharides from *Bupleurum falcatum* L. by polyclonal antibody. *Planta Medica* **62**, 341–346 (1996).

13. EFFICACY OF KAKKON-TO, A TRADITIONAL HERBAL MEDICINE, AGAINST HERPES SIMPLEX VIRUS TYPE 1 AND INFLUENZA INFECTION IN MICE

KIMIYASU SHIRAKI[1]*, MASAHIKO KUROKAWA[1], MASAMI IMAKITA[2], HITOSHI SATO[1], JUN-ICHI YAMAMURA[1], ZHI-HONG LI[1] and SEIJI KAGEYAMA[1]

[1]*Department of Virology, Toyama Medical and Pharmaceutical University, 2630 Sugitani, Toyama 930-0194, Japan;* [2]*Division of Pathology, National Cardiovascular Center, Osaka, Japan*

I. INTRODUCTION

In Asian countries many plants have long been utilized as food, traditional herbal medicines, food additives, and spices. The climate and geography of Asia are more varied than those of Europe and therefore the number of plant species in Asia is much greater than that in Europe. For example, medieval European people traveled to Asia to buy pepper or seasonings which were not native to Europe. Herbal medicines have been used and studied in Asia for many centuries. Many modern medicines have been developed from traditional herbal medicines: aspirin from willow, caffeine from *Coffea arabica*, ephedrine from *Ephedra sinica*, digitalis from *Digitalis purpurea* or *lanata*, atropine from *Atropa belladonna*, morphine from *Papaver somniferum*, quinine from *Cinchona*, cocaine from *Erythroxylon coca*, and curare from *Strychnos tomentosum*, among others.

In Japan and China, traditional herbal medicines have been used for the treatment of various diseases. The application and dosage of these medicines for each disease have been examined and established by monitoring both the efficacy and any adverse reactions in their history, and some of these medicines have a history of more than 4000 years of human use. Today the scientific bases for the use of these traditional medicines are being established. We have studied the use of traditional Chinese and Japanese herbal medicines as treatment for viral infections (Kurokawa *et al.*, 1993a, b, 1995, 1996a, b, c, 1997; Nagasaka *et al.*, 1995; Sato *et al.*, 1996; Takahara *et al.*, 1994; Yukawa *et al.*, 1996).

Kakkon-to (Ge-Gen-Tang in Chinese), a traditional herbal medicine, is one of the typical traditional formulations and is prescribed in more than 20 million doses annually in Japan. This herbal medicine has been used for the treatment of common colds including influenza infection and other diseases, since the days of ancient China. It is thought to improve the symptoms of these infectious diseases, such as fever, headache, sore throat, chills, polyarthralgia, and tension of cervical muscles (Terasawa, 1993; Jiangxu New Medical College, 1978). However, the experimental

* To whom correspondence should be addressed

studies on the efficacy of Kakkon-to and the statistical analysis of its clinical efficacy have not been extensively documented. Its pharmacological basis for the improvement of symptoms is also not clearly understood. In this chapter, we focus on the action of Kakkon-to in herpes simplex virus (HSV) and influenza infections using experimental infection systems in mice.

In the first phase of our investigation, we used a cutaneous HSV type 1 (HSV-1) infection model in mice as an experimental model for human viral infection, and demonstrated the therapeutic efficacy of Kakkon-to treatment (Nakasaka *et al.*, 1995). Treatment with Kakkon-to significantly limited the development of skin lesions and reduced the mortality of HSV-1-infected mice. Since the delayed-type hypersensitivity (DTH) response has been reported to be a major host defense system in intradermal HSV-1 infection [Larsen *et al.*, 1984; Nash *et al.*, 1980; Nash and Gell, 1983], we focused on the DTH reaction and various factors related to the natural defense system against HSV-1 infection. The first section of this chapter discusses the finding that the induction of the strong DTH reaction to HSV-1 antigen plays a major role in the treatment of cutaneous HSV-1 infection with Kakkon-to.

In the second phase of our investigation, we adopted an intranasal influenza virus infection model in mice to evaluate the therapeutic efficacy of Kakkon-to (Kurokawa *et al.*, 1996 a, b). This murine model is very useful for characterizing the development of typical pneumonia with pathologic changes similar to those in humans. We have shown that DBA/2 Cr mice are the most suitable model for analyzing the mechanism of fever production among 7 mouse strains, and demonstrated the cascade of fever production in influenza infection using the DBA/2 Cr mouse model as follows: influenza virus infection, elevated interferon (IFN) activity, interleukin (IL)-1α production, elevated cyclooxygenase (COX) activity and prostaglandin (PG)E_2 production (COX-PGE_2), then fever induction (Kurokawa *et al.*, 1996a). Kakkon-to treatment alleviated pneumonia and reduced fever in influenza infections in this model. This alleviation of pneumonia and novel antipyretic action by Kakkon-to was due mainly to the suppression of IL-1α production responsive to IFN production in influenza infection (Kurokawa *et al.*, 1996b).

II. FORMULATION OF KAKKON-TO

The hot water (HW) extract of Kakkon-to has been used orally for the treatment of viral infection for centuries in Japan and for 2000 years in China. Kakkon-to is composed of 7 medicinal herbs, *Pueraria pseudo-hirsuta* Tang et Wang (Radix), *Ephedra sinica* Stapf (Cortex), *Zizyphus jujuba* Mill. (Fruit), *Cinnamomum cassia* Blume (Cortex), *Paeonia lactiflora* Pall (Radix), *Glycyrrhiza uralensis* Fisch. (Radix), and *Zingiber officinale* Rosc. (Rhizome). The ratios of each herb differ among Kakkon-to prescriptions but they are essentially similar. The Kakkon-to prescription which we used was prepared as an HW-extract from a mixture of 7 dried herbs, *Pueraria pseudo-hirsuta* Tang et Wang (Radix, 24 g), *Ephedra sinica* Stapf (Cortex, 12 g), *Zizyphus jujuba* Mill. (Fruit, 12 g), *Cinnamomum cassia* Blume (Cortex, 9 g), *Paeonia lactiflora* Pall (Radix, 9 g), *Glycyrrhiza uralensis* Fisch. (Radix, 6 g), and *Zingiber officinale* Rosc. (Rhizome, 3 g).

III. EFFICACY OF KAKKON-TO AGAINST HSV INFECTION

A. Action of Kakkon-to Against HSV Infection

From ancient times Kakkon-to has been applied for the improvement of symptoms in the acute phase of viral infection, such as the common cold, influenza, varicella, and zoster. We examined the effect of Kakkon-to treatment on zosteriform lesions in a cutaneous HSV-1 infection murine model which presents zosteriform lesions, as shown Figure 1 (Nagasaka *et al.*, 1995). Female BALB/c mice (7 weeks old) were infected with the HSV-1 7401H strain (7 × 105 plaque-forming units [PFU]/mouse) under ether anesthesia after scarification of the shaved right midflank with 27-gauge needles. The development of skin lesions and mortality was monitored. Kakkon-to (300 mg/kg) or water was administered orally once at 4 hours before and three times (8 hour intervals) daily for successive days after HSV-1 infection. Surprisingly, Kakkon-to cured the lethal HSV infection, as shown in Figure 2. The mortality of the mice treated with Kakkon-to was 33.3 %, while that of the untreated mice was 100%. The appearance of skin lesions in Kakkon-to-treated mice was significantly retarded compared with that in water-treated mice. Similar effects were confirmed in repeated experiments using 5 different preparations of Kakkon-to. Kakkon-to exhibited significant therapeutic efficacy, i.e., the reduction of mortality and the retardation of skin lesions, in HSV-1-infected mice.

B. Mechanism of Kakkon-to Action Against HSV Infection

Kakkon-to treatment significantly reduced the mortality of HSV-infected mice and the appearance of skin lesions at a dose corresponding to clinical use. Kakkon-to did not inhibit the growth of HSV *in vitro*. The therapeutic efficacy of Kakkon-to therefore may be due to the activation of the host defense system and not to the direct anti-HSV activity of Kakkon-to. To investigate this, we first focused on the systemic reactions, i.e., cytokines in serum which activate the cellular immune system and the nonspecific cellular defense system. The effects of Kakkon-to treatment were apparent soon after infection, i.e., the retardation of the development of skin lesions were seen 3 or 4 days after infection. We therefore examined the parameters of the host defense system 2–4 days after infection. Kakkon-to treatment did not augment IFN induction, nor increase the cytokine levels such as IL-1α, IL-2, IFN-γ, and tumor necrosis factor (TNF)-α in the sera of the infected mice. The antibody to HSV-1 was not detected on day 4 after cutaneous infection. There was no significant difference in the levels of anti-HSV-1 antibody titer between Kakkon-to- and water-treated mice on days 6 and 9 after cutaneous infection. The production of anti-HSV-1 antibody in the infected mice was not affected by Kakkon-to treatment. We then examined the nonspecific cellular defense system, i.e., the natural killer (NK) and natural cytotoxic killer (NC) cell activity (Djeu *et al.*, 1983). NK and NC cell activity, and the population of T cell subsets in the spleen cells of the infected mice, were not influenced by Kakkon-to treatment. Thus, we could not find any alteration in the host defense system by Kakkon-to treatment to explain its therapeutic efficacy.

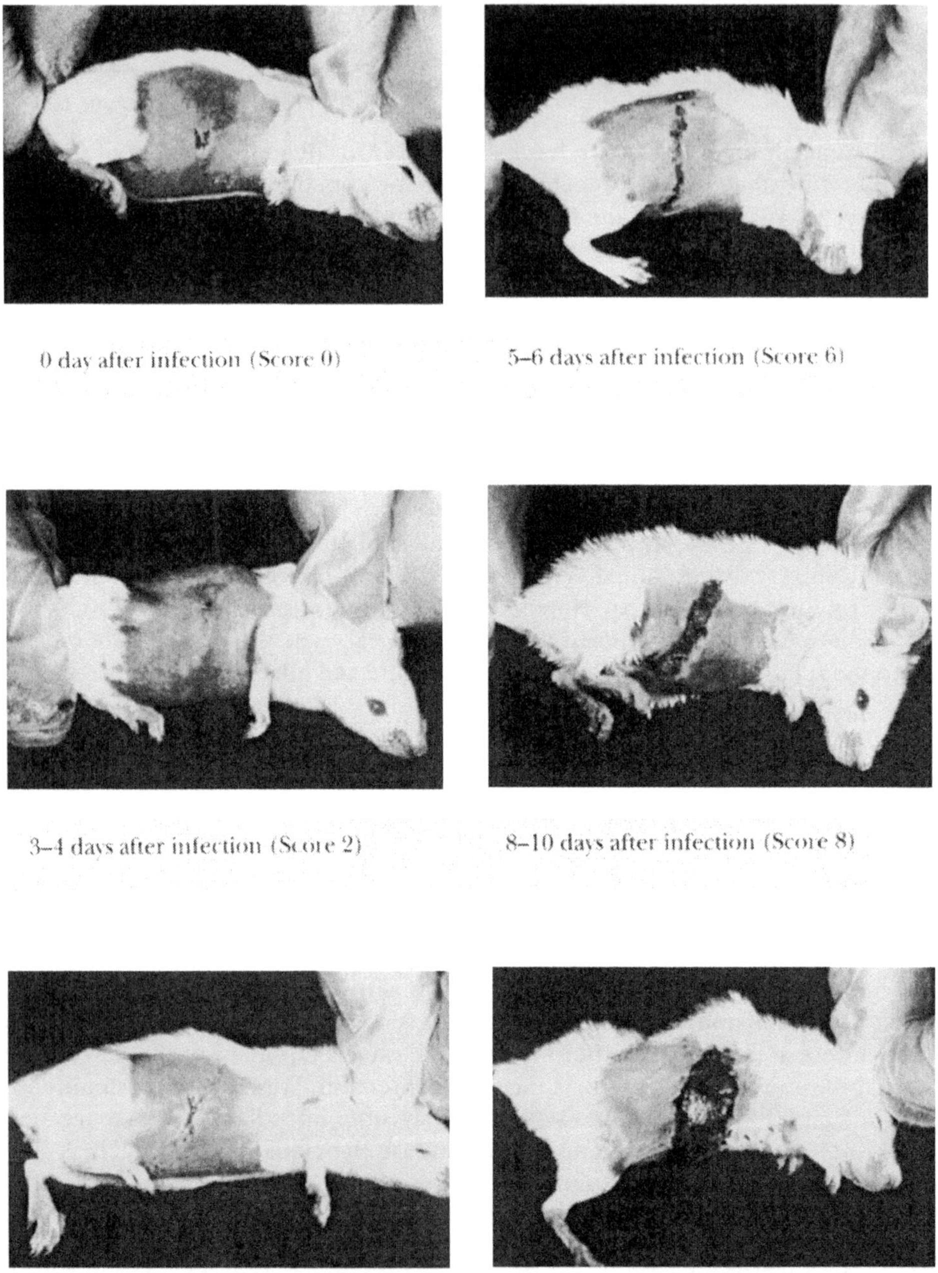

0 day after infection (Score 0)

5–6 days after infection (Score 6)

3–4 days after infection (Score 2)

8–10 days after infection (Score 8)

5–6 days after infection (Score 4)

8–10 days after infection (Score 10)

Figure 1 Development of skin lesions in mice infected with HSV-1. HSV-1 replicates in the skin and transported to the sensory ganglia. HSV-1 grown in the ganglia then spreads in the skin through neuronal transport and replicates again in the skin. Skin lesions subsequently develop from vesiculation, erosion, ulceration, to zosteriform lesion along with the dermatome(s), to severe ulcer and then death.

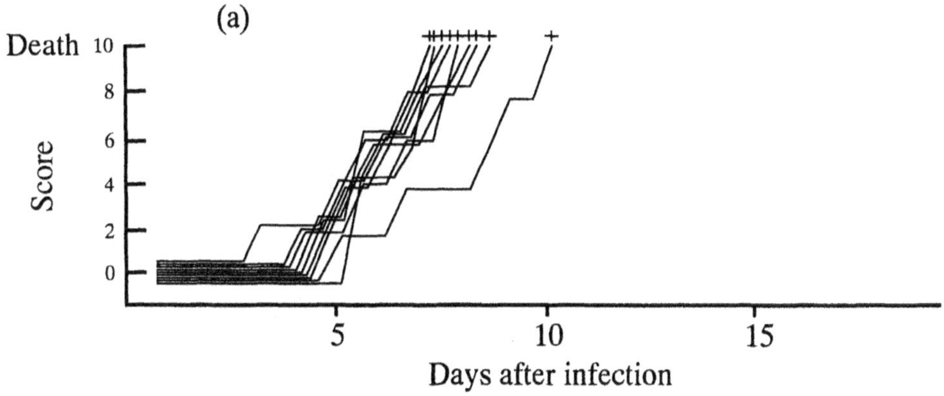

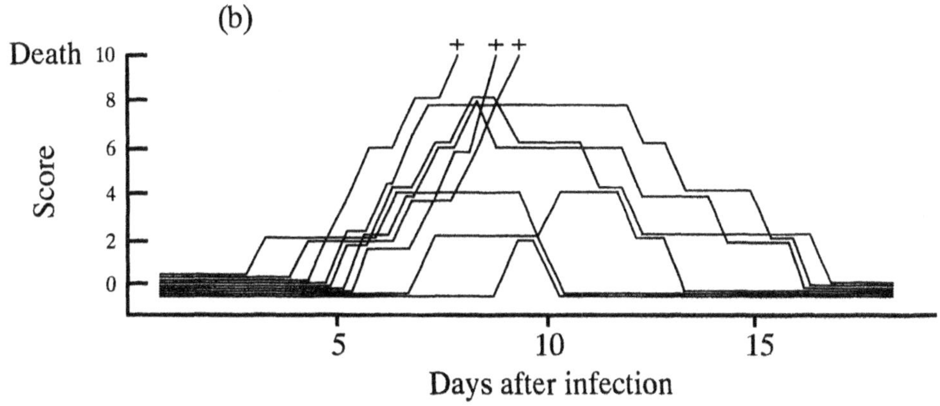

Figure 2 Effect of Kakkon-to on survival of mice infected with HSV-1 cutaneously, and retardation of their skin lesions. The nine HSV-1 infected mice in each group were administered water (a) or Kakkon-to (b) + indicates death. The mortality and the appearance of skin lesions in group (b) were significantly lower ($p < 0.01$) and retarded ($p < 0.05$), respectively, than those in group (a).

Histologically, the characteristic lesions of HSV infection were present in the skin of the Kakkon-to-treated and water-treated mice. The inflammatory skin changes in the Kakkon-to-treated and water-treated mice were similar; there was no remarkable difference in the constituents of inflammatory infiltrates, and the degree of the inflammatory response appeared to be similar between the groups except for the size of skin lesions. Inflammatory infiltrates appeared in the lesions treated with Kakkon-to earlier than in those treated with water. A local reaction to HSV was activated in the early phase of infection by Kakkon-to.

The delayed-type hypersensitivity (DTH) reaction is known to be one of the major defense systems involved in the clearance of HSV-1 and the increase in the survival rate in cutaneous and intradermal HSV-1 infection (Larsen *et al.*, 1984: Nash *et al.*, 1980; Nash and Gell, 1983). We therefore examined the effect of Kakkon-to on the DTH skin reaction to HSV-1 antigen in both cutaneous and intradermal HSV-1 infections. Figure 3 shows the relationship between the skin reaction to HSV-1 antigen and skin lesions, and the augmentation of skin reaction by Kakkon-to in the intradermal infection model. The skin reaction to mock antigen was not induced and there was less than 5% swelling at any time. Mice with skin lesions at scores 0, 2, and 4 showed strong skin reactions to HSV-1 antigen, but the skin reaction was hardly detectable in mice with skin lesions at scores 6 and 8. There was a strong relationship between skin reactivity to HSV-1 antigen and the depression of the skin lesions. Kakkon-to treatment significantly enhanced the skin reaction in mice with skin scores 0 ($p < 0.05$) or 2 ($p < 0.01$) compared with no treatment. The DTH reaction was stronger in the infected mice with mild skin lesions than in those with severe skin lesions.

Figure 4 shows the time course of the skin reaction to HSV-1 antigen in infected mice treated with Kakkon-to or water. The difference of the skin reactions to HSV-1 antigen was prominent between the mice treated with Kakkon-to and those treated with water, and was significant at 36 and 48 h ($p < 0.01$). The DTH reaction was augmented by Kakkon-to at 24 h ($p = 0.10$), 36 h ($p < 0.01$), and 48 h ($p < 0.01$) after the footpad challenge of HSV-1 antigen. The DTH reaction results are consistent with those of the histological examination.

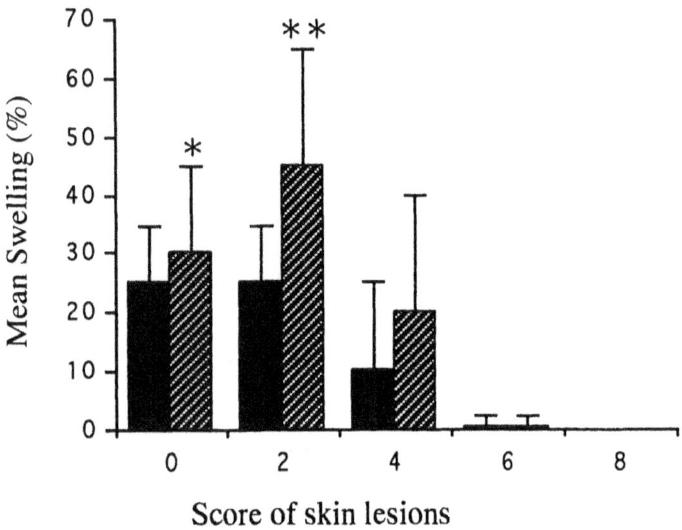

Figure 3 Relationship between skin lesions and mean skin reaction to HSV-1 antigen. Mice were infected with HSV-1 intradermally and administered Kakkon-to or water. The infected mice were challenged by HSV-1 antigen or control antigen in the footpad at 5 days after infection. The swelling of the footpad was measured and the skin lesions were scored at 36 hours after footpad challenge. Closed columns and hatched columns represent the mean percent swelling of 10 water-treated mice and 10 Kakkon-to-treated mice, respectively. The mean percent swelling represents the percentage of net increase of swelling to the thickness of the footpad before the challenge with HSV-1 antigen. Asterisks indicate a significant skin reaction compared with water-treated mice with the same score at 36 hours after the footpad challenge (*; $p < 0.05$, **; $p < 0.01$). Bars on each column indicate S.D.

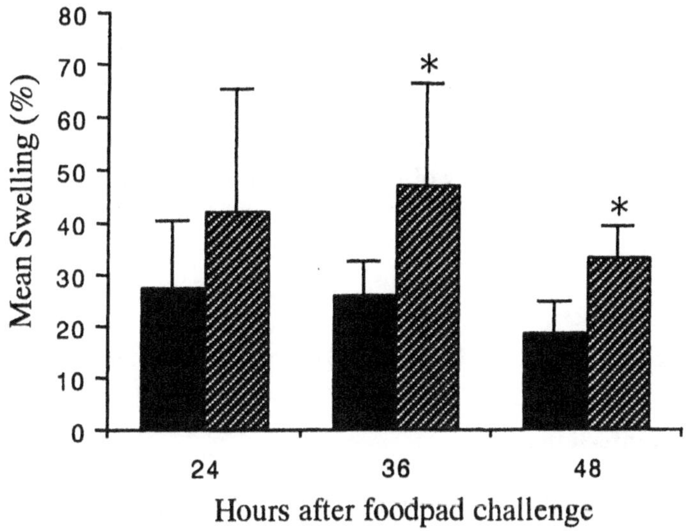

Figure 4 Time course of the skin reactions to HSV-1 antigen in mice infected intradermally with HSV-1. Hatched columns and closed columns represent the percent mean swelling of 10 mice treated with Kakkon-to and 10 treated with water, respectively. Asterisks indicate a significant skin reaction compared with water-treated mice ($p < 0.01$). The skin reaction to control antigen was less than 5% swelling. Bars on each column indicate S.D.

Kakkon-to induced strong DTH reactions to HSV antigen in infected mice, and this might have caused the retardation of skin lesions and the reduction in the mortality of the Kakkon-to-treated mice.

We used the dose of 300 mg/kg of Kakkon-to for oral administration to mice. This dose is the equivalent of the conventional dose of Kakkon-to used for humans. Information on the application of and the adverse reactions to Kakkon-to has accumulated over a very long period (Terasawa, 1993; Jiangxu New Medical College, 1978). The efficacy of Kakkon-to at this dose was equivalent to that of 10 mg/kg of oral acyclovir (ACV) treatment in our cutaneous HSV-1 infection model in mice (Kurokawa *et al.*, 1993b) and that dose of ACV also corresponds to a clinical dose for humans. Our findings indicated that Kakkon-to treatment could be as effective as oral ACV treatment for cutaneous HSV-1 infection.

The DTH reaction against HSV-1 antigen has been reported to be mediated mainly by Lyt 1^+2^- T-lymphocyte subsets in an intradermal infection model, but not by Lyt 1^-2^+T-lymphocyte subsets with CTL activity (Nagafuchi *et al.*, 1982; Nash and Gell, 1983). Macrophages activated by the former T-lymphocytes may be responsible for the DTH reaction (Roitt *et al.*, 1985). Kakkon-to might have activated Lyt 1^+2^- T-lymphocyte subsets or directly activated macrophages including Langerhans and dendritic cells in the site of skin lesions of the HSV-1-infected mice described above. We attempted to identify the specific reaction of these lymphocytes and Langerhans cells as a histological change by Kakkon-to but could not. Anyhow, Kakkon-to may activate Lyt 1^+2^- T-lymphocyte subsets or macrophages in the skin and thus retard the appearance of skin lesions.

In summary, we examined various factors related to non-specific defense mechanisms against HSV-1 infection. Kakkon-to did not modify NK cell activity, NC cell activity, or the population of T cell subsets in spleen cells, nor did it increase IFN activity or the cytokine levels of IL-1α, IL-2, IFN-γ, and TNF-α in sera of infected mice. Kakkon-to enhanced the DTH response to HSV-1 antigen and exhibited therapeutic effects. The efficacy of Kakkon-to was due mainly to the induction of the strong DTH reaction to HSV-1 antigen, and its treatment is as effective as oral ACV treatment.

IV. EFFICACY OF KAKKON-TO AGAINST INFLUENZA INFECTION

A. Influenza Infection Model

One of the most suitable applications of Kakkon-to may be for influenza virus infection. Kakkon-to is used in Japan to alleviate fever, headache, sore throat, chills, and polyarthralgia in the acute phase of influenza (Terasawa, 1993). An experimental infection model using mice has been established to study the pathogenesis of influenza pneumonia (Tashiro *et al.*, 1987a, b), but it is not a suitable model for the analysis of fever induction which is typical in humans. Therefore, we first screened 7 mouse strains to determine the most suitable model for studying fever production, and selected the DBA/2 Cr mouse strain as the most useful model (Kurokawa *et al.*, 1996a). We examined fever induction by IFN-α, which is known as an endogenous pyrogen, using the DBA/2 Cr, C57BL/6 Cr, BALB/c Cr, A/J, C3H/He, ddY, and ICR strains. Of these 7 strains, the DBA/2 Cr, C57BL/6 Cr, BALB/c Cr, C3H/He, and ICR strains significantly responded to IFN-α for fever induction. The DBA/2 Cr strain showed the most prominent response in the level of fever by rectal temperature, and therefore was used in this study. Influenza infection induces high fever in humans (38–40°C) and in DBA/2 mice (37.5–38°C). Thus influenza induces significant fever in mice but the level of the fever is lower than that in humans, possibly because of differences in the perspiration and other systems. Influenza infection and IFN-α treatment induced typical fever in DBA/2 Cr mice at 1–2 days after infection and at 10–20 hour after IFN-α treatment, respectively. Rectal temperature was therefore monitored on the second day after infection (44–48 hour postinfection) and at 14 hour after IFN-α treatment. Using these model systems, the effect of Kakkon-to on influenza infection was analyzed in parallel with the mechanism of fever induction in DBA/2 Cr and ICR mice. The antipyretic action of Kakkon-to was compared with that of aspirin. Female DBA/2 Cr or ICR mice (4 to 5 weeks old, 16–18 g) were used for the analysis of Kakkon-to efficacy. The mice were intranasally infected or mock-infected with 600–800 PFU of influenza virus [A/PR/8/34 (H1N1)] under ether anesthesia. Kakkon-to (750 mg/kg/day) or water was administered orally to the mice three times daily (approximately 8 hour intervals) on the days before and after the infection day. The development of pneumonia, rectal temperature, and body weight were monitored to evaluate the efficacy of Kakkon-to against influenza.

B. Action of Kakkon-to Against Influenza Infection

The therapeutic efficacy of Kakkon-to was thus evaluated in an intranasal influenza virus infection model in mice (Kurokawa *et al.*, 1996b). When ICR mice were infected with the influenza virus, body weight decreased markedly from 2 days post-infection, and the consolidation of lungs was clearly observed on day 4 post-infection. In contrast, Kakkon-to treatment delayed the decrease in body weight and caused an early recovery of body weight (Figure 5). The development of consolidation was also retarded in Kakkon-to-treated mice. Histopathological analysis of the lungs showed a significant retardation in the development of pneumonia in Kakkon-to-treated mice (Figure 6). As shown in Figure 7, Kakkon-to treatment significantly reduced the mortality rate of influenza-infected mice. The early recovery of body weight, the alleviation of pneumonia and the reduction in mortality caused by Kakkon-to were confirmed by repeated experiments with other preparations of Kakkon-to. However, Kakkon-to treatment did not affect the virus yields in the lungs of infected mice (Figure 8). This indicated that the alleviation of pneumonia was due not to the reduction of virus growth in the lung by the anti-influenza virus activity of Kakkon-to but to the host defense mechanism augmented by Kakkon-to. The pathologic changes in the lungs of DBA/2 Cr mice infected with the influenza virus were similar to those of ICR mice. Histologically, the epithelium of the bronchi and bronchioles showed infiltration of inflammatory cells and necrosis, and was overlaid by mucopurulent materials. These pathologic changes were milder in Kakkon-to-treated mice than in water-treated mice (Figures 6 and 9). The delay of loss of body weight produced by Kakkon-to treatment was also confirmed in the infected DBA/2 Cr mice (Figure 5). Influenza infection induced fever, and Kakkon-to treatment inhibited fever production, as shown in Figure 10. The reduction of fever was not confined to the preinfection treatment; the antipyretic activity was observed in febrile mice with influenza infection. This antipyretic activity of Kakkon-to was comparable with that of aspirin (Figure 10). Thus, Kakkon-to treatment was confirmed to have therapeutic efficacy in alleviating pneumonia and the loss of body weight, and in reducing fever production in an influenza infection model in mice.

C. Mechanism of Kakkon-to Action in Influenza

IFN is known to induce fever similar to influenza and is thought to be the major pyrogen in influenza (Dinarello, 1981; Dinarello *et al.*, 1988; Insel, 1992; Saper and Breder, 1992). IFN treatment is successful for patients with chronic hepatitis C but induces fever and influenza-like syndrome. We used both influenza infection and IFN treatment for the characterization of the mechanism of fever production and the effect of Kakkon-to treatment. Mice were infected intranasally with influenza virus or treated intraperitoneally with IFN-α to induce fever. We analyzed the activity of the cytokines IL-1α, IL-2, TNF-α, and IFN-γ, and IFN in sera of the febrile mice with influenza infection or IFN treatment, because IL-1α, IL-2, IFN-γ and TNF-α are possible pyretic mediators in influenza infection. When fever was induced 2 days

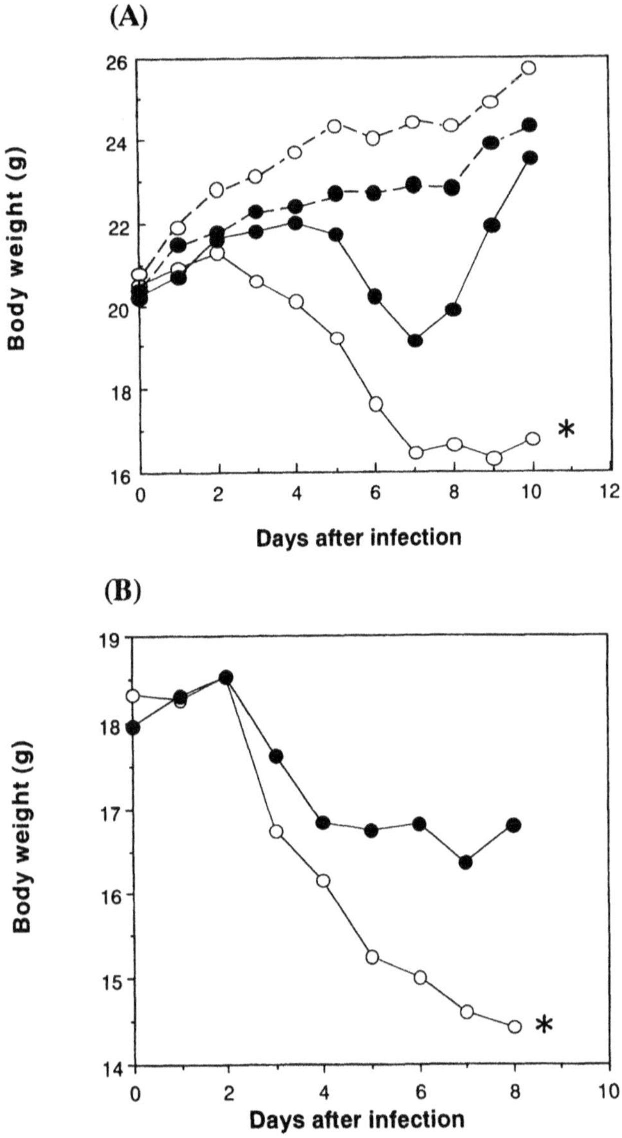

Figure 5 Effects of Kakkon-to on the body weight of mice. (A) ICR mice were mock-infected (broken lines) or infected (solid lines) intranasally with influenza virus. Kakkon-to (closed circles) or water (open circles) were orally administered as described in the text, and the body weight was measured daily. Five to 10 mice were used in each group. Asterisk indicates a significant difference from infected mice treated with Kakkon-to and uninfected mice treated with water or Kakkon-to for 1 to 10 days post-infection, $p < 0.01$ by the repeated measure ANOVA.

(B) DBA/2 Cr mice were infected intranasally with influenza virus. Kakkon-to (closed circles) and water (open circles) were orally administered as described in the text, and the body weight was measured daily. Five to 10 mice were used in each group. Asterisk indicates a significant difference from infected mice treated with Kakkon-to for 3 to 8 days post-infection, $p < 0.05$ by the repeated measure ANOVA.

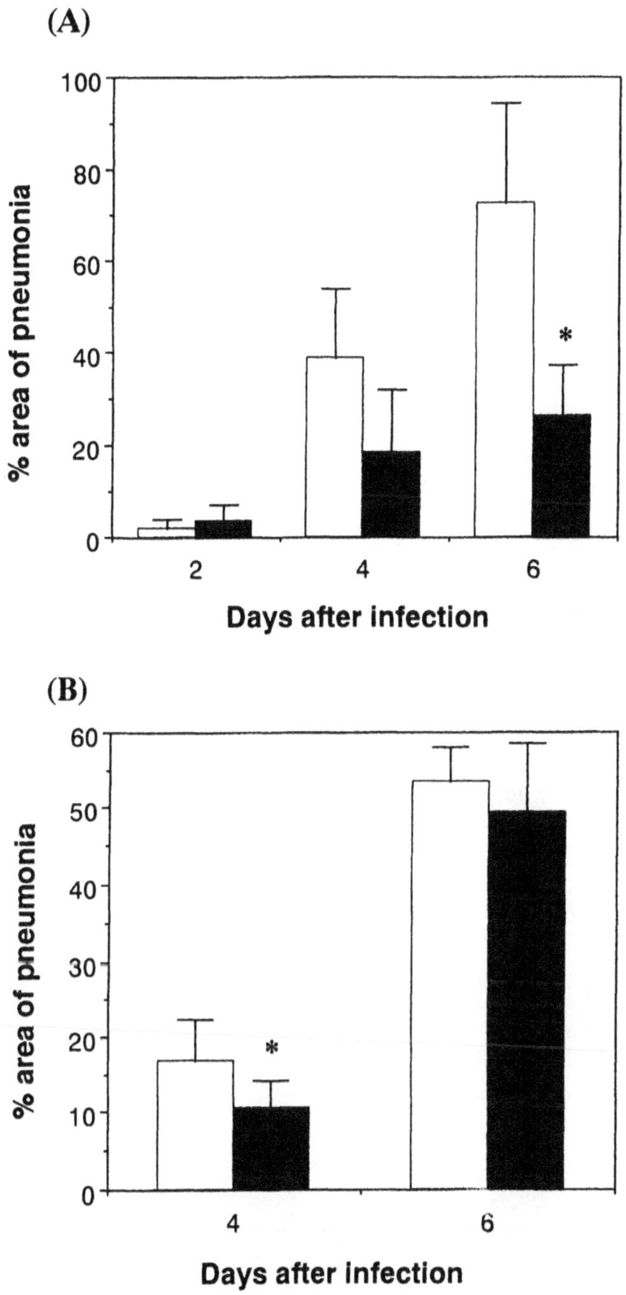

Figure 6 Effect of Kakkon-to on the development of pneumonia in mice infected with influenza virus. ICR (A) or DBA/2 Cr (B) mice were infected intranasally with influenza virus, and Kakkon-to or water was orally administered as described in the text. The lungs were removed from 3 to 4 mice in each group on days 2, 4 and 6 after infection. Open and closed columns show water- and Kakkon-to-administered groups, respectively. Asterisks indicate a significant difference from the water-administered group, $p < 0.05$ by Student's *t*-test.

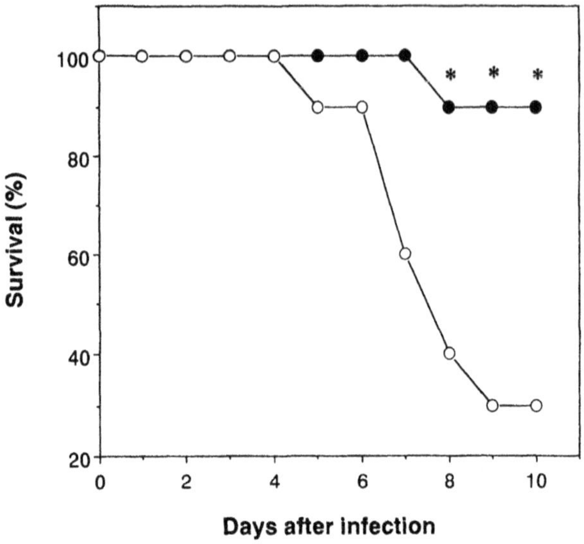

Figure 7 Survival time of influenza virus-infected mice. ICR mice were intranasally infected with influenza virus. Kakkon-to (closed circles) or water (open circles) were orally administered as described in the text. Ten mice were used in each group. Asterisks indicate a significant difference from the water-administered group, $p < 0.05$ by Fisher's exact test.

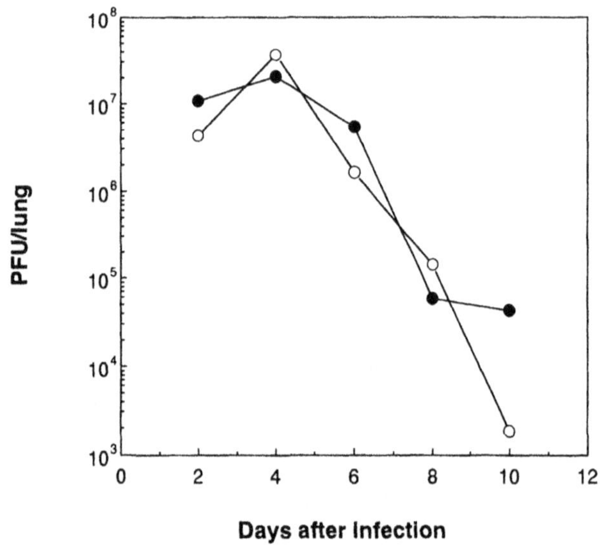

Figure 8 Effect of Kakkon-to on the growth of influenza virus in lungs. ICR mice were infected intranasally with influenza virus, and Kakkon-to (closed circles) or water (open circles) were orally administered as described in the text. The lungs were removed from 3 mice in each group on days 2, 4, 6, 8 and 10 after infection and homogenized. The virus titer in the supernatant of the homogenate was determined by the plaque assay and the mean values of 3 lungs were plotted. PFU: plaque forming unit.

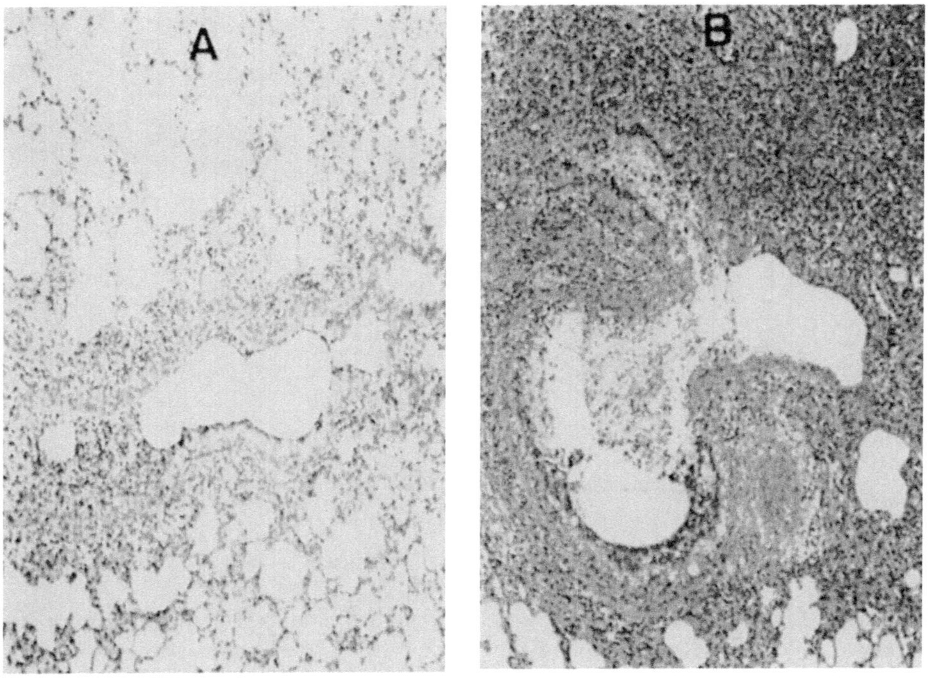

Figure 9 Histopathology samples from lungs of influenza virus-infected mice. The lungs of Kakkon-to-treated (A) and water-treated (B) mice were removed on day 4 after infection. The resected lung was fixed and stained with hematoxylin and eosin.

after infection and at 14 hr after IFN-α treatment, the serum IL-1α levels were significantly increased in the mice infected with influenza virus and treated with IFN-α (Table 1). Kakkon-to and aspirin significantly suppressed the fever. Kakkon-to suppressed the IL-1α level in both influenza-infected and IFN-α-treated mice. Comparing the IL-1α level of mock-infected or phosphate-buffered saline (PBS)-treated mice, the suppression of the IL-1α level in exogenous IFN-α treatment was weaker than that in influenza infection. As expected from the strong induction of IL-1α by IFN-α treatment, Kakkon-to was not able to suppress IL-1α production following PBS treatment because of the excessive dose of IFN-α. The IL-1α level was suppressed by Kakkon-to treatment but not by aspirin treatment.

When the levels of other cytokines (IL-2, TNF-α and IFN-γ) were examined, no significant difference was observed between infected mice treated with Kakkon-to and those treated with water. These results were confirmed by repeated experiments using three different lots of Kakkon-to. When influenza infection induced fever, IL-1α and IL-2 levels and IFN activity in serum increased significantly, but TNF-α and IFN-γ levels were not affected. Both the influenza infection and IFN-α treatment induced fever, and only the elevation of IL-1α level in sera was common between the febrile mice with influenza infection and those with IFN-α treatment. Influenza infection is known to induce IFN; IFN is known as a factor that influences the production of IL-1. Kakkon-to inhibited fever production and the elevation of

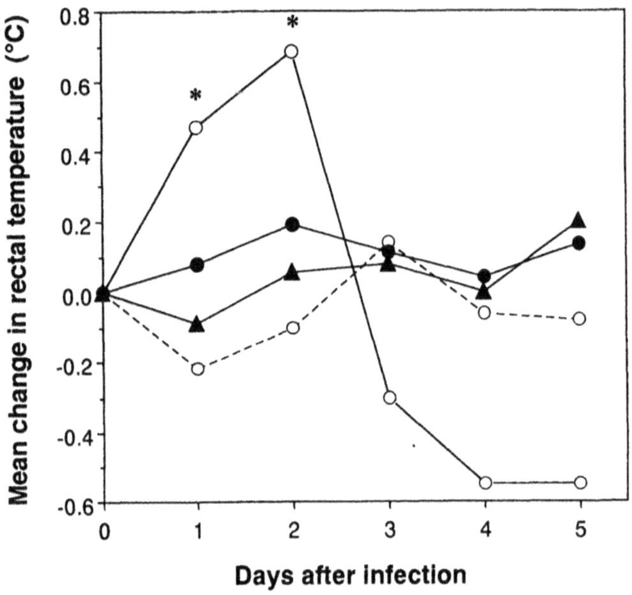

Figure 10 Effects of Kakkon-to on the mean rectal temperatures of DBA/2 Cr mice. The mice were mock-infected (broken lines) or infected (solid lines) intranasally with influenza virus. Kakkon-to (closed circles), aspirin (closed triangle) or water (open circles) was orally administered as described in the text, and the rectal temperatures of the mice were measured every morning. Five to 7 mice were used in each group. Asterisks indicate significant differences from the water-administered, aspirin-administered and mock-infected groups, $p < 0.05$ by Studentís *t*-test.

IL-1α in both infected mice and IFN-α-treated mice. These results indicate that IL-1α was a major pyrogenic factor (Hashimoto *et al.*, 1991) in febrile mice infected with influenza and treated with IFN-α.

Kakkon-to and aspirin treatments suppressed fever in both mice infected with influenza virus and those treated with IFN-α. However, the IL-1α level in the Kakkon-to-treated mice was not increased, but that in aspirin-treated mice was elevated as high as that in water-treated mice. IFN activity in the serum of the infected mice was not influenced by Kakkon-to treatment.

Anti-IFN-α/β and anti-IL-1α antibodies significantly suppressed fever production in infected mice similarly to Kakkon-to treatment (Table 2). Anti-IFN-α/β antibody as well as Kakkon-to treatment suppressed the induction of IL-1α significantly but the level of IL-1α was maintained in mice with either treatment. Cyclooxygenase (COX) is known to produce a pyrogen prostaglandin E_2 (PGE_2) in the arachidonic acid cascade which induces fever (Ballou *et al.*, 1992; Blatteis, 1992; Saper and Breder, 1992). Aspirin is one of the typical antipyretics and is known to inhibit the production of PGE_2 by inactivation of COX activity, resulting in reduction of fever (Feldberg *et al.*, 1973; Flower and Vane, 1972; Insel, 1992). As shown in Table 3, COX activity was significantly augmented by infection compared with mock-infected mice. Aspirin significantly reduced COX activity in the brain, to

Table 1 Rectal temperatures and IL-1α levels of mice infected with influenza virus or treated with IFN-α

Treatment		*Rectal temperature (°C)*	*IL-1α (pg/ml) in serum*
Influenza virus infection			
Mock-infected	Water (n = 7)[a]	37.0 ± 0.2[b]	28.2 ± 12.1[c,d]
Infected	Water (n = 10)	37.5 ± 0.1	52.8 ± 18.9
	Kakkon-to (n = 10)	37.1 ± 0.3[c]	30.0 ± 9.9[c,d]
	Aspirin (n = 7)	37.2 ± 0.3[c]	58.7 ± 24.0
IFN-α-treatment			
PBS-treated	Water (n = 5)	36.9 ± 0.2[b]	33.6 ± 10.5[d]
IFN-α-treated	Water (n = 5)	37.8 ± 0.2	93.6 ± 31.5
	Kakkon-to (n = 5)	37.1 ± 0.2[b]	46.8 ± 23.5[c,d]
	Aspirin (n = 5)	37.0 ± 0.3[e]	119.3 ± 51.8

Rectal temperatures and IL-1α levels of mice infected with influenza virus or treated with IFN-α were compared in Kakkon-to (750 mg/kg/day)- and aspirin (26 mg/kg/day)-treated mice. Rectal temperatures were measured on 2 days after infection and at 14 hour after IFN-α (5×10^4 IU/mouse)-treatment. At these times, whole blood was collected and sera were examined for IL-1α levels by ELISA. PBS: phosphate-buffered saline
[a]Number of mice used in each group.
[b]$p < 0.001$, [c]$p < 0.05$ compared to infected mice with water administration.
[d]$p < 0.05$ compared to infected mice with aspirin administration.
[e]$p < 0.01$ compared to infected mice with water administration.

Table 2 Effects of anti-IL-1α and anti-IFN-α/β antibody on rectal temperatures and IL-1α levels of mice infected with influenza virus

Treatment		*Rectal temperature (°C)*	*IL-1α (pg/ml) in serum*
Exp. 1			
Mock-infected	Water (n = 6)	37.0 ± 0.3[a]	—
Infected	Water (n = 6)	37.8 ± 0.2	—
	Kakkon-to (n = 6)	37.1 ± 0.3[a]	—
	Water + anti-IL-1α antibody (n = 6)	37.4 ± 0.2[b]	—
Exp. 2			
Mock-infected	Water (n = 5)	37.0 ± 0.4[c]	43.8 ± 15.0[c]
Infected	Water (n = 7)	37.6 ± 0.1	82.3 ± 35.6
	Kakkon-to (n = 7)	37.1 ± 0.1[a]	49.7 ± 14.9[c]
	Water + anti-IFN-α/β antibody (n = 7)	37.2 ± 0.1[a]	48.3 ± 19.2[c]

Effects of anti-IL-1α and anti-IFN-1α/β antibody on fever induction temperatures and IL-1α levels were examined in mice infected with influenza virus. Anti-IL-1α antibody or anti-IFN-α/β antibody was intravenously administered to influenza virus-infected mice, and Kakkon-to (750 mg/kg.day) or aspirin (26 mg/kg/day) was administered orally. Rectal temperatures were monitored 2 days after infection and sera were examined for IL-1α levels by ELISA. Parentheses show the number of mice used in each group.
[a]$p < 0.001$, [b]$p < 0.01$, [c]$p < 0.05$ compared to infected mice with water administration.

Table 3 COX activity in brain of influenza virus-infected mice

Treatment		*COX activity in brain (pg of PGE_2/mg of protein)*
Mock-infected	Water (n = 5)	$1.76 \pm 0.20^{a,b}$
Infected	Water (n = 5)	2.10 ± 0.04^{b}
	Kakkon-to (n = 5)	1.53 ± 0.18^{b}
	Aspirin (n = 5)	0.83 ± 0.16^{c}

Cyclooxygenase (COX) activity in the brain of influenza virus-infected mice was compared in Kakkon-to- and aspirin-treated mice. Mice were infected with influenza virus, and Kakkon-to (750 mg/kg/day) or aspirin (80 mg/kg/day) was orally administered. Parentheses show the number of mice used in each group.
$^{a}p < 0.05$, $^{c}p < 0.01$ compared to infected mice with water administration.
$^{b}p < 0.01$ compared to infected mice with aspirin administration.

less than the level of that in mock-infected mice. Kakkon-to treatment suppressed the induction of COX activity to the basal level which was observed in mock-infected mice. Therefore, aspirin inhibited the COX activity in the brain, while Kakkon-to inhibited the induction of additional COX activity. Aspirin inhibited COX activity in the brain but Kakkon-to did not. The influenza infection in the mice thus followed the following cascade for fever induction: IFN production, IL-1α production, elevated COX activity and PGE_2 production (COX-PGE_2). Figure 11 shows the cascade of fever and pneumonia in influenza and the acting sites of Kakkon-to and aspirin. We clarified that influenza infection induced IFN activity, and the production of IL-1α and COX-PGE_2, and that IFN-α injection induced IL-1α and fever. Kakkon-to treatment permitted IFN induction but suppressed IL-1α and the COX-PGE_2 and fever induction. Aspirin permitted IFN and IL-1α induction but suppressed the COX-PGE_2 and fever induction. These results were confirmed by repeated experiments using three different lots of Kakkon-to. Kakkon-to worked at the steps from IFN to IL-1α, before the COX-PGE_2 step which aspirin inhibits, and the suppression of fever by Kakkon-to was similar to the action of anti-IFN-α/β antibody. We thus have elucidated the cascade of fever induction in influenza infection and demonstrated that Kakkon-to works as a novel antipyretic agent which suppresses IL-1α production and further steps of fever induction.

The intranasal infection caused cellular infiltration in the respiratory tract of the mice and the development of pneumonia resulted from the pathological damage caused by an immune response against the influenza virus infection. The influenza pneumonia was milder in the Kakkon-to-treated mice than in the water-treated mice (Figures 6 and 9). The virus yields were similar in the lungs of the Kakkon-to- and water-treated mice (Figure 8), indicating that Kakkon-to did not inhibit the growth of influenza virus in the lungs. Thus, the alleviation of pneumonia by Kakkon-to treatment could be due to the modification of an immunopathological response against influenza infection rather than a direct cytopathic effect on viral replication. Since IFN can induce IL-1α production, the augmentation of IFN activity in an influenza infection would be able to increase IL-1α production in a

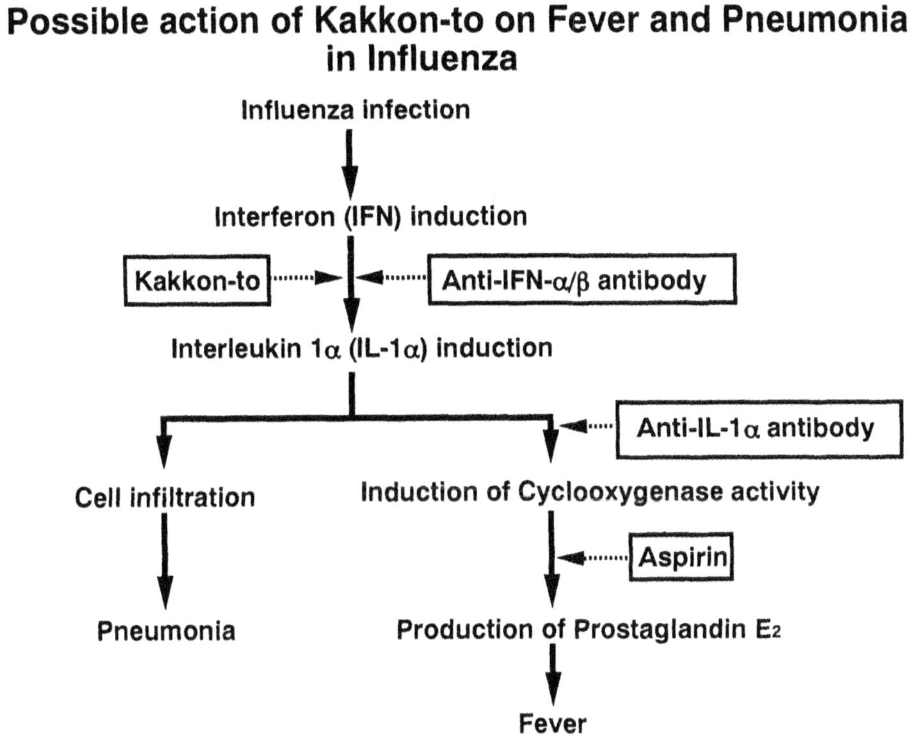

Figure 11 Fever cascade in influenza infection and the acting sites of Kakkon-to and aspirin.

murine model (Table 1 and Figure 11). In the early stage of pneumonia, however, Kakkon-to treatment did not modify the induction of IFN activity but significantly reduced IL-1α production to the level observed in the lungs of untreated mice (Table 4). IL-1α is produced from alveolar macrophages in the onset of the inflammatory response, and is produced only during the period soon after infection in the lungs of infected mice (Duff and Durum, 1983; Hennet *et al.*, 1992). The reduction of IL-1α production therefore suppressed cellular infiltration, and consequently may have resulted in the milder pneumonia in infected mice treated with Kakkon-to than in those treated with water. Kakkon-to prolonged the survival of mice rather than enhancing mortality; its suppressive action on IL-1α production therefore did not deteriorate the influenza infection. Thus, Kakkon-to suppressed IL-1α production responsive to IFN and also the reactions, including fever and the development of pneumonia, without affecting the host defense system (Figure 11). This action of Kakkon-to was beneficial to the host despite its suppression of the host response (responsive IL-1α production). This result suggests that IL-1α production induced by influenza infection could be a non-essential overreaction in the host defense system. In contrast to aspirin, Kakkon-to showed a novel antipyretic action without affecting PG metabolism. The use of Kakkon-to might be beneficial as an antipyretic agent to prevent adverse reactions caused by aspirin or aspirin-like

Table 4 Effects of Kakkon-to on the amounts of IL-1α in the bronchoalveolar lavage fluid and serum

Exp No.	*Days after infection*	*Uninfected mice*		*Infected mice*	
		Water-treated	*Kakkon-to-treated*	*Water-treated*	*Kakkon-to-treated*
1. Serum (n = 2 or 3)[a]					
	2	40.4 ± 11.6	42.8 ± 22.8	122.4 ± 18.8	62.4 ± 25.2[b]
	4	16.4 ± 10.8	22.0 ± 7.2	46.4 ± 5.2	45.2 ± 29.6
	6	24.0	8.0	30.8 ± 16.4	54.4 ± 8.0
Lavage (n = 3)					
	2	7.6 ± 13.0	4.8 ± 3.0	65.2 ± 23.6	9.2 ± 15.8[b]
	4	6.4 ± 5.8	0.8 ± 1.4	45.6 ± 40.4	10.8 ± 4.8
	6	0.0 ± 0.0	0.2 ± 0.2	16.0 ± 14.0	6.0 ± 10.4
2. Serum (n = 3)					
	2	13.3 ± 23.1	31.0 ± 37.7	191.3 ± 91.0	26.0 ± 22.5[b]
	4			26.7 ± 30.6	0.0 ± 0.0
	6			31.0 ± 37.7	26.7 ± 46.2
Lavage (n = 3)					
	2			32.6 ± 11.8	4.6 ± 7.9[b]
	4			22.8 ± 20.2	5.4 ± 2.4
	6			8.0 ± 7.0	3.0 ± 5.2

IL-1α levels (mean ± SD pg/ml) in the bronchoalveolar lavage fluid of lungs and serum prepared from ICR mice treated with Kakkon-to (750 mg/kg/day) or water.
[a]Parentheses enclose the number of mice used in each group.
[b]$p < 0.05$ *vs.* infected mice with water administration.

drugs: gastric damage, Reye's syndrome and other disturbances including those related to the suppression of the PG metabolism.

V. CONCLUSION

We have confirmed the efficacy of Kakkon-to against HSV-1 and influenza infection in murine models. Although it is not clear how much of the actions of Kakkon-to in the infection model is reproducible in human diseases, these results provide a possible scientific base for the clinical use of Kakkon-to in viral infections. Surprisingly, not all of the mouse strains examined responded to the IFN-α treatment for fever induction. This may be an example of the Kampo concept "Shou": Prescriptions should be issued according to the "Shou" of the patients and not simply by their apparent symptom(s), because the reactivity of patients to a certain prescription differs and therefore the prescription should be chosen according to the "Shou" and not uniformly according to the apparent symptom. In this context Kakkon-to is used mostly in the acute phase of influenza or the common cold. The manifestations of illness in this phase correspond to the "Shou" and Kakkon-to is prescribed at this point in Kampo medicine. The manifestations in the convalescent phase as the "Shou" do not fit the application of Kakkon-to, and our results are in accord with this idea. IL-1α is a multifunctional factor in the immune system, central nervous system, and endocrine system; it functions as an endogenous pyrogen, as an inducer of acute phase inflammatory proteins, endothelial cell and fibroblast proliferation, and PG production, and as a colony-stimulating factor, and so on. In the acute phase of influenza, the induction of IL-1α production may be an unnecessary host overreaction and should be suppressed to improve the symptoms. In the convalescent phase, appropriate IL-1α production which is not induced excessively by IFN may be beneficial to clear and repair the infected target tissues with inflammatory reactions. Therefore, IL-1α production should not be modified by Kakkon-to in this phase; other formulations are recommended at this phase.

We have concluded that the antipyretic action of Kakkon-to against pneumonia is due mainly to the suppression of IL-1α production responsive to IFN in influenza infection. Concerning the alleviation of pneumonia, IL-1α production in the lung was suppressed by Kakkon-to, and this suppression would be enough to explain the milder pneumonia. Concerning the antipyretic action, however, we have not identified the target tissue of IL-1α production which is related to fever induction. The suppression of IL-1α production seemed to be partially impaired in mice treated with an excessive dose of IFN-α, but nevertheless antipyretic action was observed. It is therefore difficult to interpret the findings in light of the possibility of the low IL-1α level related to fever or the presence of another mode of antipyretic action. The relationship between the suppression of IL-1α and fever induction in influenza was confirmed by treatments with anti-IFN and anti-IL-1α antibodies. A mode of the antipyretic action of Kakkon-to other than that by suppression of IL-1α production responsive to IFN cannot be ruled out yet.

It is interesting that the mechanism of the actions of Kakkon-to was different in the HSV-1 and influenza infections. Kakkon-to enhanced the DTH reaction in the HSV-1 infection and suppressed IL-1α production responsive to IFN in the influenza infection. These contrasting actions of Kakkon-to seem to be contradictory, but they may be accounted for by postulating an acting target of Kakkon-to as macrophage-like cells or Lyt $1^{+}2^{-}$ T-lymphocyte subsets. Macrophages or these T cells play an important role in inducing the DTH response and producing IL-1α. Therefore, if Kakkon-to works on macrophages or T cells and modulates their functions, the seemingly contradictory findings in HSV-1 and influenza infections may be reasonable. However, the precise mechanism of action by Kakkon-to in these processes is not yet clear. We have partially characterized the action of each of the seven components of Kakkon-to, and some herbs showed the activity observed in the HSV-1 and influenza infections. It may be possible to identify the single active component in each infection. Some herbal components showed the DTH induction or antipyretic action but more weakly than did Kakkon-to, although the same amount of the specific herbal extract was used as that of Kakkon-to. The combination of the seven herbal components may potentiate each biological activity to exhibit the biological activity of Kakkon-to as a whole. The combination and ratio of the seven herbal components of Kakkon-to have been tested by experience over a long history, as stated in the Introduction. As the first step in the modern scientific analysis of this typical herbal medicine, we have examined its efficacy, but further studies on the target cells of Kakkon-to, the precise mechanism of action, the identification of the active compound(s) (Kurokawa *et al.*, 1998, and the rationale for the essentiality and interactions of the seven medicinal herbal components in Kakkon-to remain to be performed.

REFERENCES

1. L.R. Ballou, C.P. Chao, M.A. Holness, S.C. Barker and R. Raghow, Interleukin-1-mediated PGE_2 production and sphingomyelin metabolism: Evidence for the regulation of cyclooxygenase gene expression by sphingosine and ceramide. *Journal of Biological Chemistry* **267**, 20044–20050 (1992).
2. C.M. Blatteis, Role of the OVLT in the febrile response to circulating pyrogens. *Progress in Brain Research* **91**, 409–412 (1992).
3. C.A. Dinarello, Demonstration of a human pyrogen-inducing factor during mixed leukocyte reactions. *Journal of Experimental Medicine* **153**, 1215–1224.
4. C.A. Dinarello, J.G. Cannon and S.M. Wolff, New concepts on the pathogenesis of fever. *Reviews of Infectious Diseases* **10**, 168–189 (1988).
5. J.Y. Djeu, L. Emanuela, P. Simonetta and A.J. Hapel, Selective growth of natural cytotoxic but not natural killer effector cells in interleukin-3. *Nature* **306**, 788–791 (1983).
6. G.W. Duff and S.K. Durum, The pyretic and mitogenic actions of interleukin-1 are related. *Nature* **304**, 449–451 (1983).
7. W. Feldberg, K.P. Gupta, A.S. Milton and S. Wendlandt, Effect of pyrogen and antipyretics on prostaglandin activity in cisternal C.S.F. of unanesthetized cats. *Journal of Physiology* **234**, 279–303 (1973).
8. R.J. Flower and J.R. Vane, Inhibition of prostaglandin synthetase in brain explains the anti-pyretic activity of paracetamol (4-acetamidophenol). *Nature* **240**, 410–411 (1972).
9. M. Hashimoto, Y. Ishikawa, S. Yokota, F. Goto, T. Bando, Y. Sakakibara and M. Iriki, Action site of circulating interleukin-1 on the rabbit brain. *Brain Research* **540**, 217–223 (1991).
10. T. Hennet, H.J. Ziltener, K. Frei and E. Peterhans, A kinetic study of immune mediators in the lungs of mice infected with influenza A virus. *Journal of Immunology* **149**, 932–39 (1992).
11. P.A. Insel, Analgesic-antipyretics and antiinflammatory agents; drugs employed in the treatment of rheumatoid arthritis and gout. In Goodman and Gilman's, *The Pharmacological Basis of Therapeutics*, Pergamon Press, New York (1992).
12. Jiangxu New Medical College, *Dictionary of Chinese Medicinal Materials*, Shanghai Science and Technology Press, Shanghai, China (in Chinese) (1978).
13. M. Kurokawa, K. Hase, H.-X. Xu, J. Yamamura, M. Koyasu, H. Sato, S. Kadota, T. Hozumi, T. Namba and K. Shiraki, A novel procedure for the identification of a fraction with anti-herpes simplex virus type 1 activity *in vivo* from hot-water extract of traditional medicines, *Geum japonicum* THUNB. *Journal of Medical and Pharmaceutical Society for WAKAN-YAKU* **10**, 195–203 (1993a).
14. M. Kurokawa, H. Ochiai, N. Kazuhiko, M. Neki, H. Xu, S. Kadota, S. Sutardjo, T. Matsumoto, T. Namba and K. Shiraki, Antiviral traditional medicines against herpes simplex virus (HSV-1), poliovirus, and measles virus in vitro and their therapeutic efficacies for HSV-1 infection in mice. *Antiviral Research* **22**, 175–188 (1993b).
15. M. Kurokawa, K. Nagasaka, T. Hirabayashi, S. Uyama, H. Sato, T. Kageyama, S. Kadota, H. Ohyama, T. Hozumi, T. Namba and K. Shiraki, Efficacy of traditional herb medicines in combination with acyclovir against herpes simplex virus type 1 infection *in vitro* and *in vivo*. *Antiviral Research* **27**, 19–37 (1995).
16. M. Kurokawa, M. Imakita, C. A. Kumeda and K. Shiraki, Cascade of fever production in mice infected with influenza virus. *Journal of Medical Virology* **50**, 152–158 (1996a).
17. M. Kurokawa, M. Imakita, C.A. Kumeda, T.A. Yukawa and K. Shiraki, Kakkon-to suppressed interleukin-1α production responsive to interferon and alleviated influenza infection in mice. *Journal of Traditional Medicines* **13**, 201–209 (1996b).

18. M. Kurokawa, H. Oyama, T. Hozumi, T. Namba, M. Nakano and K. Shiraki, Assay for antiviral activity of herbal extract using their absorbed sera. *Chemical and Pharmacological Bulletin* **44**, 1270–1272 (1996c).
19. M. Kurokawa, M. Nakano, H. Ohyama, T. Hozumi, T. Namba and K. Shiraki, Prophylactic efficacy of traditional herbal medicines against recurrent herpes simplex virus type 1 infection in mice. *Journal of Dermatological Science* **14**, 76–84 (1997).
20. H.S. Larsen, M. Feng, D.W. Horohov, R.N. Moore and B.T. Rouse, Role of T-lymphocyte subsets in recovery from herpes simplex virus infection. *Journal of Virology* **50**, 56–59 (1984).
21. S. Nagafuchi, I. Hayashida, K. Higa, T. Wada and R. Mori, Role of Lyt-1 positive immune T cells in recovery from herpes simplex virus infection in mice. *Microbiology and Immunology* **26**, 359–362 (1982).
22. K. Nagasaka, M. Kurokawa, M. Imakita, K. Terasawa and K. Shiraki, Efficacy of Kakkon-to, a traditional herb medicine, in herpes simplex virus type 1 infection in mice. *Journal of Medical Virology* **46**, 28–34 (1995).
23. A.A. Nash, H.J. Field and R. Quartey-Papafio, Cell-mediated immunity in herpes simplex virus-infected mice: Induction, characterization and antiviral effects of delayed type hypersensitivity. *Journal of General Virology* **48**, 351–357 (1980).
24. A.A. Nash and P.G.H. Gell, Membrane phenotype of murine effector and suppresser T cells involved in delayed type hypersensitivity and protective immunity to herpes simplex virus. *Cellular Immunology* **75**, 348–355 (1983).
25. I. Roitt, J. Brostoff and D.K. Male, *Immunology*, Gower Medical Publishing, London, New York, pp. 11.1–11.11 (1985).
26. C.B. Saper and C.D. Breder, Endogenous pyrogens in the CNS: role in the febrile response. *Progress in Brain Research* **93**, 419–429 (1992).
27. H. Sato, W. Goto, J. Yamamura, M. Kurokawa, S. Kageyama, T. Takahara, A. Watanabe and K. Shiraki, Therapeutic basis of glycyrrhizin on chronic hepatitis B. *Antiviral Research* **30**, 171–177 (1996).
28. T. Takahara, A. Watanabe and K. Shiraki, Effects of glycyrrhizin on hepatitis B surface antigen: A biochemical and morphological study. *Journal of Hepatology* **21**, 601–609 (1994).
29. M. Tashiro, P. Ciborowski, H.-D. Klenk, G. Pulverer and R. Rott, Role of Staphylococcus aureus in the development of influenza pneumonia. *Nature* **325**, 536–537 (1987a).
30. M. Tashiro, P. Ciborowski, M. Reinacher, G. Pulverer, H.-D. Klenk and R. Rott, Synergistic role of staphylococcal proteases in the induction of influenza virus pathogenicity. *Virology* **157**, 421–430 (1987b).
31. K. Terasawa, KAMPO, K. K. Standard McIntyre, Tokyo, pp. 101–110 (1993).
32. T.A. Yukawa, H. Sato, Y. Yoshida, S. Kageyama, T. Namba, M. Imakita, T. Hozumi and K. Shiraki, Prophylactic treatment of cytomegalovirus infection with traditional herbal extracts. *Antiviral Research* **32**, 63–70 (1996).
33. M. Kurokawa, C.A. Kumeda, J. Yamamura, T. Kamiyama and K. Shiraki, Antipyretic activity of cinnamyl derivatives and related compounds in influenza virus-infected mice. *European Journal of Pharmacology* **348**, 45–51 (1998).

14 SEARCHING FOR ANTI-HIV AGENTS AMONG TRADITIONAL MEDICINES

INES TOMOCO KUSUMOTO[1] and MASAO HATTORI[2]

[1] *Department of Biodefence Medicine, Faculty of Medicine and* [2] *Research Institute for Oriental Medicines (Wakan-Yaku), Toyama Medical and Pharmaceutical University, Sugitani 2630, Toyama 930-0194, Japan*

INTRODUCTION

The global AIDS (acquired immunodeficiency syndrome) situation has been a serious problem since this infection affects adults of working age, adolescents and children, without any effective treatment at the moment. After the first patient with AIDS was reported in 1981 in the United States, the number of AIDS patients and persons infected with its causative virus (human immunodeficiency virus: HIV) increased abruptly all over the world. According to the World Health Organization (WHO) estimation (Weekly Epidemiological Record, 15 Dec. 1995), 17 million adults were living with HIV infection at the end of 1994. The major distribution of infected people was 11.2 million in sub-Saharan Africa, followed by 3 million in South and South-East Asia, and 2 million in the Americas. WHO estimated that the number of AIDS patients in Asia will surpass that in Africa in the near future, and that in the year 2000, the total number of AIDS patients would be 10 million and that of HIV-infected people in the world 40 million. Extensive research has been conducted on the prevention and cure of this disease, but no definitively effective drugs are available at the present time. The current treatment of HIV infection is based on specific antiviral drugs in association with antimicrobial treatment and prophylaxis. However, the enormous cost of these drugs has been a barrier to AIDS patients; this therapy also requires long-term administration. AIDS is undoubtedly one of the most serious medical and social problems all over the world.

THE TARGETS AGAINST HIV

AIDS is known to be caused by HIV-1 and -2, which belong to the family of human retroviruses, subfamily of lentivirus. It is known that HIV is transmitted by sexual contact, homosexual and heterosexual, and by blood or blood products. This virus has a spike-like glycoprotein gp120 on the envelope which binds to the CD4 molecule, a protein found on a subset of T lymphocytes, at the time of infection. Since T lymphocyes are responsible for the helper or inducer function in the immune response, the replication of viruses in these cells leads to the dysfunction and depletion of CD4-positive T cells causing a profound immunodeficiency. HIV possesses some specific enzymes that work on viral replication, such as RNA-dependent DNA polymerase or reverse transcriptase (RT), RNase H, integrase and protease (PR). In the first step of replication, RT transcribes the viral RNA into a double-strand DNA. Then, this DNA is integrated into the host chromosome and the viral components are synthesized and assembled into new virions. The maturation of the virus takes place at the last step by viral PR, which cleaves the viral polyproteins at the specific amino acid sequences to give functional proteins or enzymes. The mature viruses bud from the cells and continuously infect other T-cells. The blocking of any of these steps in the viral life cycle is expected to stop the viral replication, and many RT and PR inhibitors have been studied towards the development of anti-HIV agents. Inhibitory effects on RT were first shown in the 1970s against RT of Moloney murine leukemia virus (MuLV) and that of avian myeloblastosis virus (AMV) by substances such as ethidium bromide, actinomycin, adriamycin, fagaronine and others, most of them having DNA binding properties. Later, with the outbreak of HIV, this research took on a new dimension involving natural and synthetic substances. The natural substances that inhibit HIV replication or viral enzymes will be described here. The activities of these substances are sometimes mild, but they nevertheless occasionally produce some side effects. Several medicinal plant extracts used in Chinese traditional medicine, Ayurveda (the traditional medicine of India) and herbal medicines used in Indonesia, Egypt and Panama have been tested for anti-HIV effects *in vitro* and those with some properties against the viral enzymes or replication of HIV were selected for further study. As described below, the extracts were tested for (1) inhibition of RT (which acts on the early stage of viral infection), (2) inhibition of PR (responsible for viral maturation), and (3) inhibition of HIV replication in an infected MT-4 cell assay system.

INHIBITORY EFFECTS ON REVERSE TRANSCRIPTASE

The RT activity was determined as the incorporation of $[^3H]$-dTTP or $[^3H]$-dGTP into a polymer fraction by AMV- or HIV-1-RT using $(rA)_n$-$(dT)_{12-18}$ or $(rC)_n$-$(dG)_{12-18}$ as a template-primer. The inhibitory effects differed according to the different retroviral enzymes and to the template-primer used in the assay. The differences will be discussed in the course of the text; however, a comparative assay with HIV- and AMV-RT demonstrated that the inhibitory effects of the two enzymes occurred to a similar extent.

Of the Chinese medicines tested (Table 1), a water extract of the whole plant of *Chloranthus glaber*, an ethanol extract of the leaves of *Thuja orientalis*, and water and ethanol extracts of the vine of *Spatholobus suberectus* were found to possess potent inhibitory effects against AMV-RT with an IC_{50} value (the concentration required to inhibit enzyme activity by 50%) of 10–30 µg/ml. The inhibitory property of these plants seems to be due to polyphenol constituents, although no conclusive data are available at present.

Of the plant extracts having moderate inhibitory effects, some flavonoids were isolated as the major constituents from the roots of *Scutellaria baicalensis*. Baicalein

Table 1 Inhibitory effects of various extracts of traditional medicines on reverse transcriptase from AMV and HIV

		IC_{50} (μg/ml)	
Plant name	Part used	Water ext.	MeOH/EtOH ext.
Chinese medicines[a)]			
Chloranthus glaber	whole plant	30	> 100[c)]
Scutellaria baicalensis	roots	> 100	250[c)]
Spatholobus suberectus	vine	30	10[c)]
Thuja orientalis	leaves	100	30[c)]
Ayurvedic medicines[a)]			
Areca catechu	seeds	42	30
Cinnamomum verum	bark	32	32
Coleus amboinicus	leaves	50	> 100
Eugenia jambolana	bark	n.d.	18
Ficus religiosa	bark	n.d.	36
Punica granatum	pericarp	> 100	2
Saraca indica	bark	4	n.d.
Terminalia arjuna	stem bark	34	34
Egyptian folk medicines[b)]			
Phyllanthus emblica	fruit	10	9
Quercus pedunculata	fruit	49	9
Terminalia belerica	fruit	10	9.5
Terminalia chebula	fruit	6	2
Terminalia horrida	fruit	39	3
Indonesian folk medicines[a)]			
Loranthus parasiticus	stem+branch	5	> 100
Helicteres isora	fruit	3.5	> 100
Swietenia mahagoni	bark		55
Sindora sumatrana	fruit	20	> 100
Panamanian folk medicines[b)]			
Chamaesyce hyssopifolia	whole plant	8	43
Cordia spinescens	leaves	6	36
Hyptis lantanifolia	aerial part	7	> 100
Tetrapteris macrocarpa	aerial part	> 100	8

AMV: avian myeloblastosis virus; HIV: human immunodeficiency virus.
AMV-RT (a) and HIV-RT (b) activities were determined by counting the incorporation of ^{3}H-dTTP into a polymer fraction in the presence of $(rA)_n$-$(dT)_{12-18}$ as a template-primer. c), EtOH extract. IC_{50} = a concentration required to inhibit enzyme activity by 50%.

(**1**) showed the most potent inhibitory activity (IC_{50} = 50 μM), while the main constituent of this plant, baicalin, a glucuronide of **1**, showed a lower inhibitory activity (IC_{50} = 250 μM). The effects of other flavonoids were also investigated (Kusumoto *et al.*, 1991). Of 190 flavonoids examined, 46 had over 50% inhibition at a concentration of 1 mM. 6-Hydroxyluteolin (**3**) (IC_{50} = 7 μM), pedalitin (4) (IC_{50} = 10 μM), 6-hydroxykaempferol (**5**) (IC_{50} = 8 μM) and quercetagetin (**6**) (IC_{50} = 8 μM) were the most potent inhibitory substances against AMV-RT (Figure 1). Of

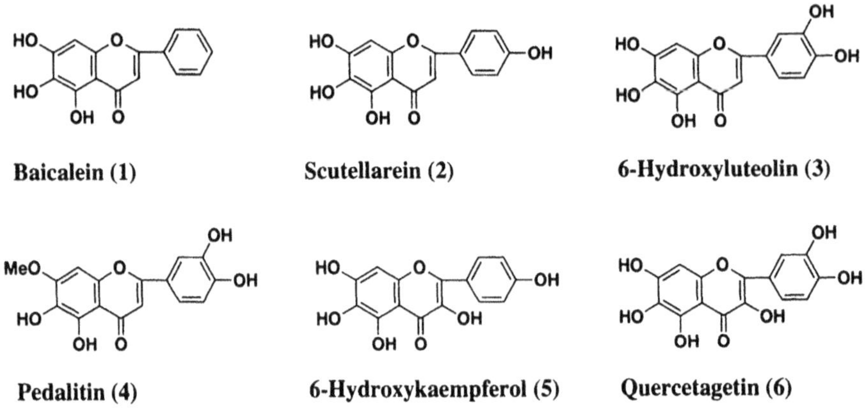

Figure 1 Flavonoids having potent inhibitory activity against AMV-reverse transcriptase.

the flavonoids having potent inhibitory effects, the number of hydroxyl groups and their positions was found to be related to the potency of inhibition. As to the number of hydroxyl groups (Table 2), all hexa-hydroxyl flavonoids showed potent inhibition but those having less than three hydroxyl groups were not effective, except for baicalein (**1**). When hydroxyl groups are arranged at the neighboring positions, such as 5, 6, 7 and 3′, 4′, and 5′, the flavonoids showed extremely strong inhibition of RT. The mode of inhibition of flavonoids such as 6-hydroxyluteolin (**3**), myricetin and gossypetin was non-competitive with regard to the substrate, and a mixed type with regard to the template-primer. The inhibitory constants (*Ki*) of 6-hydroxyluteolin (**3**), myricetin and gossypetin were 10, 94 and 90 μM, respectively, with respect to the substrate and 5,200 and 100 μM, respectively, with respect to poly(rA)-oligo(dT).

These flavonoids were also tested for DNA polymerase I. Contrary to the results with RT, the effects of the flavonoids were not strong against this enzyme, indicating, at least in part, that inhibition by these compounds is more specific to RNA-dependent DNA polymerase (RT) than to DNA-dependent DNA polymerase (DNA polymerase I).

The early reports on natural substances as RT inhibitors describe the effects of fagaronine, a benzo[c]phenanthridine alkaloid (Sethi and Sethi, 1975), and of structurally related protoberberine alkaloids on RT of RNA tumor viruses (Sethi, 1983). The former substance is a constituent of a plant *Fagara xanthoxyloides* (Rutaceae) and the latter are constituents of many plants of the family of Berberidaceae showing a wide variety of pharmacological effects. Some of the Chinese medicines, such as the rhizome of *Coptis japonica* (Ranunculaceae) and the bark of *Phellodendron amurense* (Rutaceae), contain various protoberberine-type alkaloids. Berberine (**7**) and related compounds (**8–12**) were tested to investigate the activity and structure relationship (Figure 2) (Kusumoto *et al.*, 1991; Kakiuchi *et al.*, 1987). Of the protoberberine alkaloids examined, inhibitory activity decreased in the order of coralyne (**11**), berberine (**7**), coptisine, 13-methylberberine (**9**), dehydrocorydaline and berberrubine (**12**) (Table 3). The most potent inhibitor, coralyne (**11**), has an aromatic B ring. In contrast, berberrubine (**12**), which has a

Table 2 Comparative inhibitory effects of hydroxy-substituted flavones on AMV-RT

Flavonoid	Position of OH	% Inhibition
Tri-*O*-substituted flavones		
Apigenin	5,7,4′	-9.9
Baicalein (**1**)	5,6,7	95.0
Norwogonin	5,7,8	36.9
Galangin	3,5,7	42.1
Tetra-*O*-substituted flavones		
Scutellarein (**2**)	5,6,7,4′	99.4
Luteolin	5,7,3′,4′	79.1
Isoscutellarein	5,7,8,4′	22.0
Kaempferol	3,5,7,4′	72.6
Datiscetin	3,5,7,2′	28.0
Penta-*O*-substituted flavones		
6-Hydroxyluteolin (**3**)	5,6,7,3′,4′	91.7
3′-Hydroxyisoscutellarein	5,7,8,3′,4′	31.3
6-Hydroxykaempferol (**5**)	3,5,6,7,4′	97.2
Quercetin	3,5,7,3′,4′	41.4
Robinetin	3,7,3′,4′,5′	87.5
Morin	3,5,7,2′,4′	40.5
6′-Hydroxydatiscetin	3,5,7,2′,6′	21.3
Hexa-*O*-substituted flavones		
Quercetagetin (**6**)	3,5,6,7,3′,4′	98.4
Myricetin	3,5,7,3′,4′,5′	82.5
Gossypetin	3,5,7,8,3′,4′	85.1

The concentration of flavonoid was 1.0 mM, and $(rA)_n$-$(dT)_{12-18}$ was used as a template-primer.

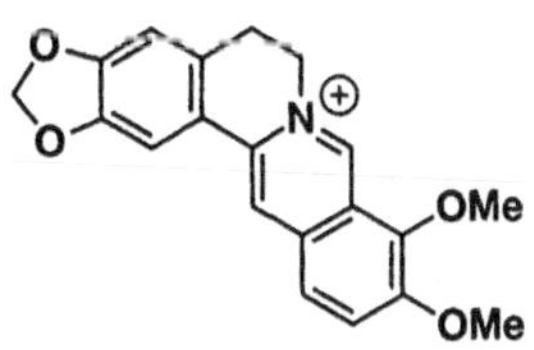

Berberine (7)

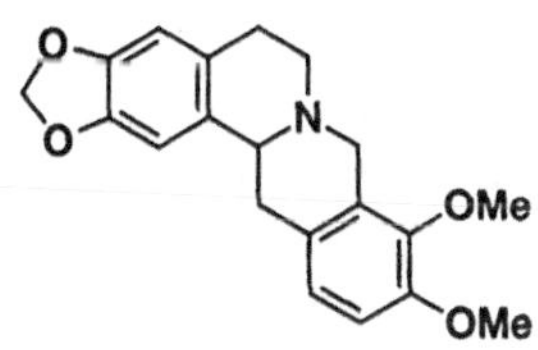

Tetrahydroberberine (8)

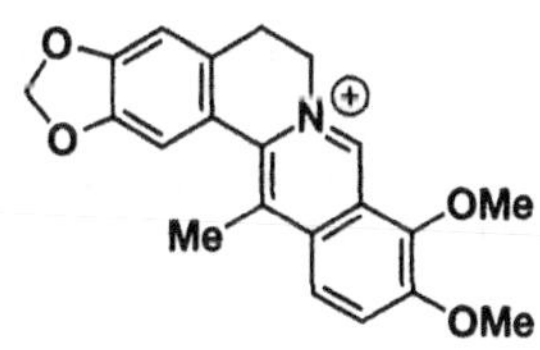

13-Methylberberine (9)

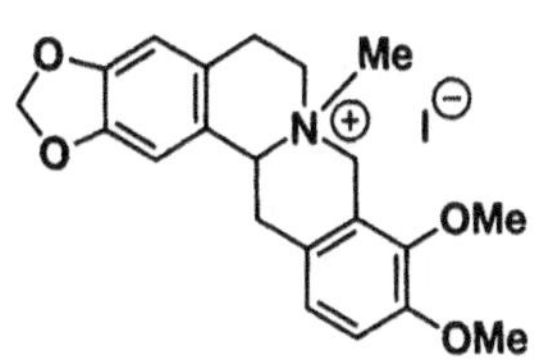

Tetrahydroberberine methyl iodide (10)

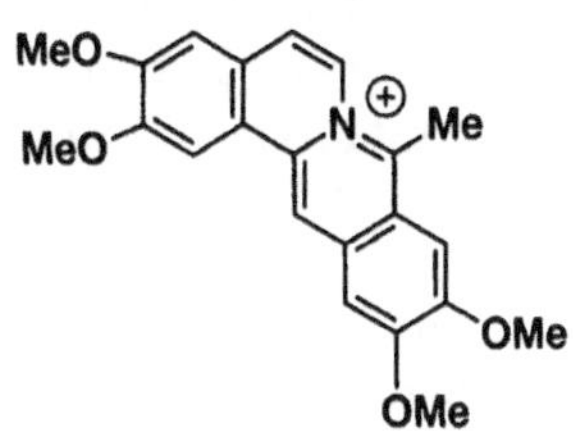

Coralyne (11)

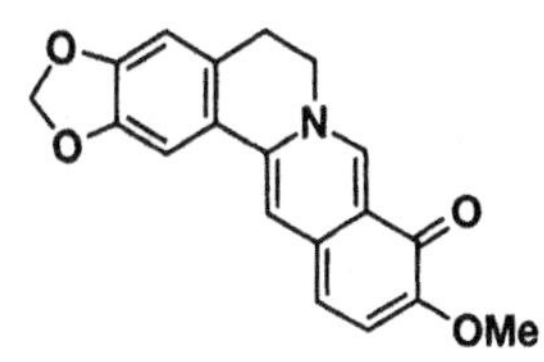

Berberrubine (12)

Figure 2 Protoberberine alkaloids tested for their inhibitory effects on AMV-RT.

Table 3 Inhibitory effects of protoberberine alkaloids on AMV-RT in the presence of $(rA)_n$-$(dT)_{12-18}$ as a template-primer.

Alkaloids	IC_{50} (μM)
Berberine (**7**)	160
Coptisine	190
13-Methylberberine (**9**)	600
Dehydrocorydaline	700
Coralyne (**11**)	3.5
Berberrubine (**12**)	750

tertiary nitrogen in its ketonic form, gave weak inhibition. Moreover, tetrahydroberberine (**8**) showed no appreciable inhibition, indicating that conversion of the quaternary nitrogen of berberine (**7**) to the tertiary nitrogen results in a significant decrease of inhibitory activity. Tetrahydroberberine methyl iodide (**10**) showed no inhibitory effect though it possesses a quaternary nitrogen. A methyl group introduced to the molecule seems to interfere with the intercalation of the compounds to a template-primer. Fagaronine was reported to show enzyme inhibition by a selective intercalation to poly(rA)-oligo(dT), suggesting that the RT-inhibition by this compound is not due to a direct interaction with the enzyme protein (Sethi and Sethi, 1975). A demonstration of the binding of benzo[c]phenanthridine alkaloids to nucleic acid was shown by CD spectra of a mixture of these alkaloids and some template-primers including poly(rA)-oligo(dT). Besides the intercalation that is the most likely mode of interaction, the electrostatic interaction between phosphate and quaternary nitrogen may also participate in the alkaloid-nucleic acid complexes (Kakiuchi *et al.*, 1987).

Of the Ayurvedic medicines (Table 1), methanolic extracts of the pericarp of *Punica granatum*, the bark of *Eugenia jambolana*, the bark of *Ficus religiosa* and the fruit of *Terminalia arjuna*, water extracts of the bark of *Saraca indica*, and water and methanol extracts of the seeds of *Areca catechu* and the bark of *Cinnamomum verum* were found to strongly inhibit RT with IC_{50} values of 2–50 μg/ml (Hattori *et al.*, 1993). Both water and methanol extracts of the pericarp of *Punica granatum* and the fruit of *Terminalia chebula* are known to contain characteristic ellagitannins: granatins A and B, punicalin and punicalagin in the former (Okuda *et al.*, 1980), and chebulagic acid and chebulinic acid in the latter (Yoshida *et al.*, 1982). It was demonstrated that these tannins strongly inhibited AMV-RT activity in a non-competitive manner as regards a template-primer (Kakiuchi *et al.*, 1985). Since the extract of the seeds of *A. catechu* contains significant amounts of condensed tannins (Nonaka *et al.*, 1981) which induce a variety of enzyme inhibitions (Hada *et al.*, 1989; Inokuchi *et al.*, 1986; Kikuchi *et al.*, 1987), we suspected that the potent inhibitory actions of both water and methanol extracts (more than 90% inhibition at 0.2–1.0 mg/ml) were primarily due to procyanidins. Polyphenols were isolated, as expected, from the *A. catechu* extracts as RT inhibitory substances by means of bio-assay directed fractionation. Of these, arecatannins A2 (**15**) and B1 (**16**) (Figure 3) showed the most potent inhibition against AMV-RT with IC_{50} values of 1.0 and 0.5 μM, respectively. Since procyanidin compounds were shown to have

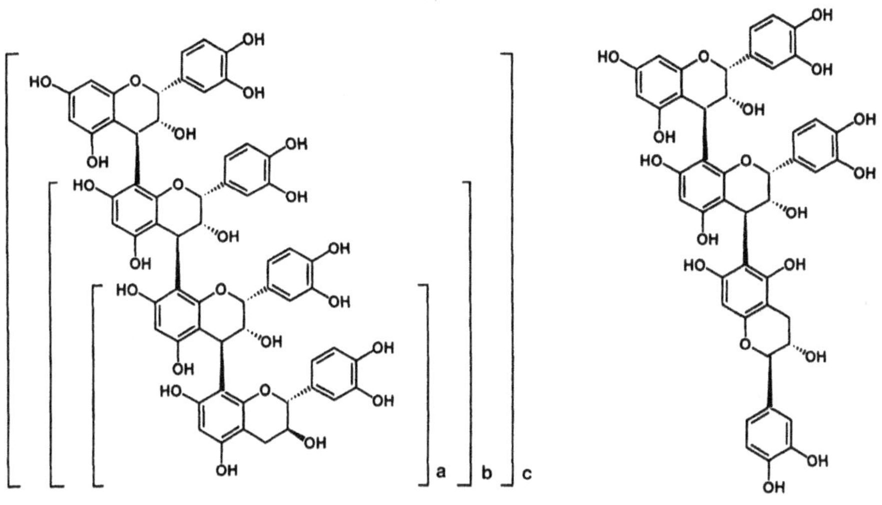

a : Procyanidin B1 (13)
b : Arecatannin A1 (14)
c : Arecatannin A2 (15)

Arecatannin B1 (16)

Figure 3 Polyphenols isolated from *Areca catechu*.

potent inhibitory effects, other condensed tannins were investigated (Kakiuchi *et al.*, 1991). The potency of inhibition increased in the order of monomer, dimer, trimer and tetramer. The reaction was inhibited to a greater extent when $(rC)_n$-$(dG)_{12\text{-}18}$ was used as a template-primer. When we compared galloylated and non-galloylated procyanidin oligomers, the galloylated compounds inhibited RT more strongly than did the non-galloylated compounds. The galloylated (-)-epicatechin trimer with 4,6 linkage showed the most potent inhibition (Table 4).

The screening of Egyptian folk medicines (Table 1) showed that methanol extracts of the fruits of *Quercus pedunculata* and *Terminalia horrida*, and water and methanol extracts of the fruits of *Phyllanthus emblica*, *Terminalia belerica* and *Terminalia chebula* had strong inhibition ability with IC_{50} values of 2–10 μg/ml. By bioassay-guided fractionation of *P. emblica*, six compounds were isolated from the most active fraction (Figure 4). These are putranjivain A (**17**), galloyl glucose, digalloyl glucose, kaempferol glucoside, quercetin glucoside and digallic acid, in which **17** has a unique structure, with spirorings. Of these compounds, putranjivain A (**17**) showed the most potent inhibition of HIV-RT. The IC_{50} was as low as 3.9 μM. From Lineweaver-Burk plots, the inhibitory mode of action was determined to be competitive with regard to a template-primer (*Ki* = 0.25 μM), and non-competitive with regard to the substrate (*Ki* = 0.89 μM). From the fruit of *T. chebula*, some ellagitannins such as chebulagic acid and chebulinic acid have been isolated and demonstrated to inhibit AMV-RT (Kakiuchi *et al.*, 1991). The inhibitory constants (*Ki*) were calculated to be 0.89 μM for the substrate and 0.25 μM for the template-primer.

Of the Indonesian folk medicines called "Jamu" (Table 1), methanolic extracts of the stems and branches of *Loranthus parasiticus* and the fruit of *Helicteres isora*

Table 4 Inhibitory effects of condensed tannins and related compounds on AMV-RT in the presence of $(rA)_n$-$(dT)_{12-18}$ or $(rC)_n$-$(dG)_{12-18}$ as a template-primer.

Compound	IC_{50} (μM)	
	$(rA)_n$-$(dT)_{12-18}$	$(rC)_n$-$(dG)_{12-18}$
Galloylated (-)-epicatechin oligomers		
4,8 Tetramer	180	94
4,8 Trimer	260	78
4,8 Dimer	350	82
4,6 Trimer	47	34
4,6 Dimer	470	260
Non-galloylated oligomers		
(-)-Epicatechin (4,8) tetramer	310	130
Procyanidin C-1	420	280
Procyanidin B-2	850	350
Procyanidin B-3	> 1000	760
Monomers		
(-)-Epigallocatechin	> 1000	> 1000
(-)-Epigallocatechin gallate	> 1000	> 1000
(-)-Epicatechin	> 1000	> 1000
(+)-Catechin	>1000	>1000
7-*O*-Galloylcatechin	590	100
Quercitrin	>1000	>1000
Quercitrin 2″-*O*-gallate	900	500

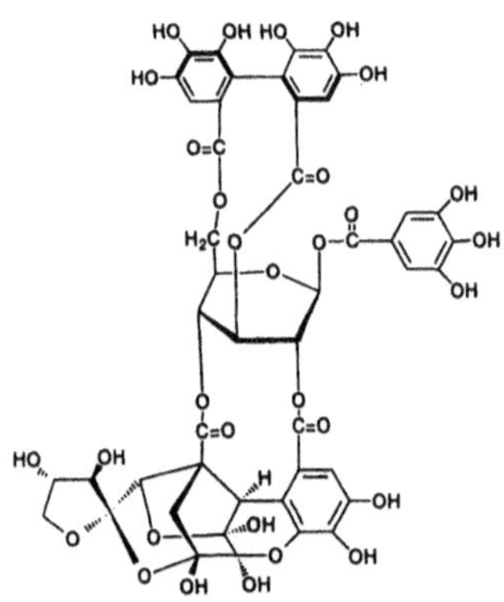

Putranjivain A (17)

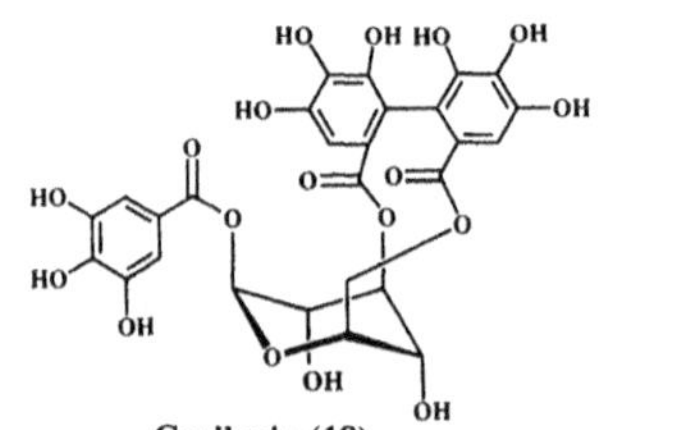

Corilagin (18)

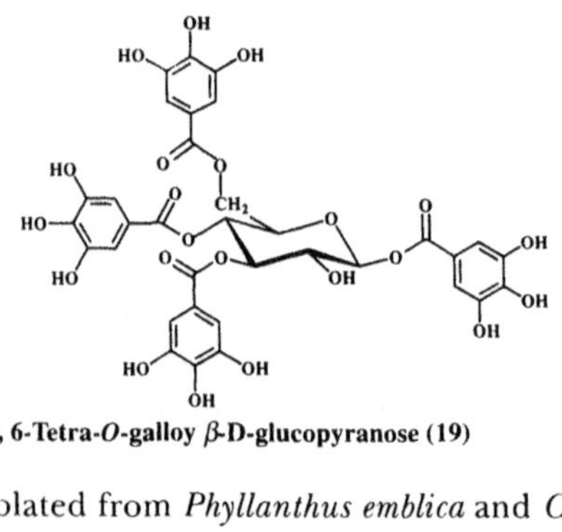

1, 3, 4, 6-Tetra-*O*-galloy β-D-glucopyranose (19)

Figure 4 HIV-RT inhibitory compounds isolated from *Phyllanthus emblica* and *Chamaesyce hyssopifolia*.

were shown to inhibit AMV-RT, with IC_{50} values of 5.0 μg/ml and 3.5 μg/ml, respectively, in the presence of poly(rA)-oligo(dT) as a template-primer (Kusumoto *et al.*, 1992b). However, more potent inhibitory activity of the extracts was observed in the presence of poly(rC)-oligo(dG); especially, water extracts of *L. parasiticus* (whole plant, stems and branches), *Curculigo orchioides* (fruit), *H. isora* (fruit), *Sindora sumarana* (fruit) and *Elephantopus scaber* (leaves), and a methanol extract of *Woodfordia floribunda* (flower and leaves) showed quite low values of IC_{50} ($\leq$ 10 μg/ml) in this assay system. These results indicated that the extracts inhibited the reaction selectively with regard to the distinct template-primer. The extracts tested for RT inhibition were also studied for their cytotoxic effects. In cell cultures containing various concentrations of the plant extracts, most of them did not affect cell viability in concentrations of 0.1–1.0 μg/ml.

Of the Panamanian plants, water extracts of the whole plant of *Chamaesyce hyssopifolia* and the leaves of *Cordia spinescens* and the aerial part of *Hyptis lantanifolia*, and a methanol extract of the aerial part of *Tetrapteris macrocarpa* had inhibitory effects with IC_{50} values of 6–10 μg/ml. Three compounds were found to be the major inhibitory substances in the extract of *C. hyssopifolia*, i.e., corilagin (**18**), quercetin 3-*O*-β-D-glucopyranoside and 1,3,4,6-tetra-*O*-galloyl-β-D-glucopyranose (**19**). Their IC_{50} values were 20, 50 and 86 μM, respectively and their mode of inhibition was non-competitive with respect to the substrate (Lim *et al.*, 1996).

INHIBITORY EFFECTS ON RTS FROM DIFFERENT RETROVIRUSES

The results described above indicate that RT-inhibitory substances have the following features: flat in structure; presence of many hydroxyl groups in its structure and/or presence of a quaternary nitrogen in the conjugated ring system. Since the early data were obtained by using AMV-RT rather than HIV-RT, a comparative experiment was performed for three flavonoids and twelve benzo[c]phenanthridine alkaloids (Figure 5) which had been shown to have potent inhibitory activity against AMV-RT (Kakiuchi *et al.*, 1985, 1987), by using three types of RTs: HIV-, MuLV- and AMV-RT (Table 5).

In the case of AMV-RT, quercetagetin (**6**), fagaronine (**24**) and nitidine (**28**) were the most potent inhibitors, with IC_{50} value < 10^{-5} M, followed by baicalein (**1**), avicine (**20**) and 6-methylnitidine (**27**). The IC_{50} values of these compounds were similar between HIV-RT and AMV-RT, except for gossypetin and chelerytrine (**21**): the former had a stronger activity against HIV-RT but the latter was stronger against AMV-RT. MuLV-RT activity was more potently inhibited by many of the samples, such as baicalein (**1**), gossypetin, avicine (**20**), chelilutine (**22**), chelirubine (**23**), 7-methoxynitidine (**25**), 10-methoxynitidine (**26**), 6-methylnitidine (**27**), sanguilutine (**29**) and sanguinarine (**30**), as compared to other RT activities. The most potent inhibitors, quercetagetin (**6**), fagaronine (**24**), and nitidine (**28**), inhibited all three RTs to a similar extent. Since the compounds tested here inhibited the AMV- and HIV-RT activities to a similar extent, it was concluded that the AMV-RT-inhibitory substances found in our previous screening may also inhibit HIV-RT.

The structure and function of virally encoded RT have been intensively studied. The basic organization in the viral polyprotein is common to RTs from different

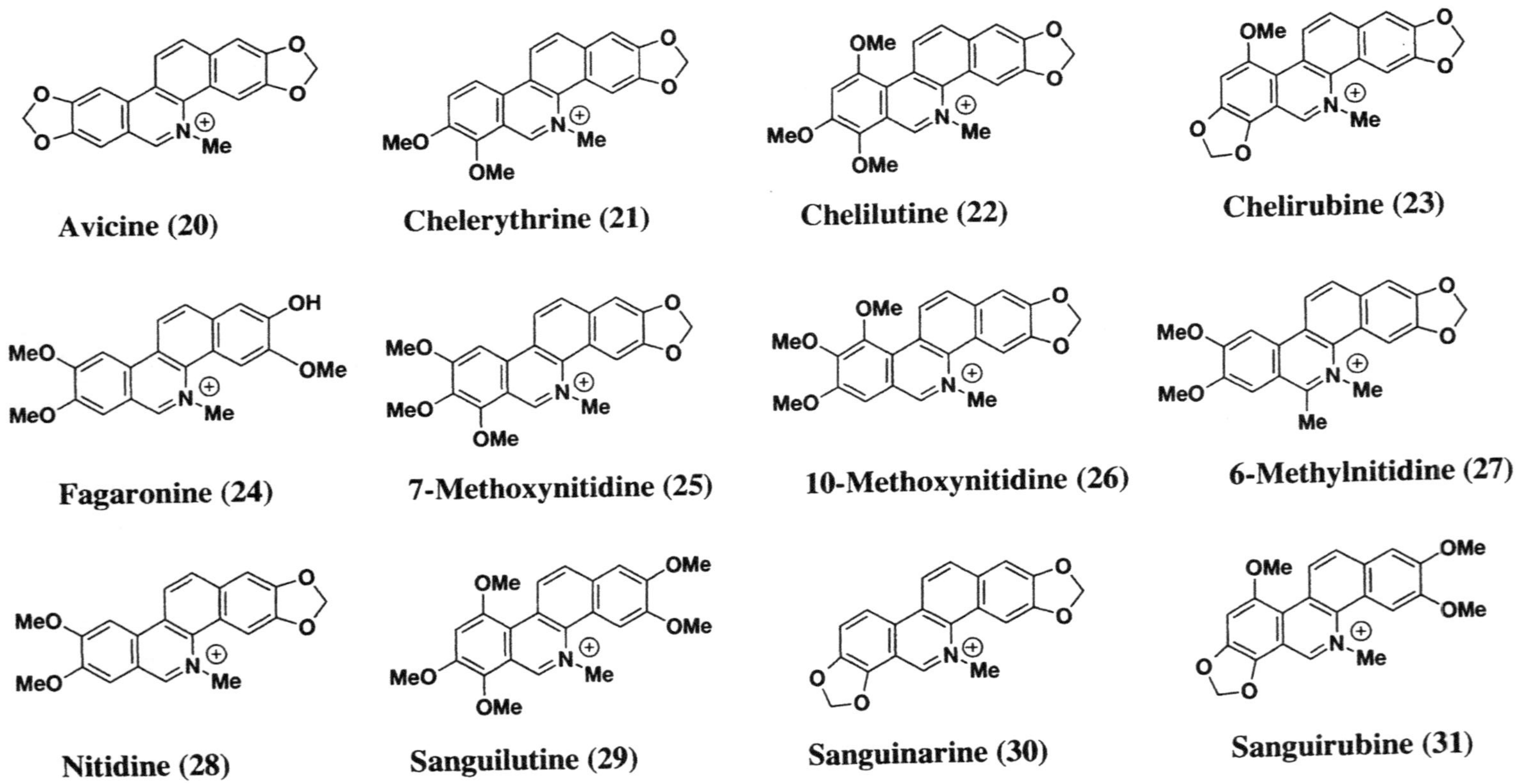

Figure 5 Benzo[c]phenanthridine alkaloids tested for their inhibitory effects on three different reverse transcriptases.

Table 5 Inhibitory effects of benzo[c]phenanthridine alkaloids on HIV, MuLV and AMV reverse transcriptase in the presence of $(rA)_n$-$(dT)_{12-18}$ as a template-primer.

Alkaloid	IC_{50} (μM)		
	HIV-RT	MuLV-RT	AMV-RT
Avicine chloride (**20**)	43	28	45
Chelerythrine chloride (**21**)	800	380	460
Chelilutine chloride (**22**)	31	20	29
Chelirubine chloride (**23**)	> 1000	500	> 1000
Fagaronine chloride (**24**)	6.4	7.3	4.8
7-Methoxynitidine chloride (**25**)	> 1000	660	> 1000
10-Methoxynitidine chloride (**26**)	> 1000	460	25
6-Methylnitidine chloride (**27**)	38	15	25
Nitidine chloride (**28**)	5.0	3.6	3.8
Sanguilutine chloride (**29**)	200	160	220
Sanguinarine chloride (**30**)	> 1000	410	> 1000
Sanguirubine chloride (**31**)	380	320	380

sources, and results of genetic analysis and amino acid sequence comparisons have been reported by Johnson *et al.* (1986), who showed that the amino acid residues were conserved in the regions of polymerization and RNase H domains of retroviral RTs, but that RTs differed in some functional and structural aspects. Functionally, they differ in their recognition of primer tRNA, i.e., HIV- and AMV-RT recognize $tRNA^{lys}$ and $tRNA^{Trp}$, respectively, as a primer, and MuLV-RT has affinity for several tRNAs (Jacobo-Molina and Arnold, 1991). Structurally, MuLV-RT is a monomer, whereas AMV- and HIV-RT are heterodimers. The three-dimensional structure of HIV-RT has been described by Kohlstaedt *et al.* (1992) in a crystal structure of RT complexed with a non-nucleotide inhibitor. In this complex, Asp^{185}, Asp^{186} and Asp^{110} residues of RT are expected to play some essential roles in the catalysis of the polymerase reaction.

The RT-inhibitory effects of benzo[c]phenanthridine alkaloids have been explained by their binding properties to nucleic acids (Kakiuchi *et al.*, 1987). In the same way, flavonoids which have flat structures also seem to interact with the template-primer, resulting in inhibition of RT. However, the different inhibitory potencies against RTs observed in the present work may have been caused by structural differences among enzymes in the formation of an inhibitor–enzyme–template-primer complex, and the exact mechanism of their inhibitory effects remains to be clarified by a study of the three-dimensional structure of each complex.

INHIBITORY EFFECTS ON HIV-1 PROTEASE

In HIV-1, the gag-pol genes are translated to the respective polyproteins which must be cleaved at specific sites by viral protease (PR) to produce essential viral enzymes, such as RT, ribonuclease, endonuclease and PR itself, and structural proteins such as p17, p9, p7 and p24. The HIV-1 PR (11,000 daltons) is an aspartic protease containing an Asp-Thr-Gly residue as its active site. Since this enzyme seems to be a promising target for the development of specific anti-viral drugs, some convenient methods for the measurement of PR activity have been established in which peptides having the amino acid sequences of the polyprotein fragments, such as gag p17/p24, gag p24/p7 and prot/RT, are used as substrates in general (Billich *et al.*, 1988; Krausslich *et al.*, 1989). To identify naturally-occurring substances which may function as a potent and specific inhibitor of HIV-PR, various medicinal plant extracts were tested for their PR inhibitory activities, using recombinant HIV-1 protease and His-Lys-Ala-Arg-Val-Leu-(pNO_2Phe)-Glu-Ala-Nle-Ser-NH_2 as a substrate (Kusumoto *et al.*, 1992a).

Some Chinese and Mongolian traditional medicines were found to have HIV-PR inhibitory activity. Water extracts of the whole plant of *Cynomorium songaricum* (Cynomoriaceae) and the rhizome of *Dryopteris crassirhizoma* (Polypodiaceae) showed 83–89% inhibition at 200 μg/ml. The bioassay-directed fractionation of *C. songaricum* led to the isolation of acetyl ursolic acid, ursolic acid and ursolic acid propanedioic acid monoester as inhibitory substances. The IC_{50} of the monoester, which showed the strongest inhibition among them, was 0.4 mM (Ma *et al.*, 1995).

Of the Ayurvedic medicines (Table 6), a methanol extract of the seeds of *Areca catechu* showed a potent inhibitory effect on HIV-PR, its IC_{50} being as low as 12 μg/ml. Of the various polyphenols isolated from this plant as described above, arecatannins A1 (**14**) and B1 (**16**) were identified as moderate inhibitory substances, their IC_{50} values being 0.41 and 0.30 mM, respectively. However, more active substances seemed to be present in the extract. Moderate inhibition was observed in the aqueous extracts of *Ficus religiosa* (bark) and *Litsea sebifera* (bark), the methanol extract of *Mimusops elengi* (bark) and both the water and the methanol extracts of *Terminalia chebula* (fruit) with IC_{50} values of 200 μg/ml.

Of the Indonesian folk medicines tested, methanol extracts of *Terminalia belerica* (fruit peel), *Swietenia mahagoni* (bark), *Garcinia mangostana* (fruit peel), and water and methanol extracts of *Woodfordia floribunda* (flower and stem) showed potent inhibitory effects (IC_{50} = 40–50 μg/ml) (Kusumoto *et al.*, 1992a). Chlorogenic acid methyl ester, (+)-catechin and (+)-gallocatechin were isolated from *S. mahagoni*. Of them, chlorogenic acid methyl ester showed a moderate inhibition, with an IC_{50} value of 60 μg/ml, suggesting the presence of other active components in the extract of this plant (Ma *et al.*, 1994).

Of the Panamanian plant extracts, methanol extracts of *Waltheria indica* (branch) and *Erythroxylum citrifolium* (trunk), and a water extract of *Xylopia frutescens* (bark) showed potent inhibitory effects (IC_{50} = 40–50 μg/ml).

INHIBITORY EFFECTS ON HIV REPLICATION AND GIANT CELL FORMATION

In an anti-HIV bioassay, the plant extracts were first investigated for their effects on viral replication in a primary screening , using MT-4 cells, a human T4-positive cell line carrying HTLV-I. When MT-4 cells are infected with HIV-1, the virus rapidly

Table 6 Inhibitory effects of various plant extracts on HIV-protease.

		IC_{50} (μg/ml)	
Plant name	Part used	Water ext.	MeOH ext.
Ayurvedic medicines			
Areca catechu	seeds	11	
Ficus religiosa	bark	200	
Lisea sebifera	bark	200	
Mimusops elengi	bark	200	
Terminalia chebula	fruit	200	200
Indonesian folk medicines			
Terminalia belerica	fruit peal	220	50
Loranthus parasiticus	stem + branche	260	100
Parameria laevigata	bark	> 500	100
Swietenia mahagoni	bark	100	40
Garcinia mangostana	peel	100	50
Woodfordia floribunda	flower + leaves	50	50
Panamanian folk medicines			
Waltheria indica	branches	96	48
Erythroxylum citrifolium	trunk	> 100	43
Aegiphila anomala	leaves	87	> 100
Cordia spinescens	leaves	100	> 100
Hyptis lantanifolia	aerial part	100	> 100
Mikania banisteriae	leaves + flower	100	> 100
Serjania mexicana	whole plant	87	> 100
Xylopia frutescens	bark	46	> 100
Egyptian folk medicines			
Boswellia carterii	resin	500	160
Croton tiglium	seeds		350
Rumex cyprius	fruit	50	
Terminalia horrida	fruit	90	

proliferates and the MT-4 cells are destroyed within 5 to 6 days (Harada *et al.*, 1985). The effects of plant extracts are evaluated as the concentration that prevents HIV-induced cytopathic effects (CPE), observed with an optical microscope. The samples showing significant anti-HIV activity in the primary screening were submitted to a more accurate test using the MT-4 cells for the inhibition of CPE. From this test, the dose that reduces the viability of uninfected cells by 50% (CC_{50}) and the dose required to inhibit HIV-1 replication by 50% (IC_{50}) are obtained and the ratio of these values (CC_{50}/IC_{50}) is given as the selectivity index (SI). Then, the test was performed on the assay system using HIV-1-infected and non-infected MOLT-4 cells in which multinucleated giant cells appear within 1 to 2 days. Formation of multinucleated giant cells or syncytia is a phenomenon observed in the co-culture of HIV-1-infected and -uninfected MOLT-4 cells. This giant cell formation may play an important role in the depletion of CD4-positive T lymphocytes in HIV-infected persons (Haseltine, 1988). Therefore, examination of this phenomenon has been a good tool to observe HIV replication *ex-vivo.*

Of the Ayurvedic medicines tested for their HIV inhibitory effects using HIV-1-infected MT-4 cells (Table 7), a water extract of *Pogostemon heynaenus* (leaves) showed appreciable inhibition of the growth of HIV, with an SI value > 48.1.

Of the Indonesian plant extracts, a methanolic extract of *Swietenia mahagoni* (bark), and water extracts of *Loranthus parasiticus* (stems and branches) and *Helicteres isora* (fruit) had appreciable inhibition, with IC_{50} = 4.2–27 μg/ml. The samples also suppressed the formation of syncytia in co-cultures of MOLT-4 and MOLT-4/HIV-1 cells. However, the SI values were not meaningful, although the methanol extract of *S. mahagoni* (bark) showed an appreciably low IC_{50}.

Table 7 Inhibitory effects of various plant extracts on HIV replication.

Plant name	Part used	Ext.	IC_{50} (μg/ml)	CC_{50} (μg/ml)	SI
Ayurvedic medicines					
Pogostemon heynaenus	leaves	W	20.8	> 1000	>48.1
Indonesian folk medicines					
Swietenia mahagoni	bark	M	4.2	10	2.4
Loranthus parasiticus	stem + branch	W	27	100	3.7
Helicteres isora	fruit	W	17	62	3.7
Sindora sumatrana	fruit	W	59	260	4.4
Andrographis paniculata	leaves	W	70	220	3.1
Curcuma aeruginosa	rhizome	W	175	240	1.4
Panamanian folk medicines					
Acalypha macrostrachya	leaves	M	> 500	190	–
Baccharis trinervis	aerial part	W	38.0	260	6.84
Bidens pilosa	aerial part	W	54.0	420	7.78
Calea jamaicensis	roots	W	62.0	140	2.26
Cordia spinescens	leaves	W	15.5	96.0	6.19
Cornutia grandifolia	trunk	W	55.0	250	4.55
Croton billbergianus	trunk	W	630	> 1000	> 1.59
Drymonia serrulata	leaves	W	24.0	> 1000	> 41.7
Jatropha curcas	branch	W	9.0	52	5.8
	leaves	M	100	94	0.94
Pereskia bleo	whole plant	M	20.8	> 1000	> 48.1
Egyptian folk medicines					
Abrus precatorius	seeds	M	2.0	43.5	21.8
Artemisia absinthium	aerial part	M	9.8	31.0	3.16
Bassia muricata	whole plant	W	135	> 1000	> 7.4
Centaurea scoparia	roots	W	46.2	> 1000	> 21.6
Croton tiglium	seeds	M	0.025	0.86	34.4
		W	2.0	100	50.0
Datura stramonium	seeds	M	4.1	6.0	1.49
		W	90.0	840	9.3

W: water extract; M: methanol extract. See text for other abbreviations.
Samples also suppressed the formation of syncytia in co-cultures of MOLT-4 and MOLT-4/HIV-1 cells. The most potent inhibition was observed for the methanolic extract from *S. mahagoni*, which also showed significant HIV-1 protease inhibition as described above.

Of the Panamanian plants examined (Otake *et al.*, 1994), methanol extracts of *Jatropha curcas* (leaves) and *Cordia spinescens* (leaves) had appreciable anti-HIV effects, with IC_{50} = 9.0–15.5 μg/ml. The extract of *C. spinescens* also inhibited the giant cell-formation, with IC_{50} = 62.5 μg/ml. SI values above 40 were shown by a water extract of *Drymonia serrulata* (leaves) and a methanol extract of *Pereskia bleo* (whole plant).

An interesting result was obtained in the Egyptian folk medicines. Six plants were shown to have an inhibitory effect below cell toxic concentrations, and the most potent inhibitory effect was shown by water and methanol extracts of *Croton tiglium* (seeds). The IC_{50} values for the methanol and water extracts were 0.025 and 2.0 μg/ml, respectively, and the CC_{50} values were 0.86 and 100 μg/ml, respectively. The SI was 34.4 for the methanol extract and 50.0 for the water extract. Other plant extracts with considerable inhibitory effects were those of methanol extracts of *Abrus precatorius* (seeds), *Artemisia absinthium* (aerial part); water extracts of *Bassia muricata* (whole plant), *Centaurea scoparia* (roots); and both water and methanol extracts of *Datura stramonium* (seeds). As mentioned above, these six plants, including *C. tiglium*, did not show inhibitory effects on HIV-1 RT and PR. To determine the precise mechanism, the possibility of these extracts interacting with the viral glycoprotein gp120 and/or CD4-positive cells was investigated, using 0.5β and anti-Leu3α monoclonal antibody (mAB), respectively. However, all of these extracts (including *C. tiglium*) did not interfere with the binding of mAB, indicating that these extracts do not directly bind to the CD4 receptor or gp120. *C. tiglium* belongs to the family of Euphorbiaceae and is known to contain tumor-promoting diterpenes such as phorbol 12-myristate 13-acetate. A member of this group of phorbol esters, prostatin, has been isolated from a Samoan plant (*Homolanthus nutans*) as an active principle with anti-HIV-1 activity (Gustafson *et al.*, 1992). Although structurally similar to phorbol 12-myristate 13-acetate, prostatin does not itself appear to have tumor-promoting activity. The extract of *C. tiglium* had quite strong anti-HIV-1 activity, suppressing the giant cell formation in co-cultures of HIV-infected and non-infected MOLT-4 cells. This suggests that these extracts can inhibit virus adsorption or fusion, since giant cell formation depends on the interaction of HIV envelope protein with virus receptors on the cell surface.

SUMMARY AND CONCLUSION

For the purpose of developing anti-human immunodeficiency virus (HIV) agents from traditional medicines, various extracts of Chinese and Ayurvedic, as well as Egyptian, Indonesian and Panamanian herbal medicines were screened for *in vitro* inhibitory effects on reverse transcriptase (RT), HIV-protease (PR), the growth of HIV and giant cell formation in HIV-infected cells. Some medicinal plant extracts were found to inhibit the viral specific enzymes and HIV replication in the above tests. Flavonoids, alkaloids and polyphenols isolated from the medicinal plants were identified as inhibitory substances against RT, and their modes of inhibition were also determined. From extracts of *Areca catechu* (seeds) and *Swietenia mahagoni* (bark), arecatannins and chlorogenic acid methyl ester were isolated as moderately inhibitory substances against PR. Extracts of *Croton tiglium* (seeds) and *Abrus precatorius* (seeds) inhibited the growth of HIV in MT-4 cells and significantly suppressed the giant cell formation in MOLT-4. Although the active substances and their mechanisms of action are still under investigation, the results achieved so far show good prospects for the exploration of traditional herbal medicines and healing by natural products.

Since HIV has a long latent period after infection in humans and no clinically effective vaccine is available at present, many medications in association with chemotherapy have been tried as treatment for AIDS. Traditional medicines whose effectiveness has been established for various diseases are expected to provide new anti-HIV agents.

REFERENCES

A. Jacobo-Molina and E. Arnold, HIV reverse transcriptase structure-function relationships. *Biochemistry* **30**, 6351–6356 (1991).

C.-M. Ma, T. Nakabayashi, H. Miyashiro, M. Hattori, S. El-Mekkawy, T. Namba and K. Shimotohno, Screening of traditional medicines for their inhibitory effects on human immunodeficiency virus protease. *J. Traditional Medicines* **11**, 416–417 (1994).

C.-M. Ma, H. Miyashiro, M. Hattori and K. Shimotohno, Screening of Chinese and Mongolian traditional medicines for the inhibitory activity on human immunodeficiency virus type 1 protease. *J. Traditional Medicines* **12**, 418–419 (1995).

F. Kikuchi, N. Miyashita, Y. Tsuda, K. Kondo and H. Yoshimura, Studies on crude drugs effective on visceral larva migrans. I. Identification of larvicidal principles in betel nuts. *Chem. Pharm. Bull.* **35**, 2880–2886 (1987).

G. Nonaka, F.L. Hsu and I. Nishioka, Structures of dimeric, trimeric and tetrameric procyanidins from *Areca catechu* L. *J. Chem. Soc. Chem. Comm.* 781–783 (1981).

H.-G. Krausslich, R.H. Ingraham, M.T. Skoog, E. Wimmer, P.V. Pallai and C.A. Carter, Activity of purified biosynthetic proteinase of human immunodeficiency virus on natural substrates and synthetic peptides. *Proc. Natl. Acad. Sci. U.S.A.* **86**, 807–811 (1989).

I.T. Kusumoto, M. Hattori, Y. Miyaichi, T. Tomimori, M. Hanaoka and T. Namba, Effects of flavonoids and alkaloids on reverse transcriptase. *Shoyakugaku Zasshi* **45**, 240–254 (1991).

I.T. Kusumoto, N. Kakiuchi, M. Hattori, T. Namba, S. Sutardjo and K. Shimotohno, Screening of some Indonesian medicinal plants for inhibitory effects on HIV-protease. *Shoyakugaku Zasshi* **46**, 190–193 (1992a).

I.T. Kusumoto, I. Shimada, N. Kakiuchi, M. Hattori, T. Namba and S. Supriyatna, Inhibitory effects of Indonesian plant extracts on reverse transcriptase of an RNA tumor virus (I). *Phytother. Res.* **6**, 241–244 (1992b).

I.T. Kusumoto, M. Hattori, H. Ishii and T. Namba, A comparative study on the inhibitory effects on flavonoids and alkaloids on reverse transcriptases of different retroviruses. *Shoyakugaku Zasshi* **47**, 291–294 (1993).

J. Inokuchi, H. Okabe, T. Yamauchi, A. Nagamatsu, G. Nonaka and I. Nishioka, Antihypertensive substance in seeds of *Areca catechu* L. *Life Sci.* **38**, 1375–1382 (1986).

K.R. Gustafson, J.H. Cardellina II, J.B. McMahon, *et al.*, A nonpromoting phorbol from the Samoan medicinal plant *Homalanthus nutans* inhibits cell killing by HIV-1. *J. Med. Chem.* **35**, 1978–1986 (1992).

L.A. Kohlstaedt, J. Wang, J.M. Friedman, P.A. Rice and T.A. Steitz, Crystal structure at 3.5Å resolution of HIV-1 reverse transcriptase complexed with an inhibitor. *Science* **256**, 1783–1790 (1992).

M. Hattori, I.T. Kusumoto, M. Soga and T. Namba, Screening of various Ayurvedic medicines for their inhibitory activities on reverse transcriptase and identification of arecatannins and embelin as major inhibitory substances from *Areca catechu* and *Embelia ribes*. *J. Med. Pharm. Soc. Wakan-Yaku* **10**, 141–148 (1993).

M.L. Sethi, Enzyme inhibition VI: inhibition of reverse transcriptase activity by protoberberine alkaloids and structure–activity relationship. *J. Pharm. Sci.* **72**, 538–541 (1983).

M.S. Johnson, M.A. McClure, D.-F. Feng, J. Gray and R.F. Doolittle, Computer analysis of retroviral pol genes: assignment of enzymatic functions to specific sequences and homologies with nonviral enzymes. *Proc. Natl. Acad. Sci. U.S.A.* **83**, 7648–7652 (1986).

N. Kakiuchi, M. Hattori, T. Namba, N. Nishizawa, T. Yamagishi and T. Okuda, Inhibitory effect of tannins on reverse transcriptase from RNA tumor virus. *J. Nat. Prod.* **48**, 614–621 (1985).

N. Kakiuchi, M. Hattori, H. Ishii and T. Namba, Effect of benzo[c]phenanthridine alkaloids on reverse transcriptase and their binding property to nucleic acids. *Planta Med.* **53**, 22–27 (1987).

N. Kakiuchi, I.T. Kusumoto, M. Hattori, T. Namba, T. Hatano and T. Okuda, Effect of condensed tannins and related compounds on reverse transcriptase. *Phytother. Res.* **5**, 270–272 (1991).

S. Billich, M.-T. Knoop, J. Hansen, P. Strop, J. Sedlacek, R. Mertz and K. Moelling, Synthetic peptides as substrates and inhibitors of human immunodeficiency virus-1 protease. *J. Biol. Chem.* **263**, 17905–17908 (1988).

S. Hada, N. Kakiuchi, M. Hattori and T. Namba, Identification of antibacterial principles against *Streptococcus mutans* and inhibitory principles against glucosyltransferase from the seed of *Areca catechu* L. *Phytother. Res.* **3**, 140–144 (1989).

S. Harada, Y. Koyanagi and N. Yamamoto, Infection of HTLV-III/LAV in HTLV-I carrying cells MT-2 and MT-4 and application in a plaque assay. *Science* **229**, 563–566 (1985).

T. Okuda, T. Hatano, H. Nitta and R. Fujii, Hydrolysable tannins having enantiomeric dehydrohexahydroxydiphenyl group: Revised structure of tercyhebin and structure of granatin B. *Tetrahedron Lett.* **21**, 4361–4364 (1980).

T. Otake, H. Mori, M. Morimoto, N. Ueba, I.T. Kusumoto, Y.A. Lim, H. Miyashiro, M. Hattori, T. Namba, M.P. Gupta and M. Correa, Anti-human immunodeficiency virus activity of some tropical medicinal plants. *J. Trad. Med.* **11**, 188–193 (1994).

T. Yoshida, T. Okuda, T. Koga and N. Toh, Absolute configuration of chebulic, chebulinic and chebulagic acid. *Chem. Pharm. Bull.* **30**, 2655–2658 (1982).

V.S. Sethi and M.L. Sethi, Inhibition of reverse transcriptase activity of RNA-tumor viruses by fagaronine. *Biochem. Biophys. Res. Commun.* **63**, 1070–1076 (1975).

W.B. Haseltine, Replication and pathogenesis of AIDS virus. *J. AIDS* **1**, 217–240 (1988).

Y.A. Lim, C.-M. Ma, I.T. Kusumoto, H. Miyashiro, M. Hattori, M.P. Gupta and M. Correa, HIV-1 reverse transcriptase inhibitory principles from *Chamaesyce hyssopifolia*. *Phytother. Res.* **11**, 22–27 (1997).

INDEX

For Product Safety Concerns and Information please contact our EU representative GPSR@taylorandfrancis.com Taylor & Francis Verlag GmbH, Kaufingerstraße 24, 80331 München, Germany

Batch number: 10731074

Printed by Printforce, the Netherlands